Swelling Concrete in Dams and Hydraulic Structures

Swelling Concrete in Dams and Hydraulic Structures

DSC 2017

Edited by

Alain Sellier
Étienne Grimal
Stéphane Multon
Éric Bourdarot

iSTE

WILEY

First published 2017 in Great Britain and the United States by ISTE Ltd and John Wiley & Sons, Inc.

ISTE Ltd
27-37 St George's Road
London SW19 4EU
UK

www.iste.co.uk

John Wiley & Sons, Inc.
111 River Street
Hoboken, NJ 07030
USA

www.wiley.com

Library of Congress Control Number: 2017940969

British Library Cataloguing-in-Publication Data
A CIP record for this book is available from the British Library
ISBN 978-1-78630-213-7

Cover image: View of Chambon dam.

Table of Contents

Table of Contents

Preface

Swelling of concrete occupies a major role in the long term concerns of owners and operators of dams and hydraulic structures. The mean age of dams in European countries, for example, is now well over 50 years.

Faced with irreversible movements of their dams or with observed cracking processes, operators need to explain the observed phenomena, justify safety conditions and in some cases plan remedial works. Underlying these concerns, the question of life duration of the structures is raised.

During the last twenty years active research has been carried out in the field, resulting in practical results in phenomena interpretation and dam modeling. An increasing number of affected dam cases have been documented, with safety reevaluations and in some cases remedial works. A small number of them have been demolished.

If it still remains difficult to establish 'a state of the art' in this domain due to the rapidly evolving context, regular international exchanges in the field appear fruitful and necessary.

Therefore, in the continuity of previous conferences in the field organized under the lead of Robin Charlwood, former President of the ICOLD Concrete Committee (Fredericton 1992 / Chattanooga 1995 / Grenada 2007 / Paris 2009 / Fontana 2013), EDF and Toulouse University-LMDC have taken the initiative to organize in Chambéry, located in the historical heart of Savoy, a workshop intending to provide a new opportunity for experience sharing and benchmarking.

The aim of the workshop is to assemble active researchers, leading engineers and experts from the practicing community and administration interested directly or indirectly in concrete swelling effects in dams and hydraulic structures. All kinds of chemical expansion phenomena, including those due to alkali aggregate reactions and those due to ettringite formation are addressed.

These proceedings include 24 papers written by renowned experts in their own field. They are divided in five chapters:

Chapter 1 provides an overview of the international context of affected hydraulic structures and a focus on Swiss dams. All continents seem to be concerned and even if temperature plays a significant kinetic role, structures in cold regions are also affected. Concrete swelling consequences strongly depend upon geometrical configurations. Interactions with electromechanical components are a main issue. A large panel of solutions are available and have already been used to minimize disorders.

Chapter 2 is devoted to physicochemical mechanisms and experimental tests. Thanks to significant work performed in the field, a real increment in the comprehension of the mechanisms can be assessed. Multiscale approaches, from aggregate to concrete scale, from chemical equations to structural modeling, enable an optimal weighting of the different mechanisms and may explain the apparent discrepancies in test results. The importance of ambient conditions (temperature, humidity, alkali content) and test procedures is highlighted. Improvements in experimental conditions and methodology are proposed. Several works focus on the importance of aggregate types in AAR but also in DEF, in relation to alkali release from granitic stone, different effects of transition zones in calcareous or silica aggregates, or to the localization of AAR inside granulates. Coupling between ASR and DEF is also now better understood through the common role played by alkalis in silica gel and ettringite formation. Thanks to available experiments related to the influence of stresses on swelling kinetics, a best estimation of confinement effects in hydraulic structures is practicable. Faced with the great diversity of tests and investigations facilitating the identification, extent and prognosis of AAR, methodological approaches and guidelines are required and proposed.

Chapter 3 presents several models dedicated to the swelling mechanisms of concrete and their structural effects. Among the main phenomena considered, all the models adopt common assumptions to consider effects of environmental conditions (temperature and humidity). A first one focuses specifically on these environmental conditions and shows the importance of rainfall periods on swelling kinetics, especially for thin structural elements. The aptitude of structural models to consider the swelling anisotropy under complex stress states is also considered as important for practitioners; it is illustrated through different applications relative to dam analysis. The possibility of using an original explicit scheme to accelerate the numerical solving of these nonlinear models is presented. Concerning the links between chemical reactions, swelling and induced damage, they are considered at different scales of analysis, with, for the finest scale, a mesoscopic model able to link ionic diffusion and swelling kinetics at the aggregate-paste interface, until different macroscopic formulations clarified in the context of classical continuum mechanics. In the latter case the swelling is either an anisotropic imposed strain depending on the stress state via empirical laws, or is based on a poro-mechanical formulation able to link the gel volume produced by the ASR, the gel pressure and the stress state via plasticity criteria controlling simultaneously the anisotropy of damage and swelling. Some analyses also show that other delayed strains of concrete (creep and shrinkage) should not be neglected compared to ASR swelling. Regarding DEF, a realistic estimation of thermal conditions at early age is a necessary step as demonstrated on a real dam case.

Chapter 4 is dedicated to the description and presentation of remedial works. It illustrates the fruitful processes including physico-mechanical approaches, experimental tests, in-situ measurements and models calibrated to monitoring results. They enable to orientate interventions, accommodate concrete swelling and estimate the impact of invasive interventions such as slot cutting. Beyond the similarities, each type of structures reveals its particularities and adaptation possibilities. The particular case of arch dams with thrust blocks illustrates the need to distinguish between external loading and internal loading such as thermal or swelling effects. For the latter, efforts are induced by the structure's confinement and can be released through adaptation processes. Bimont and Chambon dams illustrate cases where extensive remedial works programs have been carried out, for a second time in the case of Chambon dam.

Chapter 5 is devoted to the long term management of hydraulic structures. Estimation of the residual swelling potential appears to be, in this respect, a main issue. Even if long time extrapolation remains questionable, test procedures seem able to operate a distinction between concrete reactivity levels and thus to classify works sensitivity. The importance of monitoring and device redundancy is highlighted.

In conclusion, the different papers illustrate the need to progress through interdisciplinary approaches when faced with such complex engineering problems, requiring complex engineering solutions. In order to ensure an operation progresses in good conditions, a methodology based on the combined use of monitoring (irreversible evolution measurements), laboratory tests (pathology identification, swelling kinetics and potential), numerical modeling (historical behavior simulation, safety evaluation, future evolution, remedial works efficiency evaluation) and remedial works implementation (waterproofing, injections, sawing, anchoring ...) seems to bring consensus.

Finally the organizers would like to gratefully acknowledge the support of ICOLD and CFBR, Toulouse University and Electricité de France and thank the hospitality of the Hydro Engineering Centre of EDF for their organisational support.

Alain Sellier — Professor (LMDC-Toulouse University)
Étienne Grimal — Dam Specialist (EDF-CIH)
Stéphane Multon — Assistant Professor (LMDC-Toulouse University)
Éric Bourdarot — Dam Expert (EDF-CIH)

CHAPTER 1

International Context

A Review of the Effectiveness of Strategies to Manage Expansive Chemical Reactions in Dams and Hydro Projects

ROBIN CHARLWOOD* — IAN SIMS**

* *Robin Charlwood & Associates, PLLC*
Seattle, WA, USA
robincharlwood@gmail.com

** *RSK Environment Ltd*
Hemel Hempstead, UK
isims@rsk.co.uk

ABSTRACT. *This paper summarizes the authors' assessments of the impacts of alkali-aggregate and other expansive chemical reactions in a selected set of dams and hydro projects based on previously published reports at ICOLD and related meetings. Currently employed technologies to manage the expansion phenomena in existing dams are then reviewed. The effectiveness of current remediation techniques for maintaining the serviceability of the structures, frequently in the face of continuing expansion, is discussed. These measures range from short-term modifications of key items such as spillway gates, post tensioned anchors and generating equipment to longer term management by reinforcement and slot cutting of large mass concrete structures. At this time, no practical methods to terminate these chemical reactions have been identified. Consequently, owners are utilizing a variety of innovative approaches to extend the project life and avoid having to prematurely decommission and or replace many existing structures worth many billions of dollars.*

KEYWORDS: *concrete dams, expansive chemical reactions, alkali-aggregate reaction, ASR, ACR, ISA, DEF, remedial measures.*

1. Introduction

The main underlying expansive chemical reaction mechanisms affecting certain dams are: Alkali-Aggregate Reactions (AAR) including alkali-silica reactions (ASR) and Alkali-Carbonate Reactions (ACR); Sulfate related deteriorations including External Sulphate Attack (ESA) and Thaumasite Sulphate Attack (TSA), Internal Sulphate Attack (ISA) including pyrite oxidation, and Delayed Ettringite Formation (DEF). By far the most frequent process leading to the expansion of concrete in dams is ASR (alkali silica reaction) which is involved in most of the cases discussed herein.

The vulnerabilities of dams to chemical expansion of concrete are clearly specific to the dam type. Appropriate strategies to manage these effects vary substantially depending on the rate, magnitude and particular rheological behavior of the reaction and resulting concrete expansion, and the dam or hydroelectric plant structural, equipment and geological configuration.

In this paper, we present summaries of illustrative examples to provide a basis for an assessment of the effectiveness of the various management strategies adopted. Details of many of the cases may be found in proceedings of the 1992 CEA Fredericton Conference [CEA, 1992] and the 1995 USCOLD Chattanooga Conference [USCOLD, 1995], the presentations at the 2007 ICOLD/SPANCOLD Granada Workshop [ICOLD, 2007] and the 2013 ICOLD Fontana Workshop [ICOLD, 2013]. Various others are as referenced.

The opinions expressed regarding impacts, methods and effectiveness of the management strategies are those of the authors based on their interpretation of the quoted published materials and are not purported to represent those of the owners. These case histories are in the process of being more fully documented and reviewed for inclusion in the forthcoming new ICOLD Bulletin: Expansive Chemical Reactions in Concrete Dams.

2. Examples of ASR Management Strategies at Existing Dams

A set of summaries of chemical expansion cases for various dam types and associated hydraulic structures from various geographic regions are presented to provide:

– representative examples of a range of manifestations of expansive reactions in different structural configurations (dam types, geometries, etc.) that are significantly affecting safety and operations;

– identification of issues and impacts resulting from the expansive reactions;

– examples of the role of investigations, testing, monitoring, modelling;

– methods of managing the effects, do nothing, monitor, interventions, etc.;

– an assessment of the short and long term effectiveness of the management strategies adopted.

These examples have been selected based on the nature of the expansive reaction (type, expansion rate, duration), local or general extent, the epidemiology and diagnosis of the reaction, significant effects on structures or equipment, innovative instrumentation, modelling or interventions, and lessons learned regarding the effectiveness of the management strategy. They are grouped by geographic region.

2.1. *Africa*

1. ASR Case: Kariba, (Arch Dam & Spillway) Zimbabwe/Zambia [Goguel & Gurukumba, ICOLD 2007; Gurukumba, ICOLD 2013; Noret & Gurukumba, Africa 2017]

Kariba dam is a 128 m high double curvature arch dam built on the Zambezi River in the late 1950s and provides 1,830MW power generation capacity. The dam body has been known to be subject to ASR since the late 1960s. The concrete swelling is affecting the spillway sluices geometry and has damaged the upstream stoplog rollerpath upper sections. Monitoring has shown a sustained moderate vertical concrete expansion rate varying between 23 and 47 micro-strain/year but new data suggests the rate is slowing. The concrete swelling phenomenon at Kariba has been and is still being managed through continued inspection of the structure and its appurtenances, an extensive monitoring system and safety assessment studies, numerical modelling studies and remedial measures to mitigate the swelling effects. The remedial measures under consideration include repairs to the stoplog rollerpaths and provision of new upstream set of stopbeams together with an emergency gate.

2. ASR Case: Cahora Bassa, (Arch Dam & Spillway) HCB, Mozambique [Tembe, Carvalho, Hattingh, Oosthuizen, ICOLD 2014]

The Cahora Bassa Dam is a 170m high double curvature concrete arch dam located on the Zambezi river in Mozambique and was completed in 1975. The installed capacity is of the powerhouse is 2075 MW. Concrete swelling due to ASR has been observed since 1977 at a rate of approximately 25 to 40 micro-strain/year depending on the instrument type. The swelling process of the concrete is still under way and no significant decrease in swelling is evident. There is an extensive instrumentation and monitoring system with proper redundancy in place which has been used not only to support detailed finite element modelling but also to facilitate proper behaviour analysis. The development of diagonal cracks on the upper third of the downstream face parallel to dam/foundation contact is evident. The swelling has also caused small deformations of the spillway gate supports which have affected gate clearances. A spillway radial gate rehabilitation project has been completed to address the clearances. The evolution of the process is being managed by updating the monitoring system including the geodetic survey system to facilitate monitoring of absolute displacements, the in-situ measurement of stresses and strains through over coring and finite element models to follow this process.

3. ASR Case: Nalubaale (Arch Gravity Dam & PH) UEGCL, Uganda [Brueckner, Ndugga, Meri & Mahsen, Africa 2017]

The Nalubaale Hydroelectric Power Station (formerly Owen Falls Dam) is located on the Nile River in Uganda. The project consists of a powerhouse with an installed capacity of 150MW and gravity intake and an arch gravity dam and was constructed in the early 1950's. Deterioration of the powerhouse structure was first noticed in 1964 in the form of hairline cracking in concrete elements. More pronounced cracking was observed in 1977. Stabilization of the spiral casings using post-tensioned anchors was implemented between 1989 and 1991 before ASR was established as the cause of the expansion and cracking. Extensive cracking has occurred in the powerhouse generator surround concrete and floor beams and affects the crane beam alignments. No significant impacts have been identified in the intake or main dam. The expansion strain rate has been estimated to be in the range 30 to 45 micro-strain/year. A life extension program for the powerhouse is under development. The following are under considerations: structural modifications to control ASR expansion such as slot-cutting to accommodate expansion; installation of post-tensioned anchors to restrain the expansion; partial or full replacement of affected mass concrete; and, structural modifications to address the integrity of the structural frame such as installation of replacement downstream columns and strengthening of the downstream column in the region above the crane rail.

4. ASR Case: Kleinplaas Dam (Gravity Dam) DWA, South Africa [Hattingh, ICOLD 2013]

Kleinplaas Dam is located on the Jonkershoek River in South Africa and consists of a 25.5 m high uncontrolled ogee concrete gravity spillway section in the river section and flanked by a rockfill embankment with a clay core on the left flank. The dam functions primarily as a balancing dam for a scheme supplying water for domestic use to the Cape Town metropolitan area. The dam was completed in 1982. Subsequently evidence of swelling was observed including pronounced cracking and opening of horizontal construction joints of the concrete spillway section. Monitoring of the swelling has been done since 1996 with 3D-crack gauges and since 2000 with a geodetic survey system on the crest. Since 2000 vertical swelling of between 20 and 42 micro-strain/year is evident. Total vertical strain of approximately 850 micro-strain was estimated in 2014. Some decrease in vertical swelling is evident since 2009. Continued monitoring of the swelling behaviour is currently taking place. A detailed structural analysis is planned and any required rehabilitation could follow.

5. ASR Case: Kouga Dam (Arch Dam) DWA, South Africa [Hattingh & Oosthuizen, ICOLD 2012]

Kouga Dam is a 69-m double curvature concrete arch dam located on the Kougha river in South Africa and was completed in 1969. The dam has been extensively monitored during and after its completion in 1969. Initially the dam behaved as expected but since 1972, it became clear that some form of concrete swelling was occurring. In addition to this, some inelastic movement of the right flank followed a few years later that complicated the behaviour model of the structure even further. The static monitoring system of Kouga Dam showed continuing expansion since the early signs of swelling became evident. Initially only clinometers, pendulum clinometers and a geodetic network of targets on the downstream face of the dam wall were used. The static monitoring system was extended to include 3D-crack gauges, sliding micrometers and Trivecs. Subsequently real-time 3D-crack gauges, a GPS system and a permanent ambient vibration system were added. Vertical swelling of around 25 micro-strain/year was initially observed but has reduced to less than 10 micro-strain/year since 2000. Horizontal swelling has however continued unabated. Sufficient redundancy in the monitoring system has added confidence to the interpretation of unusual behaviour of some of the blocks during low water levels (3D-crack gauges and ambient vibration results). Rehabilitation of Kouga Dam is currently under consideration – possibly building a replacement structure on the downstream side. Foundation investigations have been done and detailed numerical behavioural model have been compiled using the monitoring results.

6. ASR Case: Matala Dam (Spillway) PRODEL, Angola [Casagran' Pradolin & Victor, ICOLD 2016; Bouayad, Africa 2017]

The original concrete structures of the Matala Hydroelectric Development, built in the 1950's and located in Angola, are affected by AAR. The presence of this phenomenon was, and still is, manifested in deformations of spillway piers, original pedestals for flap gates and other structures. These deformations affected the operation of the flap gates (reducing spill capacity) and resulted in important/critical bridge roller bearing rotations. The concrete structures also exhibited severe cracking. Rehabilitation works were carried out to maintain the spillway discharge capacity and to restore the integrity of the structures. Works included construction of a new gated spillway (using portions of the existing spillway), pier pinning, bridge roller support rehab, concrete repairs. Feedstock from the original quarry was used and fly ash was added to the concrete mix.

2.2. *Europe*

7. ASR Case: Chambon, (Arch-Gravity Dam & Spillway) EDF, France [Bourdarot, ICOLD 2007; Chulliat et al., *ICOLD 2012; Grimal, ICOLD 2013]*

Chambon Dam is a large curved gravity dam completed in 1934 and affected by concrete swelling due to ASR. Key issues are related to (1) the high compressive stresses developing inside the dam and the abutments and the associated potential risk of thrust towards upstream in the curved zone and shear along concrete-rock interface at the spillway located in the left part, and, (2) displacements towards downstream in the central part and right bank and potential risk of tensions in the upstream face and shear at the foundation interface.

A first remedial works campaign was performed in the 1990-1997 period (new spillway construction, decommissioning of the old one, cracks grouting, PVC geomembrane installation on the upper 60 m of the upstream face to control uplift pressures, and 8 slot cuts). All these works proved their effectiveness by the recovery of a part of the irreversible displacements of the curved part and reducing compressive stresses in the upper dam part.

In 2007, the closure of the slots and a restart of the upstream movement of the curved part justified a reassessment of the dam mechanical behavior and complementary investigations. Particular attention was paid to the vertical cracks located in the upper part which may create, under seismic events, potential unstable blocks. Taking into account the results from the last available numerical modeling, a new works campaign was carried out in 2013 and 2014 which included: seven slots-recutting with deepening of two (down to 40 m); installation of upstream to downstream post-tensioned anchors supplemented by a composite carbon-fiber net in order to reinforce the confinement of the upper part; and, replacement of the PVC membrane.

8. ASR Case: Temple-sur-Lot, (Garvity Spillway) EDF, France [Bourdarot et al., *Ejece 2010]*

The Temple-sur-Lot dam was built between 1948 and 1951 and has been in operation since 1951. It includes a gate-structure dam equipped with four double-leaf vertical lift gates (20 m wide, 10 m high).

In 1960, an inspection revealed the existence of cracks on the upstream part of spillway piers. In the following years, difficulties in the operation of the bulkhead gates led to several interventions on the mechanical parts embedded in the concrete structure. Reinforcement of the monitoring system provided a better description of the deformation of the piers and laboratory investigations revealed the existence of swelling phases inside the concrete.

During the period 1983–1988, extensive works were carried out on the piers, such as anchoring, and epoxy and polyurethane grouting. Recently (2002–2003), the guidance system of the gates was modified in order to accommodate the concrete deformations. The effects of the swelling process on the structure include general rising of the piers (1mm/year) and tilting of the lateral piers towards the gates (0.9 mm/year). Analyses, including reinforcement of the monitoring, laboratory investigations and FE modelling, were carried out in order to explain these particularities. The lateral movements of the piers towards the gates can be mainly explained by the humidity gradients between the faces.

9. DEF Case: Bimont (Arch Dam) SCPARP, France [Noret & Laliche, DSC 2017]

Bimont Dam is located near Aix-en-Provence and was first brought into service in 1952, and became a part of the Société du Canal de Provence concession in 1963. The dam is a concrete, double-curved, arch-type structure measuring 86 m in height and 180 m in length along the crest. It consists of 15 cantilevers between two abutments, and its thickness ranges from 4 m at its crest to 13 m at its foot.

The dam developed a network of cracks in some cantilevers soon after construction. Their existence was initially put down to geology, but subsequent investigations showed that specific areas of concrete were affected by Delayed Ettringite Formation (DEF). This phenomenon brought about changes in the dam's equilibrium, resulting in the formation of superficial and internal cracks. A numerical model of the dam with elastoplastic features ran in parallel with two special investigation campaigns. These measures allowed for a more in-depth understanding of how the network of cracks was formed, its spatial extent, and its probable future evolution, information which proved invaluable for the design of the dam renovation programme. The planned rehabilitation project will involve the treatment by cement grouting of the cracks and joints, waterproofing of the dam upstream face and the provision of vertical anchors to improve stability of the right abutment.

10. ASR Case: Poglia, (Hollow Gravity Dam + Wing Wall) EDISON, Italy, [Mazza, Donghi & Marcello, ICOLD 2008]

Poglia dam, located in the Lombardia Region (North of Italy) is a hollow gravity buttress structure (Marcello type) owned by Edison; its aim is hydroelectric power generation. Since the 1970s the monitoring system started to show a slow drift in elevation of the different dam blocks and, less evident, in the upstream-downstream direction. For that reason the owner carried out investigations to reach a clear explanation of the causes which gave rise to the above said drift. The full reliability of measurements and the geotechnical survey excluded the presence of problems related to the stability of foundation and abutments. Hence, the presence of possible expansive phenomena in the concrete has been explored. To this aim an on site and laboratory campaign was carried out and the presence of alkali-aggregate reaction in the concrete has been ascertained. The phenomena are moderate but, due to the non-rectilinear longitudinal dam axis and to the particular geometry of the dam blocks, a relatively severe stress-strain state, additional to the one due to operational loads, has taken place in the dam body. An advanced three-dimensional, non linear finite element model was applied to simulate the expansion AAR phenomena, and calibrated on the basis of the measurements recorded on the dam to analyse possible future scenarios, to investigate different hypothetical structural interventions and to give confidence during the work-in-progress phases. The final decision undertaken by the designer was to cut at the contraction joints in order to reduce the compression stresses in the blocks and in the right gravity shoulder. The works have been carried out during the spring 2005 in a very short stretch of time, in order to minimize the out of service time of the plant. The dam monitoring system, suitably improved during the works, has allowed us to reach a comprehensive knowledge of the actual dam safety conditions.

The joints were cut by means of a "diamond wire", and carried out on all the contraction joints with the following aims:

– the recovery of the displacements of the right wing gravity shoulder;

– re-establishment of the behaviour of each block according to the original design scheme, reducing the high compressive stresses due to AAR.

The attainment of these aims has been estimated by means of a detailed finite element model and will be confirmed with the monitoring system installed on the dam, suitably integrated during the rehabilitation works.

A safety assessment of the main block against sliding has been carried out using a limit equilibrium finite element analysis. By making the extreme assumption of the dam sub-divided in several blocks by horizontal and vertical cracks, it has been found a safety coefficient greater than 2 times the maximum hydrostatic load.

11. ASR Case: Piantelessio (Arch-Gravity Dam) Iren S.p.A., Italy [Amberg, Stucci & Brizzo, IECS 2013]

The Pian Telessio arch gravity dam is located in the Orco Valley (Piedmont, northern Italy) impounding a reservoir with a capacity of 24 mio. m3 for a normal operating level at 1'917 m a.s.l. The dam is 80 m high, with a crest length of 515 m. The crest thickness is 5.7 m, while it increases towards the base where it reaches a maximum of 35 m. The dam is equipped with a peripheral joint which separates the dam body from the foundation slab (Pulvino).

The dam began operation in 1955 approximately after 5 years of construction. In a first period of roughly 20 years, the dam presented a regular and fully reversible behavior, while since the second half of the 70s the dam is showing an upstream drift in a radial direction. The permanent displacement at crest level reached in 2008 almost 60 mm at the central pendulum, In addition to permanent displacement, horizontal cracks appeared in the upper inspection gallery, which are neither visible at upstream nor downstream faces. After excluding other causes for this observed behavior, such as for example movements of valley flanks, it was assessed that the permanent dam deformations are caused by an ongoing alkali-aggregate reaction (AAR).

In order to avoid conditions with high compressive stress at the dam heel, a limitation in the minimum water level was adopted as a temporary measure in 2003. The operational limitation in the long term was not acceptable. Rehabilitation works consisting in the execution of 16 vertical slots by means of diamond wire were proposed and finally executed in 2008 ([1]). The height of the main slots is 39 m in the central part (Figure 3), while it is limited to 31 m and 21 m towards both flanks. Between the main slots, secondary slots of 21 m height were realized.

Once the swelling stresses had been released, the slots were grouted in order to recover the arch effect required to support the pressure at full water reservoir. The rehabilitation works were carried out satisfactorily and since 2009 the dam is again under normal operation conditions.

12. ASR Case: Pracana, (Buttress Dam) Presently EDP, Portugal [Camelo, ICOLD 2007; Batista, ICOLD 2013]

Pracana dam is a buttress concrete gravity dam with 12 buttresses, 3 massive blocks in each bank, a height of 60 m and crest length of 245 m. The project was constructed in the period 1948 to 1951 for a previous owner. Since 1952, several anomalies in the dam were detected and which continuously increased. In 1971 restrictions were applied to the reservoir level. In 1972 and 1973 various unsuccessful repair works were attempted.

In 1977 the ownership of the scheme was transferred to EDP. In 1978 the reservoir was emptied due to insufficient spillway capacity; progressive deterioration of the dam; and low safety factors. The dam exhibited: intensive cracking on the upstream face; important cracks on downstream face in the transition between the web and the head of the buttresses; significant vertical cracks in the webs near the foundation; and, cracks along horizontal construction joints. This was accompanied by seepage through horizontal cracks (concrete lift joints) with excessive carbonation. Large displacements were measured by geodetic methods and analysis showed non-reversible displacements.

Mineralogical and petrographic analysis of aggregates and cement paste identified gel formation and ASR. The concrete, also suffered from insufficient fines and high w/c ratio, which led to high capillarity porosity and open "channels" along the interface with the aggregates.

Sliding along horizontal cracks was identified as a critical safety scenario.

In 1985, evaluations concluded that the expansion phenomenon in the concrete was considered the main cause of the dam deterioration. It was also suggested that this expansion only developed in the presence of infiltrated reservoir water and stability conditions should be acceptable if uplift effects into concrete cracks could be avoided. The integrity of the dam's concrete could be improved by crack treatment and grouting. A global foundation treatment should be undertaken and a careful dam monitoring program should be set up.

The dam rehabilitation program was executed in the period 1988 to 1992 and included foundation struts between buttress webs, an upstream foundation plinth, foundation treatment, concrete regeneration including cement and epoxy grouting, installation of an upstream watertight system and improvement of the monitoring system.

The upstream watertight system consisted of a PVC membrane with a HDPE geogrid for drainage. The main aims of this system were to prevent contact between the concrete and the reservoir water to limit the future potential ASR, and to limit uplift in the cracks and thereby improve the stability.

The reservoir was refilled in 1992. Reported results of observation to 2007 showed evidence that the expansion phenomenon in the concrete was still present, but significantly attenuated. The upstream membrane was suggested to be effective both in terms of limiting water access to the concrete and thereby limiting restarting of ASR expansion effects and also improving stability by limiting uplift pressures in cracks. Continuous monitoring of the expansion phenomenon evolution was recommended.

13. ASR Case: Alto Ciera (Arch Dam-replaced) EDP, Portugal [Camelo, ICOLD 2007, Batista, ICOLD 2013]

Alto Ciera was a 36 m high, 85 m creast length concrete arch dam built in 1949 with ASR and a fairly high expansion rate in the range of 120 micro-strain per year with no sign of slowing. It has been subject to progressive displacements, radial upstream and vertical upwards and cracking of dam's body.

The dam was subjected to comprehensive investigations which showed great heterogeneity of the swelling process with estimated potential expansions up to 650 micro-strain, intensive cracking special concentrated in the shoulders and on the crest; relatively great depth of the cracks in a thin arch structure; and relative intensive leakage through the body dam. It was not possible to estimate the stresses in the structure as well as to predict the future behavior of the dam. The rehabilitation of the dam was considered to be very difficult and expensive. In 2013 a new dam was been completed 200 m downstream and the old dam is under demolition.

14. ISR Case: San Esteban, (Gravity-Arch Dam) Endesa, Spain [GIL, ICOLD 2007]

San Estaban dam is a 115 m high gravity-arch dam with crest length of 295 m with adjacent 4,500 m^3/s spillway completed in 1955 on the River Sil in Spain.

The following structural anomalies were identified: wet lift joints in the high zone of the dam; misalignments of blocks at the crest; irreversible movements of joints in upper zones; wet cracks and lift joints in the upper gallery; cracking of the faces of two blocks; and progressive elevation and upstream movement of the crest.

An investigation program with drilling and permeability tests showed honeycombed concrete, deficient lift joints, connections between drillholes and shallow cracks and obtaining in-situ stresses by the overcoring method. Materials testing showed granite, diabase, gneiss and shale (presence of pyrite) aggregates with fractured structures, altered crystals and the presence of reactive quartz and ettringite and expansive gel products. The concrete had high porosity, normal mechanical characteristics, cement with a high content in CaO. A finite-element model was used for diagnosing expansion and confirmation of the phenomenon.

A remedial program of structural rehabilitation and waterproofing was carried out that included injection of epoxy resin in defective lift joints and facing of the upstream face with 12,000 m^2 of dual-reinforced resin laminate installed in 9 layers with prior cleaning (5 mm thick) and able to adjust to concrete deformations with a modulus and tensile strength capable of supporting concrete cracks of 3 mm and resistant to internal pressures due to drawing down reservoir.

The results of the actions included: filled lift joints and a rehabilitated structure, elimination of seepage but continuation of the expansion phenomenon.

Future actions include: visual inspection, numerical cartography, monitoring including in-situ stresses, additional rehabilitation based on the use of mathematical models.

15. ISA Cases: Tavascan (Gravity+Spillway w radial gates), Toran (Gravity+ Spillway) & Graus (Gravity+Spillway) Endesa, Spain [Río & Espinós, ICOLD 2007]

Tavascan dam is located on the River Noguera de Cardós in the Pyrenees, Spain. It is a gravity dam with radial gated spillway with maximum height: of 31 m commissioned in 1965. The aggregate is a metamorphic slate. Anomalous behaviors, including cracking in the spillway piers, were identified in 1981.

Toran Pont De Rei dam is located on the River Toran, in the Pyrenees, Spain. It is a gravity dam with a fixed lip spillway with maximum height: of 36 m. It is constructed with a metamorphic slate aggregate and was commissioned in 1960. Anomolous behaviors included cracking in the inspection gallery.

Graus Dam is located on the River Tavascan in the Pyrenees, Spain. It is a gravity dam with a fixed lip spillway and with a maximum height of 29 m. It is constructed with a metamorphic slate aggregate and entered service in 1971. Anomalous behaviors, including cracking in the downstream face, were detected in 1986.

Reports for all three dams indicate that movement began about 10 years after construction and some form of waterproofing was carried out in year 33, 1998. The rate of vertical movement increased slowly over at least 40 years but no changes in the rate have been observed since waterproofing was carried out. However, the rate of upstream movement stabilized in 12 to 18 years and reduced in 19 to 32 years and after waterproofing (year 33), has stabilized up to now. It is not clear if this reduction is because of waterproofing or would it have decrease naturally anyway.

16. ASR/ISR Case: Isola (Arch-Gravity Dam) AXPO, Switzerland [Otto, ICOLD 2007]

Isola dam is a concrete arch-gravity with a maximum height of 45 m, with a representative height of 30 m, and crest length of 290 m and located 1,600 m above mean sea level. This is a wide dam with crest length/height ratio of 9.7. It was completed in 1960. The mean concrete temperature is reported to be 7°C.

Pendulum readings indicate an upstream crest displacement starting 20 years after construction and reaching 35 mm in 2005 after 45 years. Vertical rise of the crest reached a maximum of nearly 5 mm in the same period, which corresponds to an accumulated vertical strain of approximately 200 micro-strain in the upper part and 50 micro-strain in the lower part. Cracking was detected in 1978 in the central downstream face and near the abutments in 1986. A crack was detected in the upstream face in 1989. In addition, cracks are apparent in the upper gallery. A horizontal compressive stress of 6 MPa has been measured.

It appears that the rate of development of expansion and associated effects, such as cracking, is slow at this dam, possibly due to the low mean annual temperature. It is suggested that most of the expansion occurs in the summer months. Laboratory tests are reported to suggest that the accumulated expansion strain could increase substantially above the present values. To our knowledge, so far, no interventions have been planned.

17. ASR/ISR Case: Illsee (Gravity & Arch Dam) AXPO, Switzerland [Otto, ICOLD 2007; Leroy & Amberg, ICOLD 2013]

Illsee dam is a concrete arch-gravity with a maximum height of 25 m, with and crest length of 270 m and located 2,360 m above mean sea level. It was built in 1926 and heightened in 1943. The horizontal layout includes two straight sections and a curved arch dam at one end.

The arch crest has been recorded to displace upstream by 45 mm in the 60-year period 1946 to 2005. The straight section near the junction with the arch section has been recorded to displace downstream by 30 mm in the same 60-year period. Cracking is visible on the downstream face at the junction between the center straight section and the arch section. The accumulated expansion strain has been calculated to be in the range 400 to 500 micro-strain in the 60-year period. Cracking is reported to have evolved in the arch section initially as horizontal cracks on the downstream face, and then horizontal cracking on the upstream face followed by opening of joints. The mean concrete temperature is reported to be 5°C.

An impervious membrane was added on the upstream dam face in 1997 in a first phase of interventions, and slot cutting and reinforcement with anchorages were added in 2012–13.

The owner concluded in 2007 that the rate of expansion of 10 to 30 micro-strain/year is increasing with time after decades with swelling observed mainly in summer. Remedial measures (membranes on the upstream face) have shown no effect on the rate of swelling after 10 years. It is noted that AAR and sulfate reactions happen at low temperatures (5° to 7°C). This dam is still in the initial phase of swelling even after 40 to 60 years of operation and the swelling potential has been suggested to be much larger than the state to date.

18. ASR Case: Salanfe (Gravity Dam) ALPIQ, Switzerland [Leroy & Amberg, ICOLD 2013]

Salanfe Dam is a gravity dam built in 1953 and now owned by ALPIQ. The dam is 616 m long, 52 m high with a crest width of 5 m. It has 42 blocks which are placed in 4 straight sections with one sharp change in alignment (concave downstream) in plan.

Vertical expansion has been measured for the last 15 years at a strain rate of 45 micro-strain/year. The geodetic network results show a significant upstream movement of the dam with a differential movement occurring at the change in alignment. Extensive surface cracking has occurred on the downstream face in the region of the corner and in the access gallery.

A program of slot cutting to reduce the stress concentrations was planned and 3 slot cuttings performed during the autumn 2012. The campaign was aimed at verifying the results of the numerical model and assessing the methodology and the performance of the work to be performed. It was decided to perform one of the cuts along a vertical joint and two others in the middle of a block. The slot cutting in the vertical joint caused slivers of concrete to fall into the slot preventing closure. The numerical model indicated an almost full closure of the 15 mm wide slot. Measurements showed a closure about halfway. Therefore, it was decided to increase the number of slot cuts to 22 and cut 20 m deep and to reduce the width of them to 11 mm. All remaining slot cuts were to be executed in the middle of blocks.

In 2013 is was too early to draw definitive conclusions regarding the global behavior of the dam but some preliminary results indicated that although most of the slot cuts closed approximately by 8 mm, some of them completely closed after a few days. This was interpreted to be a demonstration again the high heterogeneity of the ASR affected zones within the concrete mass.

It was concluded that: slot cutting seems to be an efficient mean to reduce the compressive stresses due to concrete swelling induced by ASR; the numerical modeling of the behavior of the dam affected by ASR before and after rehabilitation offers the opportunity to adjust the procedure and the strategy of rehabilitation; it is also essential to increase the monitoring efforts of the dam before performing slot cuttings: adapted and improved equipment are necessary in order to follow precisely the behavior of the structure during and after relaxation of the stresses; and simple procedures and installations are possible for performing slot cuttings. Slot cutting can then be executed efficiently.

19. ASR Case: Stolsvatn (Multiple Arch-replaced) E-CO Energi, Norway [Gunleiksrud, 2008]

Stolsvatn dam was a multiple cylindrical arch concrete dam with vertical upstream faces and pairs of concrete butresses built in 1948 on the Drammen River in Norway. The dam consisted of 13 arches in reinforced concrete, was 18 m at its highest, with total crown length of 520 m. The buttresses suffered from extensive surface cracking with concern about this allowing corrosion of the reinforcement. Epoxy coatings were applied to the external surfaces of the buttresses and additional anchors installed. It was found that AAR was present causing continual expansion of the concrete with surface cracking that facilitated freese-thaw damage.

For a long time local residents expressed concerns regarding the safety of the dam despite numerous repairs. The dam was decommissioned in 2009 and replaced by a new embankment dam downstream.

2.3. *North America/Canada & USA*

20. ASR Case: Mactaquac, (Gravity Intake, Spillway & PH) NB Power, Canada [Curtis, ICOLD 2007; Fletcher & Curtis, ICOLD 2013]

Mactaquac GS is a 670 MW hydroelectric project consisting of a zoned embankment dam and a powerhouse with a concrete gravity intake and spillways completed in 1968. In this case ASR has developed and is continuing at a rapid rate in all concrete structures. Diamond wire saw cut slots were first developed at this project in 1988 and have been used to: control the deformations and cracking of the intake and other gravity sections; maintain clearances of spillway gates; maintain turbine runner and generator air gap clearances; reduce vertical misalignment of turbine shafts and stress build up in wicket gates. Extensive post-tensioned anchors were installed in the intake to secure stability during and after initial slot cutting. A comprehensive high precision instrumentation system was installed. A special purpose finite element concrete growth modelling program was developed to use the instrumentation data to assess existing conditions and plan and monitor effectiveness of interventions and estimate future service life. In December 2016, NB Power recommended maintaining the station to its intended lifespan and is proceeding with remediation plans on that basis.

21. ASR Case: Beauharnois (PH and Gravity Abutment), Hydro Quebec, Canada [Gocevski, ICOLD 2013]

The plant has 38 generating units in three connected powerhouses with a total installed capacity of 1,903 MW. Construction of the facility began in 1930 and was completed in 1961. The effects of concrete expansion were first attributed to ASR at Beauharnois in the late 1960s. The major rehabilitation actions to date include:

– 1970–1971: Stability improvement – Strengthening (pinning to the foundation) of the right gravity dam by Post tensioning cables plus grouting of cracks;

– 1972–1975: Slot-Cutting by overlapping drillholes between the gravity dam and the water intake structure; and between the gravity dam and the administration building;

– 1980–1981: Slot-Cutting by overlapping drillholes between the units 35 and 36;

– 1995–1996: Controlled Separation: Powerhouse1 – water intake structure (slabs only);

– 2003: Controlled Separation: Powerhouse3 – water intake structure (slabs only).

Observations regarding the major interventions:

– In 1970–71 strengthening of the Right Gravity Dam by post-tensioning cables and grouting of cracks was performed. HQ previously reported that longitudinal expansion increased after the interventions at the Right Gravity Dam causing more cracking of the walls of units A, B and 1. The expansion at the middle of the dam increased.

– It was also reported previously that the slot cut that was made by overlapping 4-inch diameter drill holes between the approximately 115 m-long concrete Right Gravity Dam and the SW corner of the powerhouse in 1972 to 1975 to relieve the expansive thrust on the powerhouse structure slot closed suddenly and shook the powerhouse and that stay vanes in Units A+B were damaged and the gates experienced operational difficulties.

– It was also reported that when a 4-inch overlapping drill hole slot cut was made in 1981 between units 35 and 36 that this affected the alignment of the adjacent units and caused interruption of production of unit 36 and the gate for unit 37 jammed.

The plant continues to operate effectively with ongoing monitoring and maintenance.

22. ASR Case: R.H. Saunders (Gravity Dam, PH), Ontario Power Generation, Canada [Ho, Eastman, Adeghi, USCOLD 1995; Eastman, Ho, Adeghe, HydroVision 1998]

The R.H. Saunders Generating Station is a 16 unit, 912 MW hydroelectric project spanning the St Lawrence River and was constructed in the period 1954-58.

Equipment operating problems were first detected in 1972 when one of the generating units tripped due to distorted generator rotor-stator air gaps. Other issues include: deformation of throat ring clearances and turbine runners scrubbing the throat rings; misalignment of generator-turbine components, cracking of support beams; opening of construction joints in the water passages and generator deck; and cracking and dislocation of governor supporting walls.

A programme of slot cutting was implemented starting in 1993 to relieve stresses and stabilize the oval shaped deformations of the throat ring and air gaps. The slots were designed to also provide accommodation for future slot cuts.

OPG reported in 1995 that three 15 mm slots were successfully established in 1993 and 1994 and the equipment and structural responses were favourable. Concrete slot cutting was viewed as a better alternative to the conventional grinding of throat rings, the modification of equipment or eventual throat replacement. Consequently, slots were cut between all other units in the period 1995 and 1996. A total of 16 slots were cut in the powerhouse concrete. The slots provided stress reduction in the concrete around the throat rings, reduced and stabilized the oval-shaped throat ring and rotating equipment openings. Slot cuts between units improved runner and generator airgap clearances and provided allowance for future concrete growth.

From 2012 to 2014, a detailed condition assessment and evaluation of the generating units was carried out. Clearance reduction, discharge ring deformation (ovalling) and further structural damage due to continued concrete growth have been reported, especially in the past couple of years. Slot monitoring instruments indicate 9 out of the 16 slots have closed completely and others are about 90% closed. Longitudinal compressive stresses have been re-established, and discharge ring have deformed.

Based on the condition assessment, the OPG plans to recut the slots, starting 2017, as part of the ASR maintenance program to ensure reliable operation for the future.

23. ASR Case Otto Holden (Powerhouse and Spillway) [Khoral, Hafez, Zhao, Hong, HydroVision 2016]

Otto Holden Generating Station is an 8-unit 243 MW Powerplant located on the Ottawa River in Ontario, Canada and completed in 1952. The East and West Gravity Dams, Powerhouse, Sluice Gates and Log Sluices have been affected by ASR. Fly ash was used as partial replacement for Portland cement in the mass concrete that was used in the gravity sections. The structural instrumentation measures concrete growth rates in the range of 15 to 50 micro-strain per year.

Since the early 1970s, the station has experienced a number of structural and operational problems due to mass concrete deformation. Some of the major problems included: reduction of runner clearances to the extent that several runners were scrubbing the headcover and bottom ring seals; misalignment of the turbine/generator assembly; seizing of the wicket gates in their bushings; misalignment of the crane rails; severe cracking of carious structural components in the powerhouse; and high tensile stresses in stay vanes.

Unlike the sister plant, R.H. Saunders GS, where slot cutting was needed as one of the measures to manage AAR effects on the rotating equipment, slots were not needed at Otto Holden because of the absence of significant compressive stresses in the concrete structure. There is also little evidence of significant ovaling taking place around the rotating equipment. As a result of AAR-induced concrete movements and accompanied unit misalignment, a program of overhauls and realignment was performed from 2000 to 2008. During the overhauls, the units were modified to reduce the effects of AAR and to facilitate future realignments. A key component of the program is the modification of pivot ring so that it can be adjusted to accommodate induced movements.

In 2014, a detailed condition assessment was carried out to evaluate the current condition of the units. As a result of the condition assessment, a decision was made to overhaul the units to mitigate the unit alignment problems and extend the operating life of the generating station.

The planned overhauls will include disconnection of the top speed ring from the concrete embedded anchor bars and the pit liner so that the offset between the top and bottom speed rings can be reduced. This will improve unit alignments and reduce the frequency of unit overhauls.

24. ASR Case: Fontana (Gravity Dam & Spillway) TVA, USA [Dodson & Curtis, ICOLD 2013]

Fontana dam was completed in the early 1940s and is subject to ongoing ASR. By 1970 a significant crack had developed in the curved section of the dam. TVA drilled overlapping boreholes to develop a wide slot at that location in the dam to destress the cracked area and limit further development of the cracks. This slot closed fairly soon afterwards. In the 1990s TVA installed vertical post-tensioned anchors to improve lift joint shear strength for stability, and installed a new wide slot using the diamond wire saw cutting technique. This slot cut is being used to assist managing the ASR induced safety impacts on the dam. The build-up of loads in the anchors is monitored using load cells. The anchors are destressed periodically as necessary although TVA reported in 2013 that one of the anchors had failed due to overstress. Monitoring has subsequently been increased. The dam continues to perform safely.

25. ASR Case: Hiwassee (Gravity Dam & Spillway) TVA, USA [Dodson & Curtis, ICOLD 2013]

The dam consists of a 307-foot high gravity type concrete structure, a central spillway with seven radial gated bays, four sluices, and a 2-unit hydroelectric plant completed in 1940.

Evidence of growth discovered as early as 1939, strain meters in test cylinders indicated growth shortly after the concrete hardened. Spillway gate binding in the 1960s from mass concrete pushing into spillway. Significant cracking was found in the late 1970s in the curved sections near both abutments. Initial investigations into alkali-silica reactivity were made in 1978.

Greater than 35 mm permanent vertical expansion and more than 6 mm upstream movement since 1940. High longitudinal stresses and significant structural cracking near the abutments. Spillway gate binding and gate anchorage problems. Vertical growth has caused unit misalignment problems in the powerhouse.

Slot cuts were installed at each end of the gravity section at the start of the curved sections and at each end of the spillway in 1993 and 1994. In 2004 anchors were installed to improve the stability of gate piers. In 2006 it was confirmed that there was an adequate amount of potentially reactive material to sustain future expansion. The recalibration of the finite element model in 2007 allowed a revision in the slot cutting plan with a reduce frequency of re-cuts. The finite element model is currently being used to assess the timing of future slot cutting at the dam (narrow slots at the spillway). The finite element model is also being used to assess the dam stability for updated PMF and seismic loading.

26. ACR Case: Chickamauga, (Lock and PH) TVA/USACE, USA [Niznik & Curtis, ICOLD 2013]

Chickamauga Lock and Dam was commissioned in 1940. The lock is owned by TVA and operated by USACE. The lock chamber 18 m x 110 m with normal lift of 15 m. The evolution of ASR issues included surface cracking developed in 1943 in the lock junction portion of the lock. Extensive pattern cracking was observed in 1955. Concrete growth was identified at the powerhouse and significant cracks were found in the upper river approach wall and supporting piers in 1964/5.

Interventions from 1965 to 2012 included: installation of vertical multi-strand post tension anchors in all the chamber blocks, the upper sill, and upper and lower approach blocks along with horizontal post tensioned bars at various locations; slot cutting of upper river approach walls, post tensioning with a combination of multi-strand post tension anchors and bars; replace lower miter gate anchorages; shear pins to reinforce discharge ports in lower river approach wall, cut slot and reinforce discharge ports; chemical and cement grout used to reduce leakage; structural modifications to lower gates/anchorage connections, replace quoin blocks and gate seals; and rework of floating mooring bit tracks.

Future activities include: lock dewatering and inspections every 3 years; annual diving inspections and alignment surveys; continuous monitoring of instrumentation; lock operations personnel training to identify any changes and instructed to immediately report it to the lockmaster; update of finite-element analyses every 3 years; and pursue funding to complete construction of new lock and decommission existing lock.

27. ASR Case: Terry Lock and Dam, (Lock & Spillway) USACE-LRD, USA [Charlwood, Rutherford, Curtis & Winters, USSD, 2005]

The David D. Terry Project in Little Rock, Arkansas, is part of the McClellan-Kerr Arkansas River Navigation System and is owned and operated by the US Army Corps of Engineers. The project was completed in 1969 and consists of a 33 m wide by 183 m long navigation lock and gated spillway with eighteen 18 m wide by 8 m high radial gates. The concrete piers at the spillway are subject to alkali-silica reactions (ASR) and related expansion and some are subject to severe cracking which has affected their stability.

Testing and analysis has confirmed the hypothesis that the cracking is caused by differential expansion between pier concrete lifts constructed with 35 MPa and 20 MPa concretes. Rates of expansion have been found to vary between piers.

Repairs were completed in 2004 and 2005 for the six most seriously affected piers. These repairs included grouting and installation of inclined anchors and borehole extensometers to monitor future expansion. The condition of the concrete appears acceptable, with inferred compressive strengths above the specification values. The additional stress in the post-tensioned trunnion anchors appeared to be small at the time of the repairs. The extent of expansion of the 35 MPa concrete and the associated strain build-up in the anchors is to be re-assessed periodically.

28. ASR Case: Stewart Mountain (Arch Dam) USBR, USA [Fiedler, ICOLD 2013]

Stewart Mountain Dam is a concrete thin arch dam located in Arizona. The dam is 380 m long, 63 m high, and was built between 1928 and 1930. The dam includes a 13,000 kilowatt (kW) hydroelectric generating unit.

Evidence of concrete deterioration was first noted in 1935 as cracks, expansions and deflections; ASR was confirmed in 1943. Major dam safety modifications were initiated in 1980s mostly driven by new loads. AAR was not specifically addressed in these modifications.

Even with evidence of AAR reaction in most core specimens, overall the compressive strength (ave of 36 MPa), modulus of elasticity and Poisson's ratio have not been reduced substantially since 1946. Movement of the dam since 1968 has been less than potential indicated by lab measurements of expansions due to AAR. Potential for continued AAR still exists but is of low magnitude; 3 of 7 cores from 1977 exhibited expansions in 100% relative humidity.

Despite potentially reactive aggregate and relative humidity favourable to expansion; movements of the dam have essentially stabilized. It is concluded that production of reaction product has essentially stopped. Recent analyses found no significant issues identified as a result of ASR, the concrete in the interior of the dam is very strong (37 MPa).

ASR at Stewart Mountain Dam resulted in significant expansions of the concrete arch but strength was not significantly affected and the reaction has stopped. Modifications to dam were required for other structural issues not related directly to

ASR (unbonded lift lines were largely attributed to construction practice). Movement between dam and powerplant resulted in replacement of steel penstock and addition of expansion coupling.

29. ASR Case: Seminoe (Arch Dam) USBR, USA [FIEDLER, ICOLD 2013]

Seminoe Dam is a concrete arch dam located on the North Platte River in Wyoming. The dam was completed in 1939 with a structural height of 90 m and a crest length of 162 m. The dam is experiencing concrete expansion, cracking and deterioration due to alkali-silica reaction (ASR) and freeze-thaw damage. The mass concrete did not include entrained air and total air content is about 1% by volume.

Seminoe Dam did not exhibit the deleterious effects of ASR until after 1950. Deterioration at the top of the dam was initially attributed to freeze-thaw damage. It is likely that a slow growth from ASR was causing surficial cracking of the concrete, allowing freeze thaw damage to mask the apparent cause from internal expansion. ASR attributed to alkali-reactive quartzite was not identified until the 1970s. Compressive strength of the upper concrete has deteriorated significantly.

Instrumentation measurements indicate the crest of the dam has deflected horizontally upstream about 18 cm and vertically upward about 5 cm since 1979. Deformation causes an upstream flexure of the cantilevers of the dam and results in horizontal and diagonal cracks on the downstream face of the dam.

Repairs to reduce further damage from freeze-thaw included: resurfacing of the dam roadway; epoxy sealing of the curb/gutters/sidewalk and the top of the parapet walls; enlarging selected expansion joints through the downstream parapet wall and replacement of the concrete in the top of the outlet and penstock intake structures.

Seismic Tomography tests were performed in 2000 at three cross-sections of the dam and measured P-wave velocities in the concrete. Plots show lowest values near the crest, on the upstream face above the reservoir level and along the downstream face.

A 2004 finite element model study showed reasonable comparison to stresses and deformations measured in upper portion of dam in 2003. ASR expansions in upper portion of arch results in significant increase of load into abutments which could result in block movement along joints.

The concrete damage in the upper portion of the dam is caused by a combination of ASR, freeze-thaw damage and a lack of confinement. The concrete is continuing to deteriorate and there are no indications that the situation will stabilize. Remedial action will likely be required in the near future. Additional concrete testing is being performed and will be the basis for revisiting risks and a dam safety decision.

30. ASR Case: Friant (Gravity Dam & Spillway) USBR, USA [Fiedler, ICOLD 2013]

Friant Dam is a concrete gravity dam on the San Joaquin River in central California. It was built between 1937 and 1942. Friant dam was Reclamation's first extensive use of pozzolan (pumicite) to reduce cement content and heat generation within concrete; mass concrete contained 20% pumicate by weight of cement. Within several years after construction, concrete in portions of the dam developed abnormal cracking. Examination of cores indicated that alkali-silica reactivity involving andesite particles had developed. A separate structural mix was used for the spillway concrete that didn't include pozzolan and experienced significant ASR reaction. Worst impacts of ASR were at the dam crest parapets and curbs; blocks 367 and 42 adjacent to the spillway; the spillway piers, walls and gallery; and the river outlet works valve deck at the control house.

ASR appears to be most severe in concrete which contains high-alkali cement, does not contain pozzolan, and there is less confinement or restraint. These conditions are met for the concrete above elevation 170 m across the dam and structural concrete associated with the spillway. However, no stability issues were identified from 2004 analyses.

The spillway was originally regulated by three drum gates, which were separated laterally by concrete piers. As a result of expansion in the dam concrete, primarily at the outside ends of the two outside gates, spillway gates were replaced in 1997-98. The crest gates were designed to accommodate up to 15 cm of inward movement of the adjacent blocks and adjustable end seals were provided.

ASR damage and gate biding at Friant Dam is a function of concrete mix (whether or not pozzolan was used with high and low alkali cements) and lack of confinement at gate openings. Crest gates with adjustable seals have bought more time to deal with future expansions.

31. ACR/ASR Case: Center Hill (Gravity Dam & Spillway) USACE, USA [Hull, ICOLD 2007; Henderson 2013]

Center Hill Dam was constructed by the U.S. Army Corps of Engineers in 1948. The lake has a dual purpose: electricity production and flood control. The structure is 79 m high, and composed of concrete and earth structures, with 8 spillway gates that are 15 m wide each. Center Hill Lake is one of four major flood control reservoirs for the Cumberland River.

From 1951 to the early 1970s no problems were noted. In 1970 survey monuments were installed. In 1974 bridge expansion joints began closing, bridge support rockers began tilting and surface cracking was noted. In the 1980s, bridge joints at end spans were completely closed, more cracks appeared, and surveys show dam is increasing in elevation. In 1983 gate seals on end Tainter gates bind during operation. Multiple spillway openings were smaller than design dimension by over 2.5 cm. Petrographic examinations in 1983 indicated the potential for Alkali Carbonate Reaction due to coarse aggregates. Overcoring stress analysis showed stresses in the longitudinal direction ranged from 0.7 MPa to 5 MPa. Compressive strength was 25 MPa and tensile strength was 2.75 MPa. Petrographic Evaluation in 2001 could not confirm ACR, but found ASR gel in most of the cores.

15 mm slots were cut in the non-overflow gravity section at each end of the spillway in 2006 and 2007. Finite element stress analysis showed a 40% reduction in the stress concentrations in the end spillway openings as a result of slot cutting. The slots closed in 2007, approximately 1 year after original cut and were recut in 2009. In 2013 the slots remained open.

Future plans include: continue monitoring; maintain open monolith slot cuts; and update finite element analysis and contributing data.

32. ASR Case: Roanoke Rapids (Gravity Dam & Spillway) Dominion Energy, USA [Reinicker & Cima, ICOLD 2013]

Roanoke Rapids Dam is a 930 m long, 22 m high concrete structure completed in 1955. It includes four distinct sections: a 177 m long curved south non-overflow section (SNOS); a powerhouse with four 26 MW generating units; a 323 m long gated section with 24 radial gates; and a 228 m long fixed crest overflow section.

In 1995-1996 gate binding occurred at the gates at the ends of the spillway. 6mm shaved at each gate to maintain clearances. In 1996-1999 powerhouse units were re-aligned and lift joint cracking was observed in the SNOS and the emergency spillway. In 2006-2008 studies confirmed ASR. In 2003 leakage was observed on the downstream face of the SNOS at the gallery entrance. An unusual feature of this case is that the SNOS is a curved gravity section with the upstream face being concave, that is, the opposite of an arch dam. The longitudinal ASR thrusts in the SNOS therefore caused a “bowing” action downstream which resulted in an underwater crack opening of 25 mm on the upstream face.

In 2010 the crack openings were grouted and mild steel dowels and re-stressable post-tensioned anchors were installed in the SNOS to maintain stability. Continued monitoring of tendon stresses and deformations, coupled with finite element modelling is being used to confirm stability as the ASR continues. The most recent periodic upstream crack underwater inspections confirmed no change in the grouted cracks. New overcoring stress data is being used to update the FEM stress calibration and re-evaluate stability. The possibility of slot cuts is being considered as a possible option but is not planned at this stage.

33. ASR Case: Santeetlah (Arch Dam) Brookfield Renewable Energy, USA [Mochrie, ICOLD 2013]

Santeetlah is a concrete gravity and arch dam located on the Cheoah River in North Carolina and constructed in 1927 and originally owned by ALCOA. The structure is 320 m long; and 61 m high with a remote powerhouse with 40.4 MW licensed capacity. The dam consists of a right non-overflow gravity section, an integral intake, a right thrust block spillway, an arch section, a left thrust block spillway and a left non-overflow gravity section.

The structure has long been recognized to have used high alkali cement and reactive aggregates which led to AAR. Cracking of surface concrete has been observed throughout its history with upstream displacement of arch and thrust blocks.

Interventions included: concrete added in 1928 to address stability and seepage: flood modifications in 1938 included the lowering the crest of arch and wing wall sections raised and concrete added to thrust blocks. In 1947-1950 expansion slots were cut. In 1968 post-tensioned anchors were added. In early 1990's the slots started to close with deformations at the left side contact. In 1999-2003: the right and left side slots were re-cut and the right-side slot was re-cut in 2012. Throughout this period the dam has been regularly inspected and judged to be safe to operate. In 2013 there was no sign of a reduction in the vertical expansion rate. Presently the AAR is not causing instability or operating problems.

2.4. *South America/Brazil*

34. ASR Case: Moxoto, (PH) CHESF, Brazil [Kuperman, ICOLD 2013]

The Apolônio Sales (Moxotó) Hydropowerplant, generates 440MW, was built between 1972 and 1977 and is part of Paulo Afonso Hydroelectric Complex. Since the early stage of the commercial exploitation the units presented an abnormal performance with progressive shifting and tilting of the turbine shaft. The first visual evidences of alkali-aggregate reaction were several cracks noticed on walls and slabs in 1979. The cutting of three expansion slots between the concrete blocks, performed in 1988–1992 period, improved the performance of the generating units for some time. Meanwhile, the concrete expansion cumulated stresses and strains in the turbine parts fixed in the concrete, such as the stay vanes and the discharge rings. To counteract these effects a rehabilitation process was implemented. The concrete swelling phenomenon at Moxoto has been and is still being managed through continued inspection of the structure, an extensive monitoring system and safety assessment studies, numerical modelling studies and remedial measures to mitigate the swelling effects. It is reported that there are signs that the swelling effect is decreasing.

35. ASR Case: Pedra, (Gravity Dam & Spillway) ANEEL, Brazil [Kuperman, ICOLD 2013]

The Pedra Hydropower plant is a 60m high gravity dam located in the Contas river, Bahia state, generates 48MW and was built between 1964 and 1968. In 1991 the presence of ASR was detected in the concrete structures. Besides presenting several places with map cracking, the concrete expansion has caused a lateral thrust on the end pillars of the spillway, transmitted by the adjacent blocks, resulting in operating problems with the left end gate. In 2009, to mitigate the effects of the expansion, four slots were cut with diamond-coated cables. A monitoring system composed of multiples rods extensometers, direct pendulums, tri-orthogonal joint meters and bench marks was installed to assess the safety of the structure and to inform whether other rehabilitation procedures will be necessary in the future.

36. ASR Case: Billings-Pedras, (Gravity Dam & Spillway) ELETROPAULO, Brazil [Kuperman, ICOLD 2013]

Billings-Pedras Dam dates from 1936. The concrete structure has a maximum height of 31 m and is 145 m long, comprising three water intakes for a powerhouse that was not built, a gravity dam 35 m long and a space left in the concrete for a lock, that was also not built. AAR was identified in 1992. Instrumentation was installed consisting in tri-orthogonal joint meters, convergence meters, rod extensometers, surface marks and three-dimensional finite-element analysis was performed. In 1996, one of the floodgates began to present operational difficulties, attributed to the deformation of the guides by the expansive reaction. This problem was corrected in 2004/2005 by replacing the guides in the 3 gates. In 2002 a part of the concrete structure was coated with mortar added with micro-silica, for testing. After 4 years, the mortar fell off at many points and the cracks in the original concrete tended to surface again. In 2006 rehabilitation of the structure was carried out by installing drainage at the foundation, adding weight to improve stability conditions, cracks were sealed with injection of polyurethane, and shotcrete over steel mesh was also used in parts of the structure. The rate of expansion due to AAR is variable from 8 to 35 micro-strain/year, depending on the part of the structure that is considered.

37. ASR Case: Furnas (PH & Spillway) FURNAS, Brazil [Kuperman, ICOLD 2013]

The Furnas Hydropower plant was built in the early 1960s and has an installed capacity of 1,216 MW. The first noted signs of alkali-aggregate reaction were detected in the mid-1970s, when the following events were detected: unleveled central wall blocks and adjacent blocks; cracking on powerhouse structures;

cracking on top of spillway pillars. The first measurable displacements occurred in 1976 when an average annual rate of expansion of 22 micro-strain/year was detected.

In 1980, tri-orthogonal meters have been installed on crest joints and on the dam drainage gallery. In 1992/1993, a dragging process was identified on shoe fixed parts of spillway sluice-gates. In 1995 a significant reduction was detected in the annual expansion rate, down to 8 micro-strain/year. During tests of turbine # 6, various equally spaced diameters along the circumference of the bottom stationary wearing ring were found to be out of round. Also, the underlying bottom ring onto which the wearing ring was bolted was also found to be out of round. Further investigation revealed that all 20 wicket-gate stems were found to be out of the vertical plumb line by various amounts. A series of interventions were required in order to restore the design geometrical and dimensional conditions. A monitoring system was installed and a 3D finite-element thermo-chemo-mechanical expansion model was applied in order to analyse how the ASR is affecting the structure.

3. Observed Effects of Chemical Expansion on Dams

The general historical experience indicates that the following are typical effects of expansion in dams and hydro-projects which should be considered when planning surveillance and monitoring programs and a management and remediation strategy:

– development of map cracking at the concrete surface;

– ongoing upstream or downstream and vertical deformations;

– development of structural or macro-cracks;

– movements causing interference with gate clearances and binding of gates;

– movements affecting equipment clearances and alignment;

– development of internal expansion stresses leading to movement and leakage at horizontal lift joints;

– longitudinal stresses causing spalling of crest, parapet walls and bridge decks at vertical joints;

– separation of embedded parts, gate guides, etc.;

– changes in abutment loading and deformations affecting stability.

Our review of the above case histories presented has identified the following structure specific issues that have been addressed in various ways:

– arch dams:

- upstream movements in an "m" shape at crest,

- diagonal cracks on downstream face parallel to dam/abutment contact,

  - structural cracks in galleries,
  - signs of separation or slip on horizontal lift joints,
  - excessive leakage affecting uplift pressures and leaching of joints,
  - changes to abutment thrusts possibly affecting stability;
- gravity dams:
  - upstream crest movements (usually upstream but not always),
  - cracks at structural discontinuities, changes in geometry,
  - structural cracks in galleries,
  - separation or slip on horizontal lift joints,
  - excessive leakage affecting uplift pressures and leaching of joints;
- buttress and ambursen dams:
  - cracking at structural connections of slabs affecting water tightness,
  - cracking and deterioration of buttresses with possible impacts on reinforcing steel;
- spillways:
  - lateral movement of end piers into the openings,
  - loss of gate clearances and binding of gates,
  - cracking of piers;
- intakes:
  - lateral movement of end piers into the openings,
  - loss of gate clearances and distortion of hoists,
  - diagonal cracking in water passage piers;
- powerhouses:
  - internal expansion causing ‘ovalling’ of unit surrounds affecting runner and generator air gap clearances, distortion or cracking of stay ring, wicket gates, liners, etc.,
  - lateral displacement of structure causing loss of shaft alignment,
  - differential vertical expansion causing head cover bolts misalignment,
  - stress buildup in penstock surround concrete causing thrusts on the structure and equipment,
  - unbalanced thrusts causing diagonal cracking of intake and draft tube piers,
  - interference with intake and draft tube gate clearances,

- distortion, separation or overstressing of embedded parts,
- leakage,
- distortion of superstructure frames and misalignment of crane rails.

4. An Assessment of the Effectiveness of Management Strategies and Interventions

Table 1 presents a matrix summary of our interpretation of the selected case histories versus the structure type and main management approaches and interventions. It also shows the main reaction type in each case. This is intended to reflect a sampling of frequently affected structures and approaches among the selected cases. The types of structures among the 37 cases considered, include 23 gravity dams, 15 arch dams, 2 buttress dams, 25 spillways, 3 intakes, 7 powerhouses and 2 locks.

The management strategies and interventions considered included: anchors, grouting, slot cuts, coatings including membranes, gate and equipment modifications, monitoring and finite element modelling, and decommissioning or replacement. The authors offer the following personal opinions on their effectiveness based on our interpretations of the referenced published information. (Refer to Table 1 for specific cases of each type of action and to the case summaries and references for specifics.)

4.1. *Anchors*

Both post-tensioned high tensile and unstressed mild steel anchors have been added in many cases to: improve the integrity and stability of internally cracked concrete gravity, arch and buttress dam sections, tighten weakened horizontal lift joints and foundation contacts to improve stability, and repair damaged spillway piers.

Anchors have been used in powerhouses to repair damaged floor beams, generator enclosure walls, tighten second stage concrete around scroll cases and seek to retain downstream sections.

In all the cases presented, anchors appear to have been effective in terms of improving integrity and stability. In cases where the anchors are in expanding concrete, there has been a related stress build up in the anchors. In some cases this could significantly overstress the anchors and needs to be monitored, particularly in post-tensioned anchors. Unstressed mild steel anchors or dowels may be able to handle the additional strain plastically but this needs to be examined on a case-by-case basis. There is one known case in a gravity dam where a vertical post-tensioned tendon failed. Load cells and destressing options have been provided in several cases. The concern regarding stress build up also applies to spillway radial gate trunnion anchors. Attempts to estimate these stresses have been made in some cases.

Case Number	Project name	Reaction Type	Structure Type							Management/Intervention						
			Gravity	Arch Dam	Buttress	Spillway	Intake	Powerhouse	Lock	Anchors	Grouting	Slot Cuts	Coating	Gate/Equip	Monitor/FEM	Replace
1	Kariba	ASR		■		■								■	■	
2	Cahora Bassa	ASR		■		■								■	■	
3	Nalubaale	ASR		■		■		■		■	■				■	
4	Kleinplaas	ASR	■			■									■	
5	Kouga	ASR		■		■									■	
6	Matala	ASR		■		■				■	■			■		
7	Chambon	ASR	■			■				■	■	■	■	■	■	
8	Temple-sur-Lot	ASR	■			■				■	■		■	■	■	
9	Bimont	DEF		■						■	■		■		■	
10	Poglia	ASR			■							■				
11	Piantelessio	ASR	■	■								■				
12	Pracana	ASR			■					■	■		■		■	
13	Alto Ciera	ASR		■												■
14	San Estaban	ISA	■	■		■					■	■	■		■	
15	Tavascan/T/G	ISA	■			■					■		■			
16	Isola	ASR	■	■											■	
17	Illsee	ASR	■	■						■	■	■	■		■	
18	Salanfe	ASR	■							■	■	■			■	
19	Stolsvatn	ASR	■	■						■	■		■			■
20	Mactaquac	ASR	■			■	■	■		■	■	■	■	■	■	
21	Beauharnois	ASR	■				■	■		■	■	■			■	
22	Saunders	ASR	■			■	■					■			■	
23	Otto Holden	ASR	■			■		■							■	
24	Fontana	ASR	■			■				■	■	■			■	
25	Hiwassee	ASR	■			■						■		■	■	
26	Chickamauga	ACR	■			■		■	■	■	■	■		■	■	
27	Terry	ASR				■			■	■	■				■	
28	Stewart Mtn	ASR		■						■	■				■	
29	Seminoe	ASR		■							■		■		■	
30	Friant	ASR	■			■								■	■	
31	Center Hill	ACR	■			■						■			■	
32	Roanoke Rapid	ASR	■			■				■	■				■	
33	Santeetlah	ASR		■		■				■	■	■		■	■	
34	Moxoto	ASR	■			■		■		■	■	■		■	■	
35	Pedra	ASR	■			■						■		■	■	
36	Billings-Pedras	ASR	■			■		■				■		■	■	
37	Furnas	ASR				■								■	■	
	Total no. of cases		23	15	2	25	3	7	2	18	21	17	10	14	31	2

Table 1. *Summary of Dam Type and Interventions*

Although in theory, the compressive stress applied by anchors will reduce the rate of expansion parallel to the anchors, aside from one quoted case in a powerhouse which warrants further investigation, we have not seen any reports which show that the reduction in expansion rate by the use of anchors provides a significant long term expansion reduction benefit.

4.2. *Grouting*

Cementitious and epoxy grouting has been employed in many cases to help manage the effects of expansion by improving the structural integrity of cracked concrete and filling opened lift joints. It has been recognized that, in cases where the expansion is continuing, periodic regrouting may be necessary.

In addition, at some high-altitude projects, grouting of surface cracking has been effective in limiting the progression of freeze-thaw damage.

4.3. *Slot Cuts*

The creation of expansion joints by cutting slots has been utilized to accommodate expansion in dams, spillways and powerhouses for many years. In dams, slot cuts are usually used to help reduce stresses at structural discontinuities, changes in alignment or interfaces between arch and gravity section. The first case we are aware of was in 1947 when slots were cut in a gravity-arch dam using overlapping drill holes and which effectively reduced the stress concentration at structure interfaces.

Overlapping drill hole slot cuts were used in a few dams and powerhouses in the 1960s and 1970s. However, these slots were wide and the opening was created rapidly and, in some cases, the associated displacements have been reported to have had negative impacts on nearby concrete and equipment. The use of twisted steel wires with silicon-carbide grit was tried in one project in the late 1980s with limited success. In order to develop a more controlled approach, the use of thin (10 mm diameter) diamond wire saw cuts was introduced in 1988 in a large gravity intake. The first slot was cut slowly to allow continuous monitoring and closed after the initial cut due to the pre-existing longitudinal compressive stresses. The slot was recut to yield an open expansion joint which was the objective at that time. Subsequently, thicker (13 and 15 mm diameter) diamond wire saws were used to cut successfully at other locations. This technique has been applied subsequently at several other dams, spillways and powerhouses in the US and Europe, including an arch dam. Recutting the slots has been found to be necessary in cases where the expansion is continuing. The technique has evolved since 1988. In particular, it has been recognized that destressing the concrete invites an increase in the local expansion rate and a possible increase in the rate of deterioration of the concrete by allowing more micro-cracking. Recent projects have limited the recutting so that the slot remains closed to maintain some reduced compression in the expanding concrete and thereby maintain some restraint on subsequent expansion and reduce the frequency of recuts.

In some dam cases, several slots have been cut along the length of the dam to distribute the stress relief and minimize stress concentration buildup at the ends. In others, cuts have been localized to the area of stress concentration. Both approaches appear to be working. The need for distributed slots may increase with higher accumulated expansion strains.

Gate binding in some spillways has been addressed by cutting slots immediately adjacent to the openings. Again, the need for multiple slots to distribute the relief will depend on the magnitude of the accumulated expansion strain.

Expansion in powerhouses in the secondary concrete around the units has been shown to have caused "ovalling" of the air gaps and throat rings. The initial treatment in several cases has been to adjust stator supports and grind the throat ring steel. In cases where this option has been exhausted, thin slot cuts have been made between the units and these have partially restored clearances. This relaxation may have partially opened concrete joints with some leakage, but this appeared preferable to excavating and rebuilding the entire unit.

Wide slots have been cut in some powerhouses to separate diverging floor slabs and beams and have allowed alternative supports to be inserted.

4.4. *Coatings and Membranes*

Epoxy coatings to attempt to prevent additional water penetration has been attempted in a few cases. This has not been successful as the epoxy cracks with ongoing expansion. It may have some short-term benefit in limiting freeze-thaw penetration.

PVC membranes, and one with a resin laminate, with drainage curtains have been used on the upstream face of several large dams. These have been successful in limiting water pressure in cracks and lift joints, thereby improving stability. The membranes also limit leakage and associated leaching on joints. Of the cases considered herein, only one appears to show any success in controlling the reaction. That case is Pracana which is an extraordinary case for many reasons where the reservoir was dewatered for 12 years while extensive rehabilitations took place and it appears that the expansive reaction has been significantly attenuated as a result. It also appears that since refilling the reservoir, the expansive reactions have not so far restarted significantly in the concrete structures, quite probably due to the presence of the upstream membrane.

4.5. *Gate and Equipment Adjustments*

Spillway gate binding has occurred at a number of cases due to pier deformations into the spillway openings. The initial approach in most cases has been to shave the gates to increase clearances. When that has been exhausted, the seals and/or roller bearings have been modified. In most cases, there have been designed to accommodate forecasts of future expansion. In one case, a new rubber gate was installed with the capacity to absorb future expansion movements.

Powerhouse superstructures and crane rails have been found to become deformed in several cases. These appear to have been effectively handled by structural modifications.

Modifications to generator equipment have also been made and are usually complex and site specific. These include turbine shaft re-alignments, modifications to stator supports, wicket gate adjustments.

4.6. *Monitoring and Finite Element Modelling*

Precise monitoring is playing a key role in most projects, firstly to provide reliable data to confirm the existing behavior, and when interventions are made to monitor the response. In addition, the data are used to calibrate finite element models to help understand the existing behavior and condition, and then to forecast future behavior and, when appropriate, to simulate possible remedial strategies and interventions.

In a number of cases this has been sufficient so far, no structural or equipment modifications have been implemented.

In most cases where significant interventions have been made, elaborate monitoring and modelling have been an integral part of the assessment and design process. Various special purpose finite element models, adapted to model the expansive non-linear time and stress material behavior have been used.

4.7. *Decommissioning or Replacement*

Decommissioning and replacement has occurred in a small number of cases. This includes two of the cases listed. To the authors' knowledge there are perhaps four or five others. In most cases these decisions have been precipitated by seriously declining material properties. In one it was based on economic analysis considering the cost of ongoing maintenance and repairs, as well as the indirect costs of maintaining public confidence.

In several projects, estimates of remaining economic service life are being made. These are very dependent of forecasts of future behavior including estimates of the future expansion and on damage accumulation.

5. Conclusions

There are fundamentally two possible approaches to managing existing reactions in dams and hydro projects:

– attempt to control the rate and duration of the expansive reactions; and

– options to live with the expansion and treat the consequences of the expansion to extend the useful service life for as long as possible.

Both approaches typically involve ongoing monitoring and management.

5.1. *Controlling the Reaction*

There are a number of instances where membranes or coatings have been added to the upstream face, with good benefits regarding leakage and uplift control, but we see no documented cases where this has measurably controlled the reaction and associated expansion rates, except for the Pracana case which is a special situation which is unlikely to be considered as a management option at other dams. In our opinion, due to the massive sizes of the structures and the low permeability of the concrete, if it can ever be achieved the time required to dry the interior concrete to a degree that would affect the reaction, probably is in many hundreds of years, rendering this option impractical. This is the case even without water ingress, the residual water in the interior concrete after hydration will usually be sufficient in most cases to drive the reaction for many years.

Similarly, post-tensioned anchors have been installed in several cases, with good benefits regarding stability and structural integrity, but we see no documented cases where they have reduced the expansion rates. Numerous laboratory and modelling efforts have shown that the expansion rate is reduced by compressive stress. However, for this to significantly reduce the rate of expansion, compressive stresses in excess of about 5 MPa are required to be present in the bulk of the structure. Except in smaller components, e.g. bridge beams or thin buttresses perhaps, the amount of post-tensioning steel required is impractical.

Although we see some limited success in concrete pavements, we see no successes in dams using chemical means, such as lithium salts (the effectiveness of which is dependent on reaction type) or carbon dioxide, due to difficulties in achieving extensive uniform penetration in massive structures.

In summary, despite the problems, the one approach to control ASR that may warrant further research may be to seek ways to accelerate the drying process by comprehensive sealing and long-term treatment.

5.2. *Living with the Expansion*

The objective in most cases has no doubt been to maximize the remaining economic service of the project. The rate, magnitude and duration of the residual expansion will have been a key factor in managing the remaining service life of a dam or hydro project and is very difficult to estimate in the cases where expansion is continuing. The earlier notion that with AAR the alkali source is the cement is now realized often to be only part of the long-term story; in many cases alkalis become available from certain aggregates or SCMs, or a recycling process with time. This will clearly affect the duration of the reactions and in many cases, could cause the reaction effectively to continue indefinitely.

A number of effective strategies to extend the life of projects with existing expansive reactions have been discussed. Only in cases where the reaction is ceasing or has ceased, can these be considered final solutions. In most of the cases herein, the reaction is continuing and the strategy has to be one of ongoing management. In some cases, no immediate interventions are planned, the strategy is to monitor the evolution of the phenomena and the condition and safety of the project. In other cases, significant interventions have been made and in many of these, repeated interventions will be required.

The use of anchors and grouting to help maintain the integrity and stability of the concrete structures has been found to be effective option in many cases.

Slot cutting has been effectively used in many cases in dams and powerhouses to control stresses, concrete cracking and equipment interferences. Current techniques, combined with comprehensive monitoring and modelling, allow this to be done in a controlled way to minimize undesirable side effects. Multiple slot recutting has been demonstrated to be feasible and effective although there are no doubt limits to the number of times this can be done before associated deformations in adjacent concrete and equipment become unacceptable.

The use of membranes to minimize leakage and deterioration of the concrete by leaching, and to control uplift pressures in cracked dams has also been effective. There are cases which show that such membranes can operate effectively for long periods and effectively replaced if and when necessary.

A wide range of spillway gate and powerhouse equipment adjustments and modifications have been developed and successfully implemented. A number of projects are in a monitoring stage, where they can be safety operated without major interventions at this time. These also rely on comprehensive monitoring and modelling to assist in understanding the current status and forecast future requirements.

In the case of sensitive structures where the expansive behavior will continue indefinitely, possibly due to supplementary alkali supplies from aggregates, SCMs or recycling, there may be limits to sustainable life extension.

6. Acknowledgements

The authors would like to thank the authors and presenters at the referenced CEA, USCOLD, SPANCOLD, ICOLD and other workshops, for the availability of their materials and trust that we have fairly reported on them. We also thank members of the ICOLD Committee on Concrete Dams, particularly Mario Berra, and Francesco Amberg, and members of the RILEM Committees on ASR for their cooperation.

However, we reiterate, that the opinions expressed in this paper are those of the authors and are not purported to represent those of the contributors or authors of the referenced publications or owners of the projects.

7. References

Amberg, F., Stucchi, R., Brizzo, N., "The effect of temperature on the development of the Alkali Aggregate Reaction at the Pian Telessio dam", *9th ICOLD European Club Symposium*, Venice, April 2013.

Bouayad A., "Reconstruction de l'évacuateur de crues de l'aménagement hydroélectrique de Matala, Aspects de la sécurité des barrages", *Africa 2017 – Water Storage and Hydro Deevelopment for Africa, International Journal for Hydropower and Dams*, Marrakech, Morocco, March 2017.

Bourdarot, E., Sellier, A., Multon, S., Grimal, E., "A review of continuum damage modelling for dam analysis", *European Journal of Environmental and Civil Engineering*, published Online on 05 Oct. 2011.

Brueckner, R., Ndugga, N., Meri, T.C., Mahsen, M., “Alkali-aggregate reaction at Nalubaale hydropower station, Uganda”, *Africa 2017 – Water Storage and Hydro Development for Africa, International Journal for Hydropower and Dams*, Marrakech, Morocco, March 2017.

Casagran, G., Pradolin L., Victor, J., “Dam Rehabilitation in the presence of Aar – The Matala Hep”, *84th ICOLD Annual Meeting of the International Commission Large Dams*, Johannesburg, South Africa, May 2016.

Charlwood, R.G., Rutherford, J.H., Feldman, D.C., Winters, L., Terry, D.D., “Lock and Dam ASR Repairs”, *USSD, Annual Meeting*, Salt Lake City, April 2005.

Chulliat, O., Grimal, E., Bourdarot, E., Boutet, J.-M., Taquet, B., “Q95-R15: Le Gonflement des Barrages en Béton. Apports des Recherches Scientifiques: Application au Barrage du Chambon et à son Confortement”, *80th ICOLD Annual Meeting of the International Commission Large Dams*, Kyoto, Japan, May 2012.

Eastman, K.T., Ho, M.S., Adeghe, L.N., “Innovative Techniques to Mitigate Concrete Expansion – R.H. Saunders GS”, *HydroVision Conference*, Reno, Nevada, 1998.

Gunleiksrud, O., E-CO Vannkraft, Personal Communication, 2008.

Hattingh, L.C., Oosthuizen, C., “Q95-R33: Unusual Behaviour of a Large Arch Dam in South Africa as a result of Swelling due to Chemical Reaction”, *80th ICOLD Annual Meeting of the International Commission Large Dams*, Kyoto, Japan, May 2012.

International Conference on Concrete Alkali-Aggregate Reactions in Hydroelectric Plants and Dams, October 1992, Fredericton, NB, Canada, Canadian Electrical Association (CEA) & the Canadian National Committee on Large Dams (CANCOLD).

International Workshop on Alkali Aggregate Reactions, Fontana, TN, USA, *81st Annual Meeting of the International Commission on Large Dams*, August 2013.

Khoral, P., Hafez, A., Zhao, D., Hong, E., “Operating Hydro Turbines with AAR Induced Growth in Concrete Dams – Otto Holden Generating Station”, *HydroVision International Conference*, July 2016.

Mazzà, G., Donghi, G., Marcello, A.C., “The rehabilitation of Poglia dam: the support of experimental investigations and numerical modelling in the design and during the work-in-progress stages”, *76th ICOLD Annual Meeting of the International Commission Large Dams*, Sofia (Bulgaria), June 2008.

Noret, C., Clave, O., Gurukumba K., Chibvura, C., “Kariba Dam, more than 50 years of operation and surveillance”, *Africa 2017 – Water Storage and Hydro Development for Africa, International Journal for Hydropower and Dams*, Marrakech, Morocco, March 2017.

Noret, C., Laliche, K., “Bimont Dam Case: Studies and investigations inside the dam body”, *International Workshop, Dam Swelling Concrete 2017*, Chambéry, Savoie, France, June 2017.

Second International Conference on Concrete Alkali-Aggregate Reactions in Hydroelectric Plants and Dams, October 1995, Chattanooga, TN, USA, United States Committee on Large Dams (USCOLD) now United States Society on Dams (USSD).

Special Workshop on Chemical Expansion Of Concrete In Dams & Hydro-Electric Projects, October 2007, Granada, Spain, ICOLD Committee on Concrete Dams, Spanish Committee on Concrete for Dams and The International Journal on Hydropower & Dams.

Tembe, I., Carvalho, E., Hattingh, L., Oosthuizen, C., "The importance of redundancy in the surveillance of aging dams – the Cahora Bassa experience", *International Symposium, 82nd ICOLD Annual Meeting of the International Commission Large Dams*, Bali, Indonesia, June 2014.

Swelling Dams in Switzerland

FRANCESCO AMBERG* — ROGER BREMEN* — PATRICE DROZ** — RAPHAËL LEROY*** — JOHANNES MAIER**** — BASTIAN OTTO*****

** Lombardi Engineering Ltd*
Via R. Simen 19, 6648 Minusio, Switzerland
francesco.amberg@lombardi.ch, roger.bremen@lombardi.ch

*** Stucky SA*
Rue du Lac 33, 1020 Renens, Switzerland
pdroz@stucky.ch

**** Alpiq SA*
Ch. de Mornex 10, 1001 Lausanne, Switzerland
raphael.leroy@alpiq.com

***** Swiss Federal Office of Energy SFOE*
Supervision of Dams, 3003 Bern, Switzerland
johannes.maier@bfe.admin.ch

****** Axpo Power AG, Hydroenergie*
Parkstrasse 23, 5401 Baden, Switzerland
bastian.otto@axpo.com

ABSTRACT. *The paper exposes the results of a study carried out by the ad hoc Working Group of the Swiss Committee on Dams aiming to investigate the current situation in Switzerland related with expansive phenomena in concrete dams. The behaviour of a dam affected by this type of chemical reaction is characterized by trends, with both horizontal and vertical non-reversible displacements. The presence of trends have therefore been identified, analysed and compared on various structures. Among 154 dams, a relevant number, roughly 50%, is concerned by the problem. The expansions are quite low with average rates up to 30 μm/m/year and until now repairing work have been performed only in few cases.*

KEYWORDS: *concrete dams, alkali-aggregate reaction, chemical expansion, monitoring, Swiss dams.*

1. Introduction

In 2013 the Swiss Committee on Dams created the Working Group "Alkali-Aggregate Reactions" with the aim to investigate the current situation on the Swiss dams in relation with this problem. One of the results provided by the Working Group is presented in this paper. It consists in the identification of concrete dams potentially concerned by expansive phenomena based on the analysis of their behaviour. In particular, the presence of a non-reversible drift, both in horizontal and in vertical directions, is identified and characterized. The members of the Working Group are listed as author of the present paper. The entire work is also exposed in a specific publication of the Swiss Committee of Dams.

The alkali-aggregate reaction is the most common chemical reaction that causes swelling of concrete. Dams are particularly sensitive to this type of phenomenon for the following reasons:

– the structures are permanently in contact with water, one of the required condition for the reaction;

– dams are large and massive structures, where small expansion might produce already relevant displacements;

– dams are not reinforced with steel and are therefore sensitive to cracking.

Chapter 2 of the present paper gives a synthesis of the Swiss dams studied, while Chapter 3 offer an overview the situation regarding the monitoring devices available in dams for the assessment of their behaviour pertinent to a swelling phenomenon. The behaviour of the Swiss dams is statistically exposed in Chapter 4 and the analysis of the observed tendencies is discussed in Chapter 5. The collected data have mainly been provided directly by the members of the Working Group, while the dam operators and owners have been contacted only in particular cases. The main conclusions are finally summarized in Chapter 6.

2. Swiss Dams

The database on the Swiss dams has been provided by the Federal Office of Energy, Section Dams, in charge of the dam safety in Switzerland. The database contains the name of each dam, the height, the reservoir volume, the dam class, the type as well as the year of construction. Switzerland has 154 concrete dams and 85 embankment dams under federal supervision. Small dams, basically with a height less than 10 m and a reservoir less than 500,000 m^3, are under the responsibility of the regional authorities. In the present paper only the 154 concrete dams under federal supervision are considered. According to their height H and the reservoir volume V, these are classified as follows:

– Class I: 74 dams (50%), H>40 m, or H>10 m and V>1 mio m^3;

– Class II: 35 dams (25%), H>25 m, or H>15 m and V>50,000 m^3, or H>10 m and V>0.1 mio m^3, or H>5 m and V>0.5 mio m^3;

– Class III: 35 dams (25%), smaller dams, in particular up to 25 m high but with small reservoir volumes.

In the database 11 dams, all gate-structure, are not classified. Figure 1 shows the type and height of the dams in function of the year of construction. The Swiss concrete dams are subdivided in:

– 50% gravity dams (77 dams);

– 35% arch dams (54 dams, Sera dam is considered twice, the original dam built in 1952 and the new one built in 2010 (Leroy *et al.*, 2010));

– 12% gate-structure dams (18 dams);

– 3% others (2 buttress, 2 multiple arches and 2 arch-gravity dams).

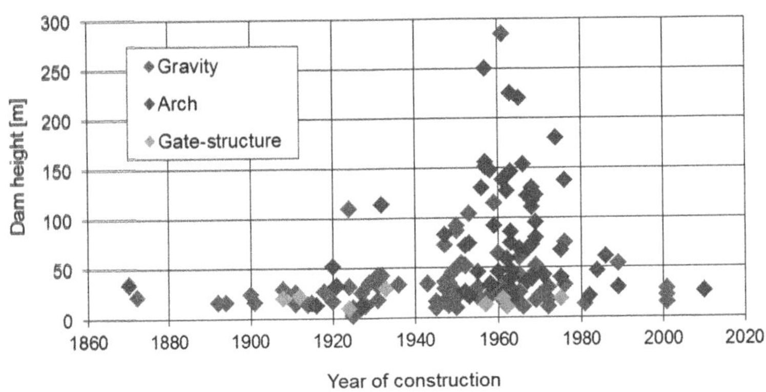

Figure 1. *Type and height of the dams in function of the year of construction*

3. Monitoring Devices

Dams are monitored structures. This is an essential characteristic allowing an early identification of any anomalous behaviour before a swelling phenomenon becomes clearly visible and eventually problematic for the dam safety. The information on the type of monitoring devices available for the assessment of the behaviour has a certain relevance and has therefore been provided by the members of the Working Group. Altogether, four categories of instrumentation have been considered, two for the horizontal displacements and two for the vertical ones:

– horizontal displacements:

- pendulum: as commonly used monitoring device, pendulum are very precise and have a sufficient reading frequency to measure the seasonal variations,

- others devices, typically geodetic measurements. geodetic nets are also widely used, but the reading frequencies are often not sufficient to measure the seasonal variations. in many dams the measure interval is 5 years;

– vertical displacements:

- levelling: levelling together with geodetic measurements is the most commonly used monitoring device to record the vertical displacements. the measurement frequency is typically not sufficient to determine accurately the seasonal variation,

- invar wires, extensometers and other devices: in some cases other instruments have been installed to monitor the vertical displacements within dams. these types of devices are rarely present since the beginning of the operation, in particular in older dams. therefore, the monitoring of the behaviour over the entire service live of a structure is generally missing.

The availability of the abovementioned four categories of monitoring devices for the different dam classes is presented in Figure 2.

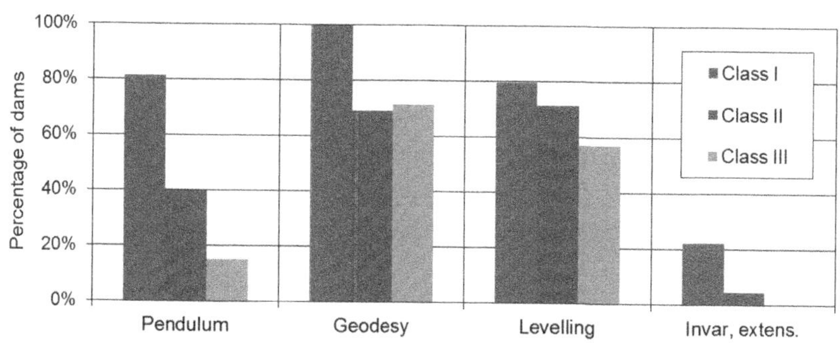

Figure 2. *Monitoring devices available in concrete dams (subdivision in dam classes)*

According to Figure 2, it can be stated that:

– dams of class I are relatively well monitored. 80% of the structures have pendulums and for all a geodetic survey is performed;

– the monitoring of smaller structures reduces progressively. Roughly 70% of the dams of classes II and III have a geodetic survey including levelling measures, but pendulums are less frequent. For example, in only 15% of class III dams pendulums are installed;

– invar wire or extensometer placed vertically within the dam body are quite unusual in Swiss dams.

The values illustrated in Figure 2 do not include all the 154 concrete dams in Switzerland. At the time of this publication the database is completed for around 100 dams (65–70% of the structures). This percentage does not change significantly with dam class.

Finally, it can be affirmed that, even if it does not result from Figure 2, gate-structures are less monitored than gravity or arch dams.

4. Behaviour of Swiss Concrete Dams

One main objective of the Working Group is the identification of Swiss concrete dams exhibiting a behaviour, which is characterized by the presence of non-reversible tendencies that are compatible with a concrete expansion. In arch dams an upstream drift and a progressive crest rising is expected, while in gravity dams the horizontal drift might also be in downstream direction. The first step of the study focusses therefore on the identification of the presence of such tendencies. Hereby, the following 5 answers were possible to the question if trends are observed:

– yes, the dam shows a long-term trend in horizontal or in vertical direction (even if said trend is not related to a concrete expansion);

– probably yes, but the readings are still insufficient or the trend is so small that it cannot be fully confirmed;

– situation unknown, basically due to a lack of monitoring;

– probably not, but the readings are not fully sufficient to confirm this conclusion;

– no, the dam shows a fully reversible behaviour without any trend. An eventual permanent displacement at an early age due to creep is not to be considered as trend in the current evaluation.

Figure 3 shows the situation for all 74 concrete dams of class I.

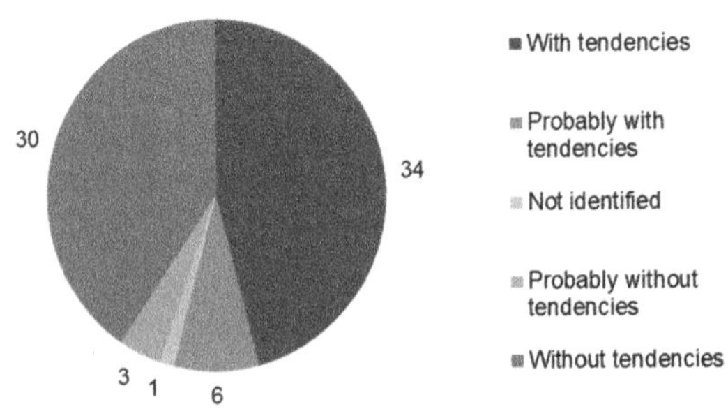

Figure 3. *Presence of trends in class I dams (the values around the pie chart indicate the number of dams)*

From Figure 3 it can be concluded that the knowledge on the behaviour of the relevant dams in Switzerland is satisfactory. Only one dam could not be clearly assessed and also the number of structures with uncertainties is limited. It appears that the number of dams affected by trends on the long term is quite relevant. 34 dams have a confirmed trend and 6 dams a probable one.

Figure 4 shows the summary for all concrete dams in Switzerland including also the dams of class II and III.

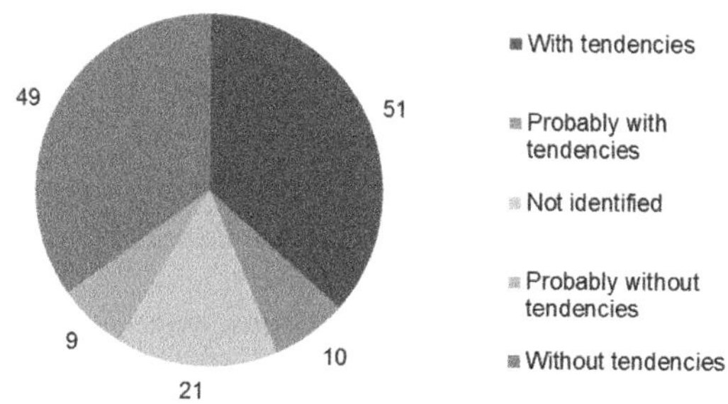

Figure 4. *Presence of trends in the Swiss concrete dams (90% of the dams) (the values around the pie chart indicates the number of dams)*

Compared with Figure 3, the number of dams with uncertainties increased. Figure 4 also considers only 90% of all the concrete dams, since the database is still not fully completed. The increase of dams with an unknown behaviour reflects the situation of an insufficient monitoring of the smaller structures. From Figure 4 it can be roughly summarized that around half of the concrete dams in Switzerland have a certain trend. In Total, up to 60 dams are more or less concerned by these phenomena.

The Working Group also verified an eventual effect of the construction period, concluding that the number of dams with trends remain more or less between 40 and 60% independently from the age.

A further statistical result is presented in Figure 5, where the presence of trends is exposed in function of the structure type. Gravity and arch dams are nearly equally affected by non-reversible phenomena, while the situation of gate-structure dams based on the analysis of their behaviour is less well identified. This is probably due to the complexity of gate-structures as well as due to a reduced monitoring available in this type of dam. This does not necessarily mean that gate-structures are less controlled than others dams. Gate-structures are generally complex structures with many galleries and passages, where the visual inspection might have a relevant role in the assessment of the dam behaviour. The presence of large gates can also provide relevant information for the assessment of the structural conditions. This type of information is however less accessible to the Working Group than just the simple analysis of trends performed for the others massive dams.

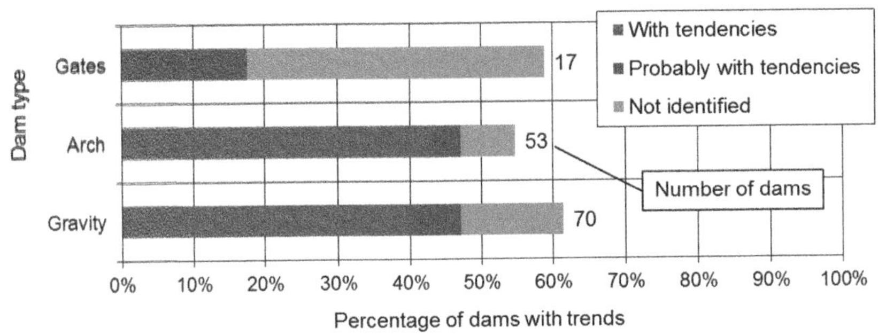

Figure 5. *Presence of trends as function of the dam type*

The distribution of dams showing certain trends in the Swiss territory together with the geology is illustrated in Figure 6. The higher reactivity in the Alps (Austroalpine, Penninic and Autocthon) is related mainly with deformed and strained quartz while the dams in Helvetic, Jura and Molasse contain more Limestones and Sandstones.

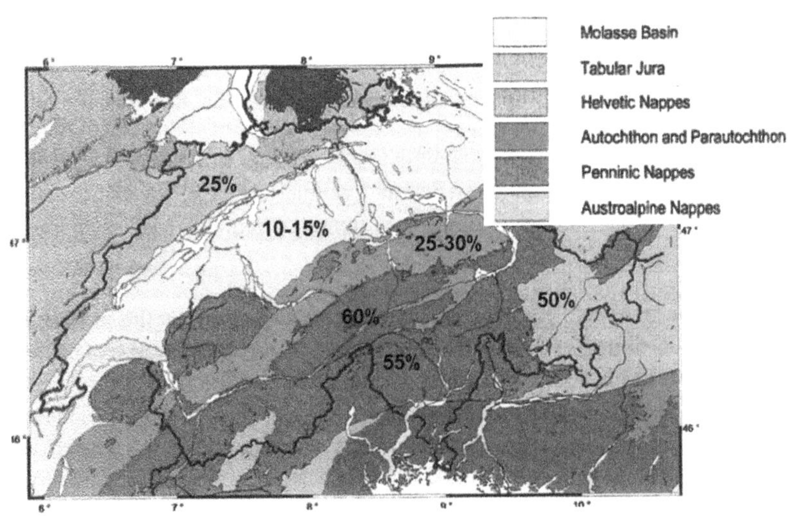

Figure 6. *Percentage of dams with trends in relation with the geological zones of Switzerland*

A last interesting analysis is related to the causes of the identified trends. Hereby, the Working Group considered the following 5 possibilities:

– concrete expansion due to chemical reaction already confirmed, eventually also by laboratory investigations;

– concrete expansion probable, i.e. not confirmed by laboratory investigations, but behaviour is compatible with a concrete expansion (crest rising, horizontal drift, cracks);

– cause unknown and to be identified;

– probably other cause: observed trend is less compatible with a concrete swelling, but the causes are still not clear;

– other cause: reason of observed trend is well known and is not related to a swelling reaction within the concrete.

Figure 7 shows the situation of the 61 dams with trends (confirmed or probable) previously exposed in Figure 4. Apart of few particular known cases, the large majority of the dams showing non-reversible trends are potentially affected by concrete swelling. In around 1/3 of the dams the cause is clearly related to this chemical phenomenon.

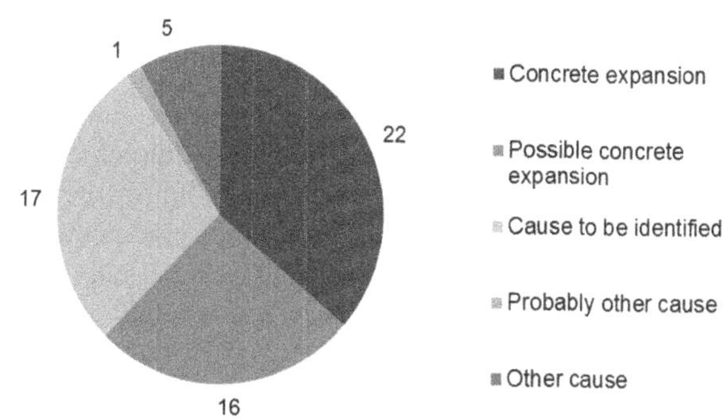

Figure 7. *The assessment of the causes for dams exhibiting trends*

5. Analysis of the Tendencies by Expanding Dams

5.1. *Initiation Time*

Section 5 aims to present the behaviour of a certain number of dams potentially concerned by a concrete expansion. Typically, dams affected by concrete expansion are characterized by an initial period of regular behaviour followed by a phase where horizontal drift and crest rising can be observed. The initial period of regular behaviour is defined as initiation time and is presented in Figure 8.

The inclined line corresponds to the current date at the time of data collection. It is therefore not possible to have points above this line. However, it cannot be excluded that dams today not showing any expansion will continue to behave regularly in the future. From the figure it can be clearly seen that in some cases the initiation time may be quite long, up to 50 or even 80 years. This observation is not unique, since similar results have already been published in the past (Charlwood and Solymar 1994). In other cases, on the other hand, the expansion started nearly immediately after construction.

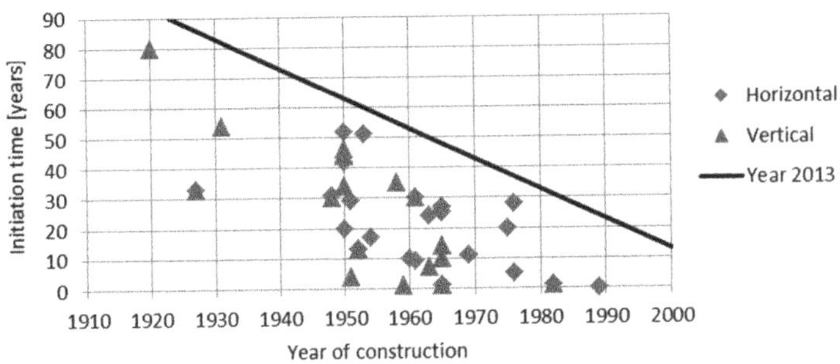

Figure 8. *Time of initiation of the observed concrete expansion in various dams represented in function of the year of construction*

Another interesting observation is that the initiation time for the vertical direction seems to be shorter than that for the horizontal direction. In other words, the expansion in vertical direction seems to occur slightly before the drift in horizontal direction. The average initiation time of a selected number of dams is 20 years in vertical direction and 26 years in horizontal. Here, only 11 dams with an initiation time defined for both direction are taken into account.

5.2. *Drift in Horizontal Direction*

A certain horizontal drift and permanent displacement does not have the same relevance for a small rigid gravity dam as it would have on a large and flexible arch dam. To compare the behaviour observed on different structures, it is therefore necessary to find other parameters. Ideally, one should determine the average expansion, but based on the measured displacements in horizontal direction this is not straightforward. For this an appropriate structural analysis should be at hand. Therefore, to relativize the observed behaviour with the dimension of the structure, it was decided to compare the permanent displacement with the annual variation induced by the regular operating conditions (water level and seasonal temperature variation). The permanent displacement is therefore divided by the typical annual variation of the displacement measured in the same point. The obtained parameter is illustrated in Figure 9. Arch and gravity dams are represented distinctly. A value of 1.0 means that the permanent displacement accumulated over a period of 10 years corresponds to the annual variation. A value of 2.0 means that the permanent displacement is the double of the annual variation.

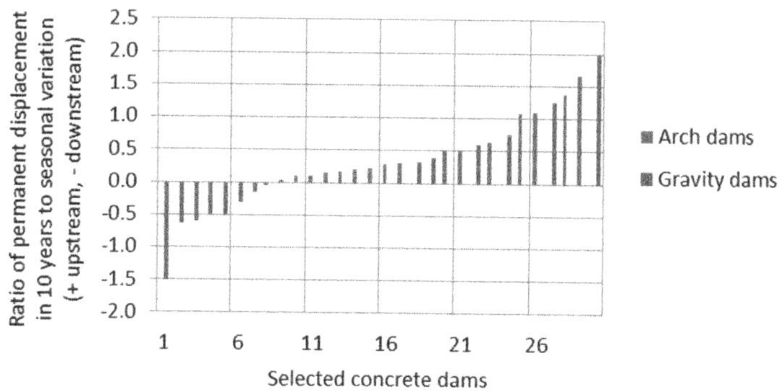

Figure 9. *Horizontal drift rate compared with the seasonal elastic variation (30 dams)*

Negative values indicate that the horizontal drift is directed in downstream direction. This might appear unusual for a dam with concrete expansion, but according to Figure 9 there are more than one gravity dams exhibiting a drift towards downstream. No arch dam shows a drift in downstream direction. The relevance of the temperature on the expansion development has already been commented and presented (Amberg, 2011 and 2012), showing that the horizontal drift of gravity dams is mainly determined by a differential expansion between downstream and upstream face. In the Alpine region the temperature distribution is largely influenced by sunshine, and therefore by the orientation of the dam. Figure 10 shows the rate of expansion in the gravity dams considered in Figure 9 in function of the orientation of each dam.

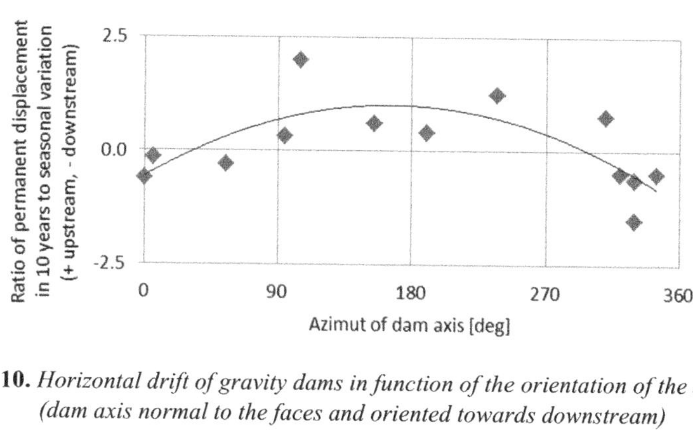

Figure 10. *Horizontal drift of gravity dams in function of the orientation of the dam axis (dam axis normal to the faces and oriented towards downstream)*

The result presented in Figure 10 seems to rather confirm the effect of the sunshine: gravity dams with the upstream face towards the south (azimuth of dam axis between -45°, i.e. 315°, and +45°) exhibit a permanent displacement towards downstream, while for gravity dams with the downstream face towards the south (azimuth of dam axis between 90° and 270°) a permanent displacement exclusively towards upstream is observed.

By the end of 2013, the total permanent displacement exceeds in some cases 8 times the regular seasonal variation. Repairing works have been undertaken in certain cases where the permanent displacements exceeded 5 times the seasonal variation. For lower displacements still no works have been performed.

5.3. *Vertical Displacements*

To assess the vertical displacement the levelling readings at crest elevation are generally taken into account. The non-reversible component of displacement is then divided by the dam height, to obtain an average expansion in vertical direction. The corresponding values for a selection of arch and gravity dams are illustrated in Figure 11.

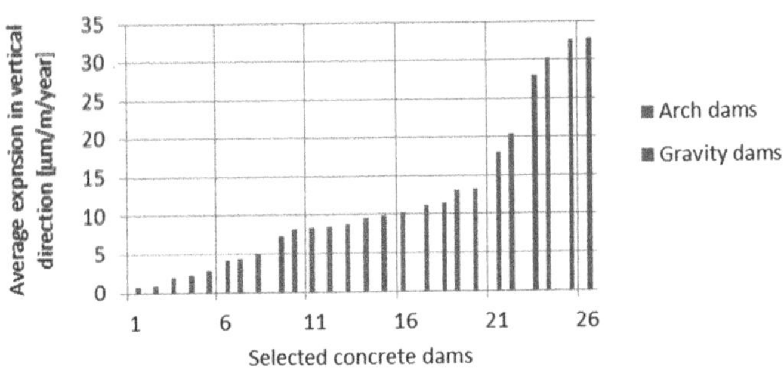

Figure 11. *Average vertical expansion rate over the whole dam height (26 dams)*

The expansion rate reaches at maximum 30-35 µm/m per year which can be considered as still moderate. At Mactaquac an expansion of up to 140 µm/m per year has been estimated (Hayward *et al.*, 1991). The values illustrated in Figure 11 are quite similar to the ones presented in a recent study performed by Electricité de France (Sausse and Fabre, 2013).

Figure 12 shows finally the total expansion in vertical direction developed until the end of 2013. Like before, it corresponds to the average value, which is determined dividing the permanent rising of the crest by the dam height. The two dams with values higher than 800 µm/m have been rehabilitated.

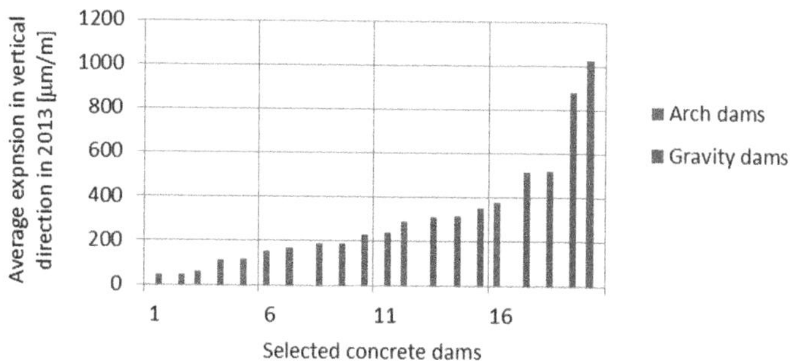

Figure 12. *Average vertical expansion over the whole dam height (20 dams)*

6. Conclusions

The paper exposes the analysis of the behaviour of 154 Swiss dams with the aim to identify the impact of the concrete expansion induced by chemical reactions on this type of structures. The results show that around 50% of these dams exhibit trends and permanent displacements generally compatible with a concrete expansion. In 22 dams the presence of a chemical reaction is confirmed by laboratory investigations. This result clearly confirms the relevance of the phenomenon.

The swelling dams are generally characterized by an initial period of operation with a regular and reversible behaviour. This initiation time might vary between 0 and many years, reaching up to 50 or even 80 years in some cases. After this initial period the effect of the concrete expansion becomes visible with crest liftings and horizontal drifts. Such tendencies are nearly linear with time. The mean expansion rate estimated in vertical direction varies between 1 and 30 µm/m/year. In horizontal direction the arch dams move toward upstream, while gravity dams can also move in downstream direction, depending on their orientation. In fact, gravity dams tend to drift towards north, possibly due to the greater expansion at the face exposed to south and thus to the sunshine.

Repairing works have been executed in few cases, such as Illsee, Sera and Salanfe. In these dams the total concrete expansion reached roughly 500 to 1,000 µm/m. In the future also repairing works on further dams have to be expected.

7. References

Amberg F., "Performance of Dams affected by Expanding Concrete", *Dams in Switzerland, Source for Worldwide Swiss Dam Engineering*, Baden-Dättwil, buag Grafisches Unternehmen AG, 2011, 309-314.

Amberg F., "A review of expanding concrete cases and consequences on dam performance", *Proceedings of the Hydro International Conference*, Bilbao, 2012.

Charlwood R.G., Solymar Z.V., "A review of alkali aggregate reactions in dams", *Dam Engineering*, Canada, Vol. 5, Issue 2, July 1994, 31-62.

Hayward, D.G., Thompson, G.A., Charlwood, R.G., Steele, R.R., "Remedial measures at the Mactaquac generating station", *17th ICOLD Congress on large dams*, Vienna, Q. 65 R. 47, 1991.

Leroy, R., Micoulet, G., Tognola, F., "Rehabilitation of Serra dam (Switzerland) affected by ASR", *Dam maintenance and rehabilitation, Proceedings 2nd International Conference*, Zaragoza, 2010.

Sausse, J., Fabre, J.P., "Diagnosis of Dams Affected by Swelling Reactions: Lesson Learned from 150 Monitored Concrete Dams in France", *Dam Engineering*, Vol. XXIII, Issue 1, 2013, 5-17.

CHAPTER 2

Physico-Chemical Mechanisms. Experimental Test

Behavior of Concrete Deteriorated by ASR under Triaxial Load

ADRIEN HILAIRE* — ALAIN B. GIORLA** — CYRIL DUNANT*** — LIONEL SOFIA* — KAREN SCRIVENER*

* *Laboratory of construction materials*
EPFL-STI-IMX, Lausanne 1015, Switzerland
adrien.hilaire@epfl.ch, lionel.sofia@epfl.ch, karen.scrivener@epfl.ch

** *Oak Ridge National Laboratory*
One Bethel Valley Road, Oak Ridge, TN 37831, USA
giorlaab@ornl.gov

*** *Engineering Department*
Trumpington Street, Cambridge CB2 1PZ, United Kingdom
cfd30@cam.ac.uk

ABSTRACT. *The ASR-induced expansion is the result of complex mechanical and chemical processes at the microscale. The effect of load on the expansion and degradation mechanism is significant. Despite several studies done on this topic, it is still an open question. In dams, a particular type of load is also present: pressurized water infiltrates the pore network of the cement paste and the micro-cracks. In this work, the influence of mixed load is studied. Concrete samples affected by ASR are submitted to a fluid pressure and an external via a triaxial setup. According to the results, the pressure of the fluid may increase the amplitude of the expansion.*

KEYWORDS: *triaxial test, pore pressure, damage, dam, ASR.*

1. Introduction

In Switzerland, hydroelectric dams produce a large part of the electricity. Most of these structures were built between 1950 and 1970. ASR affects an estimated 20% of them. Even if the degradation caused by ASR is a slow process, some structures are starting to require active monitoring and repairs. To establish an appropriate action plan for the maintenance, a reliable prognosis of the effects of the reaction on the dams is necessary.

Though this reaction was discovered 75 years ago (Stanton, 1940), ASR is still extensively studied and remains poorly understood. Several experimental studies have been published to analyse the effects of a mechanical load or restraint on the ASR induced expansion (Dunant, 2009) (Berra *et al.*, 2010) (Multon *et al.*, 2006) (Larive, 1997).

A common observation emerges from these works: under compression, the expansion is reduced (or even negated) in the direction of the load. This feature is positive from the structural engineering point of view, but the question of the possible redistribution of damage or load remains unsolved.

An experimental campaign was launched in the LMC laboratory by Giorla (Giorla, 2013). The initial purpose was to investigate the delayed mechanical behaviour of concrete subjected to various multi-axial stresses representative of the loads applied to a dam. An experimental device based on tri-axial cells used for rocks mechanics was developed.

However, the poromechanical behaviour of concrete was not considered during the design of this set-up, and the presented results show an important impact of this factor. The pore pressure of concrete rapidly reaches equilibrium with the confining pressure of the fluid. Consequently, the concrete samples were not only subjected to classical external loadings: the internal pore pressure also affects their behaviours.

Despite the unexpected presence of this internal loading, this experimental set-up is useful for the understanding of the behaviour of massive dams. This specific kind of load is also relevant; water is pressurised at the bottom of dams. Moreover, to our knowledge, the influence of the fluid pressure on the behaviour of concrete deteriorated by ASR has never been studied. The whole experimental set-up is described, and results are presented in this paper. Finally, the main post-mortem observations of the concrete samples are reported.

2. Materials and Methods

2.1. *Materials and Mix Designs*

The effects of applied stress and internal pressure on ASR induced expansion are investigated on two concrete mixes. The first is made with reactive aggregate, which were previously characterised (Ben Haha, 2006) (Dunant, 2009) (Chappex, 2012). The other is a reference mix with non-reactive aggregates. Both aggregates come from the Swiss Alps and have similar mechanical properties.

The mix design is comparable to the one used in the experimental campaign of Dunant (Dunant, 2009). The water/cement ratio and the cement content are respectively equal to 0.43 and 364 kg.m^{-3}. The aggregates were crushed and sieved to best approach Bolomey reference curve. Additional sodium hydroxide was dissolved in the mixing water; the Na_2O_{eq} content is equal to 0.9% of the mass of the cement. The detailed composition of the mixes is shown in Table 1.

Cement CEM I 42.5 R	364
Water	156
Aggregate (0.2-2.5 mm)	470
Aggregate (2.5-4 mm)	282
Aggregate (4-8 mm)	470
Aggregate (8-12.5 mm)	376
Aggregate (12.5-16 mm)	282
Sodium hydroxide	2
Super-plasticizer	4

Table 1. *Mix designs for the reactive and the non-reactive concretes*

2.2. *Experimental Program*

2.2.1. *Description of the Triaxial Devices*

Six cells are used to apply a tri-axial load: the concrete specimens are subjected to an axial force and to a fluid pressure. A hydraulic jack similar to the ones used for creep tests applies the axial stress. Concrete samples are not sealed; they come into direct contact with the pressurised fluid. The samples are thus kept saturated for the duration of the experiments. To minimise leaching (Chappex, 2016), an alkaline solution is used (0.320 mol/l NaOH). Three concrete cylinders are held in each cell: two are reactive (R1 and R2) and one is non-reactive (NR).

The expansion of the samples is monitored with fibre optics sensors. A Bragg network is carved on the fibre and measures the length. Two gauges are used; they record respectively the radial and the longitudinal strain. Dunant validated this method of measurement (Dunant, 2009). Special care was taken to keep the fibres sealed from contact with the water. A temperature sensor is also embedded in each sample. A schematic view of the setup is presented in Figure 1.

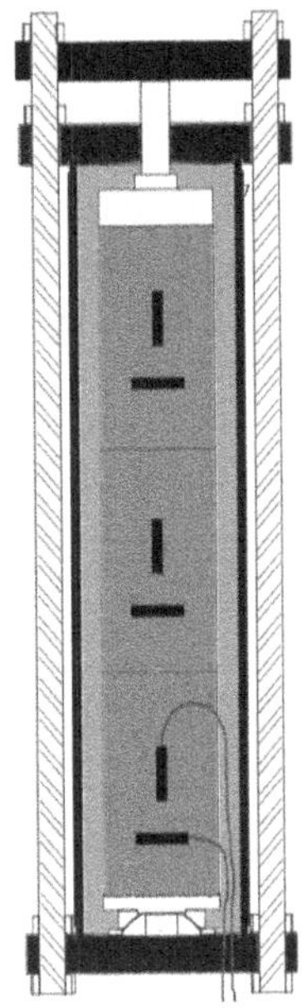

Figure 1. *Schematic view of the triaxial setup*

2.2.2. *Environmental and Mechanical Boundaries Conditions*

The cylindrical samples (diameter=16cm, height=32cm) were cast and stored in a fog room at 20°C for 52 days. After the cure, they were placed in the tri-axial cells, which were filled with the NaOH solution. The whole set-up was kept in a temperature-controlled room at 22°C for one week. Then, the temperature was increased to 38°C at a constant rate equal to 0.85°C/h. Unfortunately, ten months after the mechanical loading, the temperature-control system experienced some problems; consequently, the temperature was not well regulated during a three months' period.

The specimens were then loaded: they are first subjected to this isotropic pressure, then the axial pressure is applied using a jack. During the first days of loading, the pressure of the alkaline solution equilibrates with the internal pore pressure of the concrete. The fluid pressure was adjusted to counterbalance this pressure loss. Each cell is associated to one specific load level. The loading configurations are described in Table 2. The first column of Table 2 describes three loading cases; the axial load is relatively low (σ_a = 2MPa). Therefore, these configurations are mainly related to the pressurized alkaline solution; they are quasi-isotropic and characterize the effect of the internal pore pressure on the ASR induced expansion. With the three others configurations (2nd and 3rd columns), concrete is subjected to the combined loading of the pore pressure and the uniaxial external force.

		Axial Load, σ_a	
	2 MPa	6 MPa	11 MPa
P_{NaOH} 5 MPa	•	•	•
P_{NaOH} 10 MPa	•	•	
P_{NaOH} 15 MPa	•		

Table 2. *Loading configurations used in the tri-axial cells*

3. Results and Discussion

Figure 3 shows the evolution of the measured data. Each plot is associated to one loading condition. The longitudinal and the radial strain are plotted versus the equivalent time for the reactive samples (R1 and R2) and for the non-reactive one (NR). The equivalent time is calculated by assuming an Arrhenius relation between the expansion rate and the temperature (the activation energy is supposed equal to 50 $kJ.mol^{-1}$). An offset is applied to the raw data to not consider the strains observed at the moment of loading. Similarly, the values are corrected for thermal strains. The strain evolutions recorded when the room temperature was unregulated are not presented. The strain evolutions presented in this paper characterize the delayed behaviour of concrete deteriorated by ASR and subjected to pore pressure and uniaxial mechanical loads.

Under quasi-isotropic loading conditions, the non-reactive samples do not exhibit a delayed behavior. Non-deteriorated concrete only exhibits a significant creep if a significant external mechanical load is applied (σ_a = 6MPa or σ_a = 11MPa). From these results, if unjacketed tri-axial tests conditions are applied, concrete is characterized by a non-viscous behavior. This characteristic was also noticed for cement pastes (Ghabezloo *et al.*, 2008).

Free-expanding reference experiments were not performed during this experimental campaign. However, Dunant recorded the free-expansion of similar concrete specimens (same mix composition and same origin of the aggregates) immersed in a solution with the same alkaline content of the present study. After 250 days, the rate of expansion of these unloaded samples is largely slowed down (see Figure 2), the level of the final plateau is between 1.2 and 1.6 millistrain depending on the strain considered (radial or longitudinal).

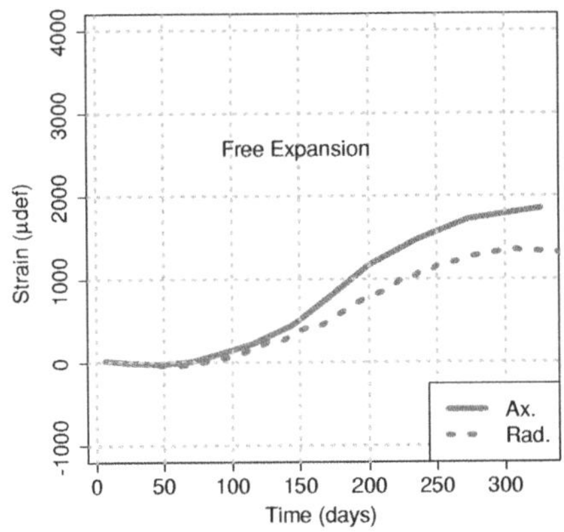

Figure 2. *Longitudinal and radial free expansion observed by Dunant*

Immersed in a pressurized fluid, the behavior of concrete affected by ASR is different:

– If the pressure, P_{NaOH}, of the alkaline solution is superior to 10MPa, higher expansion values are recorded for concrete specimens. The highest expansion is observed when the fluid pressure is equal to 10 MPa. The radial strain reaches 10 millistrain for the R1 sample. However, for P_{NaOH} = 5MPa, the recorded expansions are similar to the free-expansions measured by Dunant.

– After eleven months of immersion, the expansion slowly reduces. However, after the period during which the temperature was misregulated, the rate of expansion increases again. Even after one and a half year, the kinetic of the expansion is not notably decreased.

– As regards to the dispersion, two regimes emerge. Below 0.5 millistrain, the dispersion is quite low and typical of this kind of test. However, above this threshold the dispersion increases significantly.

Although the pressurized fluid does not cause any significant delayed strain in a non-deteriorated concrete, the ASR induced expansion is strongly influenced by this pressure. Under a uniaxial mechanical load, for similar mixes, Dunant showed that the expansion in the loading direction is significantly reduced compared to free-expansion. This observation is in accordance with previous experimental work (Multon, 2006) (Larive, 1997): the expansion decreases as the load increases. The present results lead to a similar conclusion. An external load induces a confinement effect; the expansion is reduced in the loading direction.

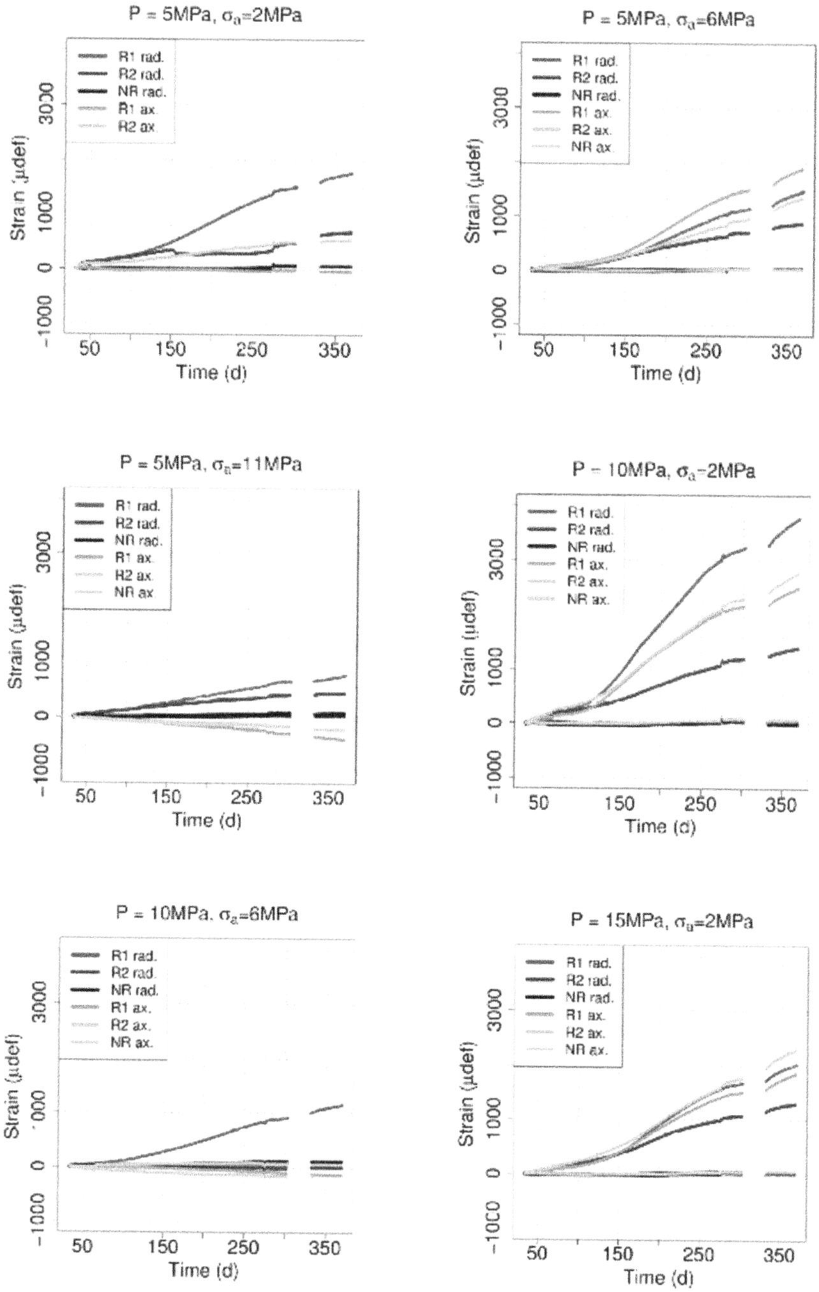

Figure 3. *Radial and axial delayed strain of concrete samples due to ASR and fluid pressure submitted to the different loading configurations*

The presented results seem paradoxical. On the one hand, non-reactive specimens do not exhibit any delayed behavior under a fluid pressure. On the other hand, the pressurized fluid largely impacts the expansion caused by the development of the ASR reaction. Even if the influence of multi-axial stress on ASR was not intensively analyzed in the past, the existing studies show that the mechanism through which the ASR is affected by the mechanical boundary conditions is essentially micro-mechanical. The chemical process and the production of the gel are not linked to the stress state applied to the concrete. Therefore, the increase of the ASR induced expansion induced by the fluid pressure has its explanation in the mechanical problem.

Micro-mechanical modelling has successfully been used to demonstrate the role played by external loads on the ASR [6][9]. In this approach, the expansion is related to the formation of cracks in the aggregates and in the cement paste. This cracking pattern is the consequence of the presence of ASR gel pockets in the aggregates. Then, the propagation of the damage occurs in two phases. First, damage is confined to the aggregates; second, cracks propagate in the cement paste. The viscoelastic behavior of cement paste links this latter cracking pattern to the kinetics of the reaction. The observations in this study are consistent with this hypothesis: problems linked to cracking and percolation present a high level of dispersion.

One factor may explain why the expansion rate of concrete affected by ASR is dependent on the pressure of the alkaline solution. The effect of the fluid pressure on the lips of the cracks can explain this additional swelling. This effect is expected to be particularly important when cracks propagate in the cement paste.

4. Post-Mortem Characterization of the Samples

4.1. *Cracks and Damage Features of Concrete Samples*

4.1.1. *SEM Analysis of the Microscopic Damage*

At the end of the triaxial tests, backscattered electron (BSE) images of the polished sections are acquired. Aggregates present some thin and dense cracking facies (see Figure 4); it is characteristic of alpine aggregates degraded by ASR, it corresponds to the reactive sites.

For some aggregates, this cracking pattern is accompanied by the presence of cracks propagating through the aggregate. These larger cracks can be partially filled with gel (see Figure 4).

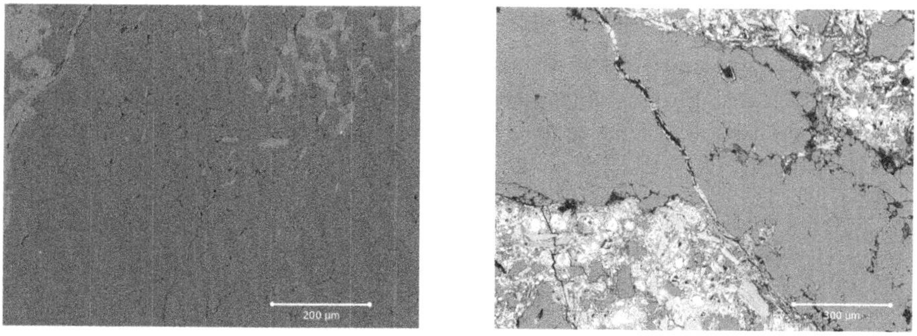

Figure 4. *Micro-cracking pattern induced by ASR in aggregates*

The picture's size presented on Figure 4 is too low to be representative of the mesoscale of concrete. To overcome this limitation, the polished sections of concrete are automatically scanned and a grid of micrographs is created. This process allows to us get a more representative image of the concrete at the mesoscale while maintaining a high resolution. An example is showed on Figure 5.

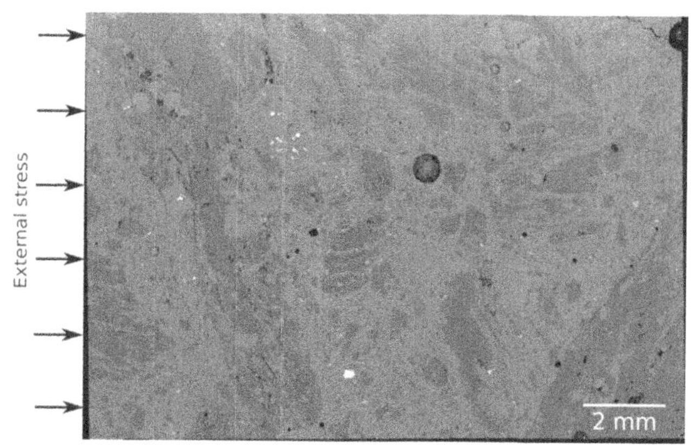

Figure 5. *Meso-scale crack pattern induced by ASR in aggregates*

The slice from the concrete sample submitted to the loading $\sigma = 6MPa$, $P_{NaOH} = 5MPa$ was cut along the axial direction to determine if the mechanical stress has an influence on the crack propagation. Large cracks percolate through the aggregates and the cement paste. The percolation and the propagation of the cracks seem to be mainly controlled by the mesostructure of the concrete. Cracks in the aggregates preferentially propagate according to the largest dimension of the aggregate. According to Figure 5, the load is not able to govern the cracking process, a large and long crack propagate perpendicularly to the loading direction.

4.1.2. *Macro-Cracking Pattern at the Surface of the Samples*

A visual examination of the surface of concrete samples is done. The cracking pattern is revealed by the impregnation of concrete by spraying ethanol on the surface of the samples. This method enables to highlight the presence of a large cracking pattern.

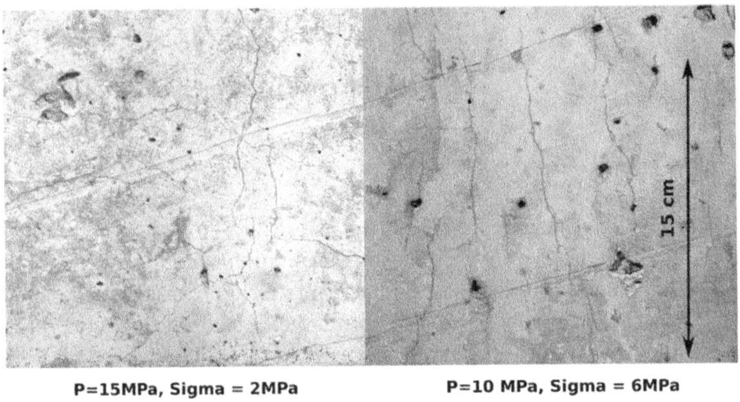

Figure 6. *Maps of the macroscopic cracks pattern (the stress is applied along the vertical direction)*

These large cracks are linked to the combined effects of ASR and mechanical loads. Two maps of the samples' surface are compared on Figure 6. On the right side of the figure, the studied case is associated to the loading configuration, $\sigma = 2MPa$, $P_{NaOH} = 15MPa$; it is a quasi-isotropic stress state. The crack pattern does not present a preferential orientation. On the left side, it is associated to $\sigma = 6MPa$, $P_{NaOH} = 10MPa$; the amplitude of the stress is higher and therefore the deviatoric component of the stress tensor cannot be neglected. The cracks mainly propagate along the vertical direction which is the direction of the load applied via the hydraulic jack.

The stage of the ASR process associated to these macro-cracks is when cracks inside the aggregates percolate in the cement paste. From the previous observations, this process is partly governed by the deviatoric component of the stress tensor. This type of oriented cracking pattern is characteristic of the failure mode in compression for concrete (Carpinteri, 2001). According to previous modeling studies, this alignment of cracks when uniaxial loads are acting on ASR affected concrete has been explained by the tensile failure criterion of the cement paste (Dunant, 2006) (Charpin, 2013) (Cuba Ramos, 2017).

These observations suggest the macroscopic elastic tensor of the ASR sample becomes anisotropic when uniaxial loads are acting on the sample.

4.2. *Residual Mechanical Properties*

The formation of cracks induces a loss of the concrete stiffness. This loss of stiffness is quantified through tensile tests along the axial direction. This mode of solicitation is more appropriated to the assessment of the mechanical damage.

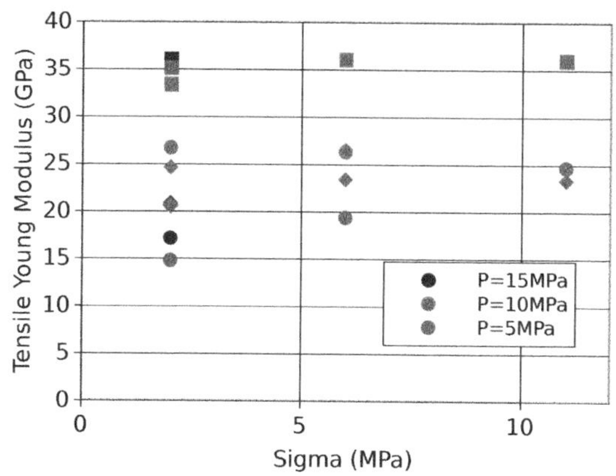

Figure 7. *Pseudo Young modulus tests vs axial stress* σ_a

The residual Young modulus of the concrete affected by ASR have been plotted versus the amplitude of the axial load on Figure 6. As expected, the stiffness of the non-reactive specimens has not been reduced by the loading configurations and it is equal to 35 GPa. In contrast, the stiffness of the reactive specimens is reduced, this decrease may be strongly pronounced. The highest loss of stiffness rises up to approximatively 60%, it is observed for the specimen with the highest expansion level (σ = 2MPa, P_{NaOH} = 10MPa). The dispersion of the results is significant for the reactive specimens with the highest expansion; on the contrary, the two cases associated with a negative expansion (σ = 11MPa, P_{NaOH} = 5MPa & σ = 6MPa, P_{NaOH}=10MPa) exhibit similar results between the two reactive specimens. This observation confirms that a high advancement of the degradation caused by the ASR may limit the representativeness of the results. However, there is not direct correlation between the axial load and the residual stiffness.

4.3. *Discussion*

The influence of the load on the degradation process differs according to the associated scales and mechanisms. The chemical reaction induces a thin and dense cracking pattern inside the aggregates. The characteristics of these cracks are not dependent on the load, they are only linked to the aggregates' feature (mineralogy, sites of amorphous silica). With the advancement of the reaction, larger cracks may propagate in the aggregates. Once again, the path of propagation inside the aggregates is not mainly affected by the loading configuration. The main influencing factors seem to be the shape and the mineralogy of the aggregates. Finally, cracks percolate in the cement paste and macroscopic cracks appear. In contrast, this mechanism of the damage process is strongly affected by the presence of an external load. The failure criterion of the cement paste forces crack to propagate along the loading direction.

This damage induced by ASR have significant influence on the residual stiffness of the reactive specimens. Initial elastic modulus may be reduced by 50%. Observation confirms that a high advancement of the degradation caused by the ASR may limit the representativeness of the results.

5. Conclusion

In this paper, the influence of the fluid pressure is studied via an experimental campaign. The experimental set up is presented and results are presented. Contrary to the effects of an external mechanical loading in which expansion is reduced in the direction of loading, the pressure of the fluid may increase the expansion of concrete. This factor could have a significant effect on the structural behaviour of hydroelectric dams.

The final mechanical degradation of concrete is quantified via the tensile pseudo young modulus. This damage is significant and rises up to 60%. The damage features are qualitatively assessed via the observation of the cracking patterns. Only the propagation of cracks in cement paste seems to be affected by an external load. At the mesoscale level the ASR mechanical problem is mainly linked to the characteristics of the aggregates.

However, it has to be noted that the load amplitudes of this study are much higher as compared to the real structures (if the height of the dam is 200m, the fluid pressure will be only equal to 2MPa). Therefore, direct conclusions about the dam safety cannot be drawn from these experimental results without thorough studies.

6. Acknowledgements

The authors would like to thank the Swiss Federal Office for Energy (OFEN) and Swiss-electric Research for their financial support.

7. References

Ben Haha, M. "Mechanical effects of alkali silica reaction in concrete studied by SEM image Analysis". PhD thesis, École Polytechnique Fédérale de Lausanne, 2006.

Carpinteri, A., Ciola, F., Pugno, N., Ferrara, G., and Gobbi, M. E., "Size-scale and slenderness influence on the compressive strain-softening behavior of concrete". *Fatigue & Fracture of Engineering Materials & Structures*, vol. 24, No. 7, 441-450, 2001.

Chappex, T., "The role of aluminum from supplementary cementitious materials in controlling alkali-silica reaction". PhD thesis, École Polytechnique fédérale de Lausanne, 2012.

Chappex, T., Sofia, L., Dunant, C., Scrivener K., "A robust testing protocol for the assessment of ASR reactivity of concrete". *ICAAR Proceedings*, Sao Paulo, 2016.

Charpin, L., "Modèle micromécanique pour l'étude de l'anisotropie de la réaction alcali-silice", PhD thesis, Université Paris-Est, 2013.

Cuba Ramos, A., "Multi-scale modeling of the alkali-silica reaction in concrete". PhD thesis, École Polytechnique fédérale de Lausanne, 2017.

Dunant, C., "Experimental and modelling study of the alkali-silica-reaction in concrete". PhD thesis, École Polytechnique fédérale de Lausanne, 2009.

Ghabezloo, S., Sulem, J., Guédon, S., Martineau, F., and Saint-Marc, J., "Poromechanical behaviour of hardened cement paste under isotropic loading". *Cement and Concrete research*, vol. 38, No. 12, 1424-1437, 2008.

Giorla, A., Scrivener, K., and Dunant, C., "Influence of viscoelasticity on the stress development induced by alkali-silica reaction". *Cement and Concrete Research*, vol. 70, No. 18, 2015.

Larive, C., "Apports combinés de l'expérimentation et de la modélisation à la compréhension de l'alcali-réaction et de ses effets mécaniques". PhD thesis, École Nationale des Ponts et Chaussées, 1997.

Multon, S. and Toutlemonde, F., "Effect of applied stresses on alkali-silica reaction-induced expansions". *Cement and Concrete Research*, vol. 36, No. 5, 912-920, 2006.

Stanton, T. E., "Expansion of concrete through reaction between cement and aggregate". *Proceedings of the American Society of Civil Engineering*, vol. 66, No. 10, 1781-1811, 1940.

Yurtdas, I., Burlion, N., Shao, J.-F., and Li, A., "Evolution of the mechanical behavior of a high performance self-compacting concrete under drying". *Cement and Concrete Composites*, vol. 33, No. 3, 380-388, 2011.

Importance of Alkali-Wrapping for CPT

KAZUO YAMADA* — YUICHIRO KAWABATA** — SHOICHI OGAWA*** — KENGO HAGA*** — YASUTAKA SAGAWA**** — T. OCHIAI*****

** National Institute for Environmental Studies*
10-2, Fukasaku, Miharu, Tamura, Fukushima 963-7700, Japan
Yamada.kazuo@nies.go.jp

*** Port and Airport Research Institute*
3-1-1, Nagase, Yokosuka, Kanagawa 239-0826, Japan
kawabata-y@pari.go.jp

**** Taiheiyo Consultant, Co., Ltd.*
2-27-8, Higashi-Nihonbashi, Chuo-ku, Tokyo 103-0004, Japan
Shoichi_Ogawa@taiheiyo-c.co.jp, Kazuko_Haga@taiheiyo-c.co.jp

***** Kyushu University*
744, Motooka, Nishi-ku, Fukuoka 819-0395, Japan
sagawa@doc.kyushu-u.ac.jp

****** Mitsubishi Research Institute*
10-3, Nagata-cho, Chiyoda-ku, Tokyo 100-8141, Japan
takamasa@mri.co.jp

ABSTRACT. *Concrete prism test (CPT) has been thought to be a reliable laboratory test evaluating the reactivity of aggregate or concrete mixture. Problems pointed are the alkali loss during keeping in a humid chamber called as alkali leaching and limited moisture supply or drying even in a humid chamber. These problems can be avoided by a method wrapping a concrete prism with wet paper containing alkaline solution called as alkali-wrapping (AW). Depending on the reactivity of aggregate, the effect of AW is different but for late-expansive aggregate, AW increased the expansion several times more for a long term. Chemical analysis of paper of wrapping cloth after test has been carried out in order to evaluate the ion balance between concrete prism and wrapping cloth. Of course wrapping with pure water, some amounts of alkali are moved to wrapping cloth. However, when alkali solution estimated having the similar concentration with pore solution, significant amounts of alkalis were moved from wrapping paper to concrete prism even in the case of non-reactive pure limestone and the reason is estimated to be caused by carbonation of wrapping solution. In the case of highly-reactive andesite, significant amounts of alkali silica gel were generated in the concrete prism and inevitably moved to wrapping cloth. There should be an optimum alkali concentration for wrapping paper compensate the movement of alkalis between a concrete prism and wrapping paper but more study is required.*

KEYWORDS: *alkali-silica reaction (ASR), alkali-wrapping, concrete prism test, alkali leaching, moisture supply.*

1. Introduction

Concrete prism test (CPT) has been thought as the most basic evaluating method for the alkali reactivity of aggregate or job mixture of concrete. However, CPT is not perfect as shown by Ideker *et al.*, 2012. For some aggregate at least, expansion by CPT is smaller than that of field expansion of concrete block. Lingdard *et al.*, 2012 estimated the reason as loss of alkalis from specimens called as alkali leaching during CPT.

The alkali leaching can be avoided by wrapping a concrete prism with wet cloth containing alkaline solution called as alkali-wrapping (AW) (Yamada *et al.*, 2016a). For highly reactive andesite, the expansion behaviors were similar despite of the methods of curing. However, for less reactive aggregate or concrete mixture containing fly ash, the expansion by AW-CPT was much more than without AW that is RILEM AAR-4.

The difficult point of traditional CPT is not only in controlling alkali leaching but the fluctuation of the amount of leached alkalis depending on various factors such as the place and batch of CPT, aggregate type, alkali content and temperature. For regulated condition such as RILEM AAR-3 or AAR-4, it is possible to compare the test results of CPT with field experiences in order to judge the alkali reactivity of aggregate. However, if once the expansion is tried to be estimated quantitatively considering various factors affecting ASR, it will be difficult by the ambiguous alkali contents in concrete. Therefore, AW-CPT will be a promising procedure as a performance test for the quantitative expansion estimation by ASR.

Recently, the authors have pointed out another important factor affecting ASR expansion. As is well known, moisture supply significantly affects ASR expansion as shown in Figure 1 (Yamada *et al.*, 2016a). The moisture supply might have been thought enough in a humid chamber. However, there were significant differences between simple curing in a humid chamber and active water supply by wrapping with wet cloth. In the case of rapid reactive aggregate TO (andesite), AW had no serious effect on expansion. However, in the case of late expansive aggregate WI (hornfels), AW increased the expansion triple.

For both cases, there were quite big differences in mass gain. Without AW, the specimen lost mass meaning drying. On the other hand, by AW, the mass increased with time continuously. For the expansion prediction, the importance has been also shown by the simulation results that the modelling of moisture was a critical factor (Kawabata *et al.*, 2016).

One not decided point of AW-CPT is the optimum alkaline concentration for wrapping cloth. As pointed by Lingdard *et al.*, 2012, some amounts of alkalis in wrapping cloth may move into concrete in some conditions. There should be an optimum value. By the analysis of chemical composition of wrapping cloth through experiments changing alkali concentration of wrapping cloth, the phenomena happened between concrete prism and wrapping cloth are discussed in this paper.

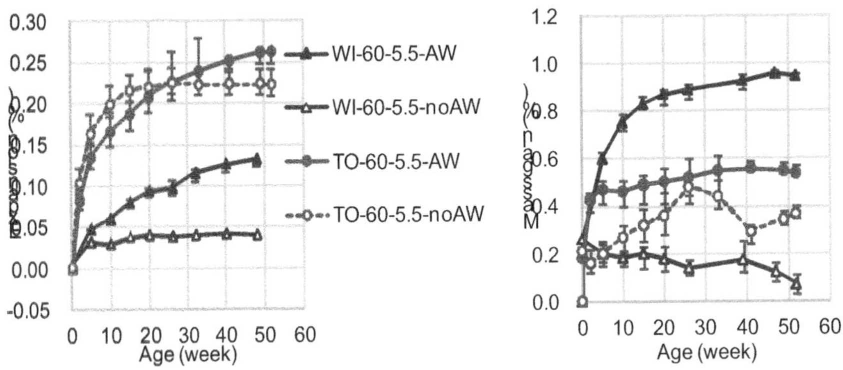

(a) Expansion (b) Mass gain

Figure 1. *Comparison of CPT with and without AW (Yamada et al., 2016a). Aggregate type: curing temperature (°C); alkali content: AW; WI: late expansive hornfels; TO: rapid reactive andesite in the pessimum proportion*

2. AW-CPT

2.1. *Procedure of AW-CPT*

In order to avoid alkali leaching and drying, AW-CPT, a method to wrap concrete specimen with wet cloth containing alkali solution was applied. The concentration of alkali hydroxide solution imitates pore solution in concrete. The testing procedure for AW-CPT are summarized below (Figure 2).

Although the general procedure is almost the same as RILEM AAR-3 (38°C)/ AAR-4 (60°C), the concrete specimens should be wrapped with wet cloth containing alkaline solution. After de-moulding concrete specimens (75×75×250 mm or 100×100×400 mm), they are submerged in water for 30 minutes.

During this submersion, limited amount of alkalis will be leached out but the amount is limited. In order to obtain a homogeneous condition of water content of concrete specimens, submersion in water will be preferable. The reference readings are performed after submersion of the specimens.

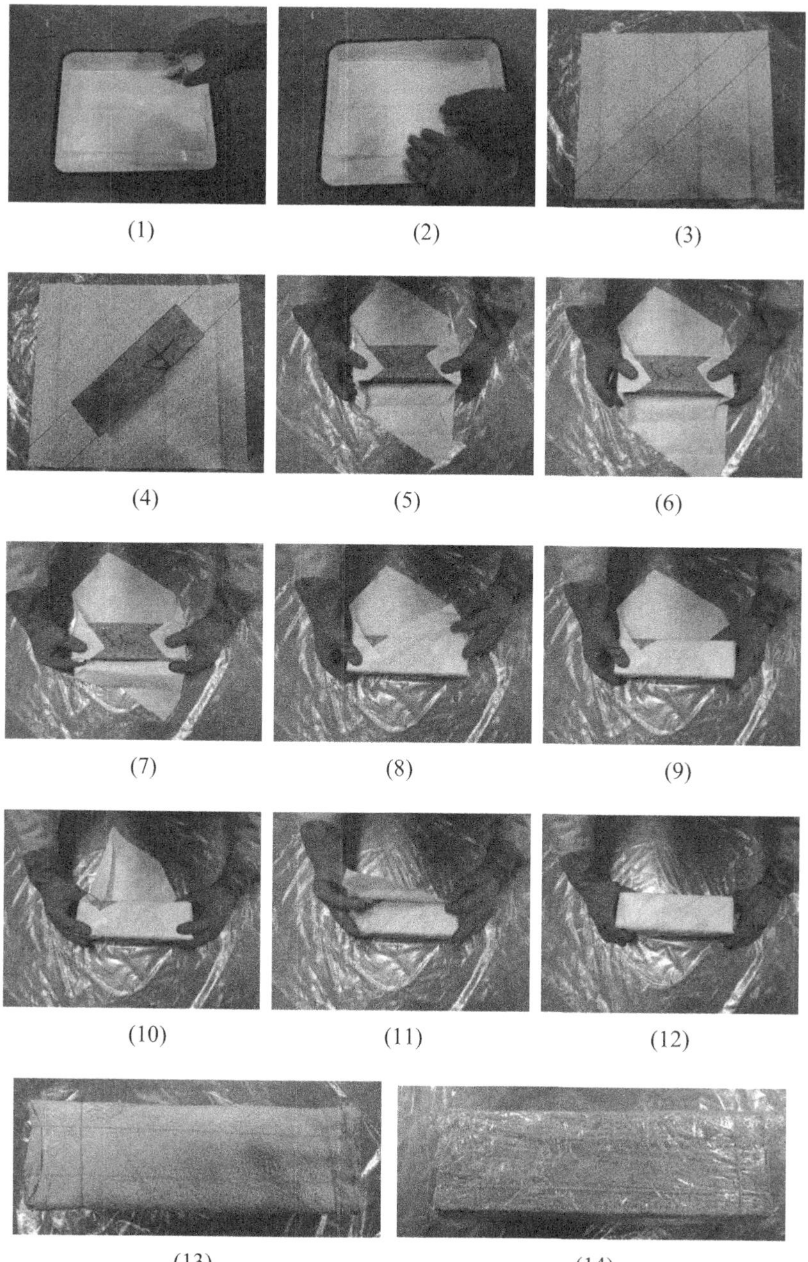

Figure 2. *Wrapping procedures for AW-CPT*

50 g of alkali hydroxide solution imitating pore solution of concrete is prepared in a flat tray of appropriate size (Figure 2 (1)). Alkali concentration is discussed in the section 2.2. 50 g of solution corresponds to the maximum amount which can be absorbed by one piece of cloth. The concrete specimen is wrapped with a piece of polypropylene non-woven cloth (355 mm × 425 mm, Crecia kimtex pops up white, Nippon Paper Crecia, Co., Ltd.). In the case of 75×75×250 mm, one piece of cloth is placed in the container (Figure 2(2)). For 100×100×400 mm, two pieces are used. The wet cloth is placed over thin plastic film on a table (Figure 2 (3)). When the wet cloth is picked up, there is a case that some droplets can fall from the cloth. Therefore, to minimize loss of the solution, the cloth should be placed on thin plastic films in advance so that the droplets can be contained in the cloth and plastic film. Then the specimen is wrapped. One pieces of cloth is sufficient to cover all the surface of the specimen if the specimen is placed appropriately as shown in Figure 2 (4)-(12). The cloth was tied by rubber bands for the cloth not to be detached from concrete specimen (Figure 2 (13)). Then the specimen is overlapped by plastic film in order to avoid drying (Figure (14)). Afterwards three wrapped specimens are stored in a moisture tight storage container over water. The container was also stored in a temperature-controlled chamber. The minimum reading of the comparator is 0.001 mm. The electronic balances to weigh the specimen and wet cloth have the minimum reading of 1 g and 0.1 g, respectively.

Before the later measurements, the storage container is moved to a room kept at 20±2°C to cool the specimens for 1 day. Special caution should be taken for the specimens not to contact with water at the bottom when the containers are moved to the room. Then the length and mass of the specimen as well as the mass of the cloth are measured after taking the cloth off slowly from the specimen. After measuring the weight of the cloth, water is added to the cloth, for keeping 50 g of solution in the cloth. This process is necessary to supply sufficient amount of moisture for ASR. On the contrary it should be avoided to supply excess amount of alkalis from the wrapping cloth to concrete specimen. This is why pure water is used during the test. The same cloth is used during the test unless the cloth is heavily damaged.

2.2. *Decision of Alkali Concentration of Wrapping Cloth*

One of the important issues is the method to decide the concentration of the solution in AW. Lower concentration of the solution will induce alkali leaching. In fact, wet cloth with pure water induces alkali leaching while the degree of alkali leaching is less than unwrapped one (Yamada *et al.*, 2014). On the contrary, higher concentration will supply alkali to the concrete, resulting in more expansion.

Therefore, the method to decide the hydroxide ion concentration of wrapping cloth is important. The best approach to decide the concentration seems to balance with concrete pore solution. Nevertheless, a certain amount of alkali can be supplied by water uptake accompanying alkalis from the cloth with increasing hydration and ASR over time. This alkali uptake from the cloth is inevitable in early ages while it can be estimated by calculating mass balance between the cloth and the concrete. However, in longer ages, hydration is saturated and additional water is supplied at the timing of measurement and finally the water content in the wrapping cloth leaches 50 g the same amount of water at the begging. If the alkali distribution is in equilibrium between concrete specimen and wrapping cloth, the alkali movement become limited. This might be correct for the case of unreactive and no-alkali-absorbing or releasing aggregate. For reactive aggregate, extra alkalis by wrapping cloth will be consumed by ASR. Contrary, alkali-silica gel (ASG) will be extruded from concrete specimens to wrapping cloth (Yamada *et al.*, 2016b). This consumption of alkalis will be more for more reactive aggregate but the introduction of alkalis balancing for no-reactive aggregate is in the safer side of experiments estimating the alkali-reactivity.

The best way may be the extraction of pore solution and analyse the hydroxide ion concentration. However, the experiments strongly depend on the procedures which have not been authorized and thus they have considerably large variations. For instance, the concentration of extracted pore solution is influenced by the pressure applied. Therefore, the calculation of hydroxide ion concentration is an alternative way. Detailed discussion will be made in future. In this study, OH^- concentration of pore solution in concrete having total alkali content of 5.5 kg/m^3 was assumed to be 1.33 M and used as NaOH solution for wrapping based on Kawabata and Yamada (2015).

3. Experiments

3.1. *Materials and Measurements*

Two series of tests were carried out. In the first series (S1), three types of aggregate were used. One is no reactive pure limestone LS for fine and some part of coarse aggregate. Second is highly reactive andesite TO as 30 mass% of coarse aggregate with 70 mass% of LS. The last is late expansive hornfels WI as 100% of coarse aggregate. In the second series (S2), two types of aggregate were used. One is LS. The other is very highly reactive andesite N as 30 mass% of coarse aggregate. A popular mixture in Japan, Japanese normal Portland cement with a water to cement ratio of 0.50 and unit water content of 160 kg/m^3, was used. The total alkali content of concrete was adjusted to be 5.5 kg/m^3 by adding NaOH. Alkali reductions evaluated by chemical method ASTM C 289 were 8, 27, 100 and 207 mmol for LS, WI, TO and N, respectively.

The prism size was 10×10×40 cm for S1 and 7.5×7.5×25 cm for S 2. The outline of procedure is described in section 2.1. The detailed conditions are as follows. NaOH solution of AW was 0.0 M for S1 and those were 1.06, 1.33, and 1.60 M for S2. The humid containers having three concrete prisms were stored in a larger humid chamber under a temperature-controlled condition at 40°C for S1 and at 60°C for S2.

3.2. *Analysis of Wrapping Cloth*

At some ages, wrapping cloths were removed from specimens and used for element analysis. One piece of cloth was cut in small pieces and putted in a plastic bottle. Then, 0.5 L of water was added in the bottle and the bottled was healed boiling water bath for 6 hours. The solution was filtrated and water was added to be 1 L in a volumetric flask. This procedure was repeated twice and 2 L of extracted solution was used for analysis. Around 90% of water soluble alkalis can be recovered by this process. Then, the cloth was applied for extraction process again but in 1 M-HCl. The cloth after triple extractions was dried at 105°C for one night and weighed. Obtained solutions were analyzed by atomic adsorption for Na and K and by ICP for Ca and Si.

4. Results and Discussions

4.1. *Reaction Products Generated by ASR in Wrapping Cloths*

Before discussing the optimum alkali concentration of AW, reaction products generated by ASR in wrapping cloths are reviewed after Yamada *et al.*, 2016a. The estimated phases generated and the distribution of elements are shown in Table 1. One important fact is the movement is not only alkalis but also calcium and silica. Moreover, carbonation took place more at higher temperature. From this result, the following mechanism can be estimated:

– ASR generates ASG and ASG is extruded from concrete prism to wrapping cloth.

– Not only alkalis but also calcium is accompanied by ASG.

– Silica is distributed in ASG, C-S-H, and silica.

– Carbonation changes the calcium containing phases from C-S-H or portlandite to calcite and silica.

– Expansions are not proportional the amount of ASG generated.

Temp. (°C)	Si (mmol)			Ca (mmol)		Na (mmol)	K (mmol)		Expan. (%)
	ASG	C-S-H	Silica	C-S-H	Calcite	ASG	ASG	K-S-H	
60	**7**	2	28	0	25	**14**	1	4	0.11
40	**15**	0.5	10	0	11	**24**	3	0	0.14
20	**8**	7	0.4	9	0	**17**	2	0	0.10

Table 1. *Generated phases in wrapping cloths (Yamada et al., 2016a) Aggregate – highly reactive andesite, alkali content – 3.0 kg/m^3, age – 45 weeks*

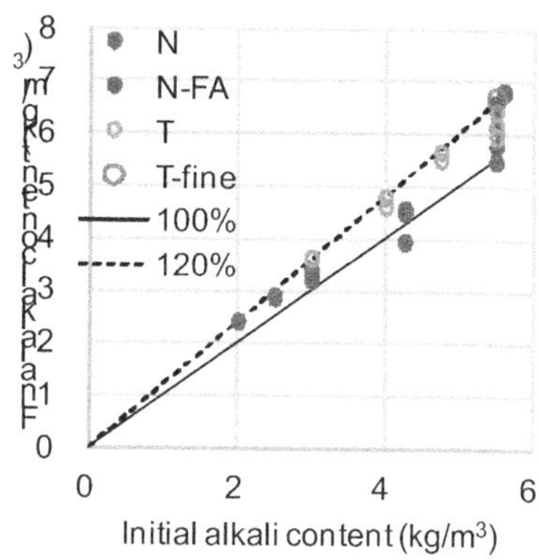

Figure 3. *Initial and final alkali contents (Yamada* et al.*, 2016a). N: very highly reactive andesite (expansion at 45 weeks – 0.06-0.17%); N-FA: fly ash contained (exp. – 0.04-0.09%); T: chert (exp. – 0.02-0.07%); T-fine: chert as fine aggregate (exp. – 0.07%)*

In Figure 3, alkali balance between initial and final alkali content is shown (Yamada *et al.*, 2016a). By AW, depending on the type of aggregates, alkali increased from 0 to 20%. In the case of less expansion, alkali supplies from wrapping cloths seem more.

4.2. *Alkali Concentration in AW*

In Table 2, alkali concentrations in wrapping cloths of S1 where pure water was used as wrapping solution are shown. Alkali concentrations in wrapping cloth after test for 26 weeks were 0.81 and 0.29 M for aggregates TO and WI, respectively. The expansions of aggregate TO and WI was 0.15 and 0.00%, respectively. In the case of aggregate TO, significant ASG is expected to be generated but in the case of aggregate WI, almost no ASG is expected. Therefore, the big difference in alkali concentration is expected to be related with ASG extrusion in wrapping cloth. At this age, most of alkalis in wrapping cloths are water soluble. Significant amounts of extracted alkalis seem to be accompanied by ASG in the case of aggregate TO. The alkali hydroxide concentration was estimated 1.0 M based on the estimation of 1.33 M for 5.5 kg/m^3 of total alkalis. Alkali concentration for the case of aggregate WI is lower significantly than the estimated value. There is a possibility of alkali absorption by aggregate decreasing alkali concentration. Although the effect should larger in the case of TO, the results were reverse. There should be other mechanisms discussed in the next section.

Starding condition				Na			K			Na_2Oeq			Water	$NaOH_{eq}$	Exp.
Agg.	Total alkali	Age	Alkali in AW	Hot water	HCl	Total	Hot water	HCl	Total	in AW/ piece	original in specimen	Leaching ratio	in AW	in AW	
	kg/m^3	weeks	mol/L	mg	mg	mg	mg	mg	mg	g	g	%	g	mol/L	%
TO	4.19	26	0.00	1210	4.1	1214	213.0	1.6	214.6	1.8	16.8	10.8	71.8	0.81	0.15
WI	4.19	26	0.00	540	1.6	542	52.5	0.0	52.5	0.8	16.8	4.6	84.4	0.29	0.00

Table 2. *Alkalis in water-wrapping cloth after curing at 40°C for 26 weeks*

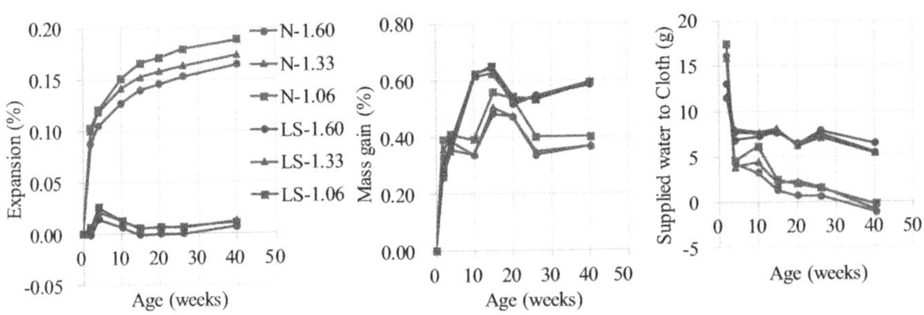

Figure 4. *Expansion and mass change for aggregates N and LS with various NaOH concentration in WA (60°C, total alkali content – 5.5 kg/m^3)*

In Figure 4, expansion, mass change of concrete specimen and amount of supplied water to wrapping cloth at each measurement of S2 are shown. LS did not show expansion. N showed rapid expansion only in 2 weeks and gradually saturated after 10 weeks. The mass gain of LS leached 0.6% that is 20 g at 10 weeks and saturated. That of N leached 0.4% that is 13 g and saturated at 4 weeks. The amount of water supply to cloth was more in early ages. It decreased gradually for N but saturated for LS around 7 g. Total amounts of supplied water were 29 and 54 g for N and LS, respectively. There are big differences between mass gain of concrete specimen and cloth. Around 7 g water was lost in the case of LS during succeeding two measurements. On the other hand, in the case of N, ASG was added to cloth and there was no room to add water after 40 weeks. This may be one reason of saturation of expansion.

Initial condition				Na			K			Ca			Si			*Est. Silica	Water	Est. NaOHeq	Exp.
Agg.	Total alkali	Age	Alkali in AW	Hot water	HCl	Total	Hot water	HCl	Total	Hot water	HCl	Total	Hot water	HCl	Total				
	kg/m^3	weeks	mol/L	mg	mg	mg	mg	mg	mg	mg	mg	mg	mg	mg	mg	g	g	mol/L	%
N	5.5	4	1.06	1210	11	1221	124	6	130	5	141	146					45.0	1.2	0.12
			1.33	1470	13	1484	135	8	143	4	127	132					45.0	1.5	0.12
			1.60	1576	17	1592	138	8	146	3	141	144					45.0	1.6	0.10
		50	1.06	927	354	1281	34	82	116	6	885	891	494	26	520	4.9	42.4	0.97	0.19
			1.33	960	346	1306	33	69	102	8	876	884	510	27	537	4.4	45.2	0.94	0.17
			1.60	1013	442	1455	30	77	107	6	1053	1059	529	28	557	5.4	44.4	1.01	0.16
LS		4	1.06	625	2	627	100	0	101	396	128	524					45.0	0.66	0.03
			1.33	715	2	717	104	0	104	570	115	685					45.0	0.75	0.02
			1.60	778	2	781	103	0	103	169	96	265					45.0	0.81	0.01
		40	1.06	545	3	548	117	1	118	2	426	428	24	4	29	-0.1	43.3	0.62	0.01
			1.33	582	6	588	120	1	121	3	414	417	30	17	47	-0.9	43.3	0.65	0.01
			1.60	596	3	599	116	1	117	3	425	428	39	15	54	0.0	42.1	0.69	0.01

Table 3. *Elements in water-wrapping cloth after curing at 60°C for 4 or 40 weeks*
** After HCl extraction, something was still remained in wrapping cloth and the amount is assumed to be silica*

In Table 3, alkali, calcium and silicate in wrapping cloth after tests at 60°C for 4 or 40 weeks of S2 are shown. Pure limestone did not show expansion and no ASG is expected. Aggregate N showed significant expansion even in 4 weeks and significant amounts of ASG are expected. At 4 weeks, in the case of LS, detected elements are water soluble Na, K, and Ca and acid soluble Ca. The most characteristic point is in large amount of water soluble Ca corresponding to portlandite that is thought to be moved from concrete. Some parts of portlandite seems to be carbonated to calcite even in this short period of time.

For longer age 40 weeks, water soluble Ca disappeared and changed to acid soluble calcite. Accompanying this carbonation, water soluble Na deceased. Alkalis are still in water soluble form. Estimated alkali concentration is significantly lower than NaOH concentration 1.33 M in the beginning.

The difference in NaOH concentration in initial wrapping cloth affected in some degree at 4 weeks but the degree of difference decreased at 40 weeks. Alkali absorption by aggregate is limited for LS. Judging from these behaviours, carbonation may affect the ion solubility equilibrium but the detailed quantitative explanation should be evaluated by phase equilibrium calculation.

In the case of aggregate N, from the age of 4 weeks, majority of calcium is in acid soluble form. Compared to LS, roughly double amounts of Na were detected. This Na is corresponded to NaOH and water soluble silicate but unfortunately it was not measured at 4 weeks. The amount of water soluble K was more than the case of aggregate LS and this may be caused by the generation of ASG. The counter ions of silica gel are expected to be Na and K. Because pH was maintained high, water soluble Ca seems small. But some amounts of calcite were generated.

After longer age 40 weeks, water soluble Na decreased and acid soluble Na increased. This tendency is same for K. Regarding to Ca, the water soluble one was still limited but acid soluble one increased significantly almost seven times more. However, still majority of silicate was water soluble and acid soluble was limited. Remained substance after acid extraction is expected to be silica according to former study (Yamada *et al.*, 2016). ASG in this age may be more stable than that at early ages and some part of alkalis have characters of acid soluble. By HCl, only alkalis are extracted but silica was remained as indissoluble phase for acid. ASG seems to accelerate the transfer of Ca from concrete to wrapping paper and the formation of calcite. Again this carbonation may have affected the less alkali concentration than early age.

4.3. *Optimum NaOH Concentration in AW*

Judging from the experiment of no-reactive LS and almost no-reactive WI, the estimated NaOH concentration 1.33 M for 5.5 kg/m^3 of total alkalis in concrete seems too high although carbonation may decrease it in some degree. As a nature of alkali concentration in pore solution, it is not easy to estimate accurate value as shown by large scattering of many previous studies. However, it is sure that water wrapping resulted in loss of alkalis. Therefore, there should be an optimum value. A next series of experiment are going by changing NaOH concentration in wrapping for 5.5 kg/m^3 of total alkalis in concrete and the conclusion will be obtained in a few months.

Besides from the alkali balance in no-reactive concrete between wrapping cloth and pore solution, alkalis supplied by wrapping cloth is used for ASR and contained in ASG extruded form concrete. This is inevitable movement of alkalis for reactive aggregate. However, if the water-wrapping is applied for marginal reactive aggregate, some alkalis will be lost and the reactivity of aggregate will be underestimated.

The target of AW-CPT is not judgment test of alkali reactivity of aggregate or concrete but is the performance test. Therefore, alkali wrapping should be designed based on the case of no-reactive aggregate.

5. Conclusions

Concrete prism test for alkali-reactivity of aggregate or concrete mixture with wrapping by wet cloth with alkaline solution imitating pore solution in concrete, AW-CPT was discussed from the viewpoint of reactions generated in wrapping cloth. By water-wrapping, some parts of alkalis were lost but higher alkali concentration wrapping increased alkalis in concrete. Therefore, optimization should be made in the future study. By the generation of alkali-silica-gel, ASG in concrete, significant amount of ASG extruded to wrapping paper accompanying alkalis and calcium. For longer ages, ASG becomes stable and water-insoluble but acid soluble forms. Carbonation started from early ages and may decrease alkali concentration by changing ion equilibrium.

6. Acknowledgements

Some parts of this study have been carried out as a part of the Nuclear Regulation Authorities, NRA project for advanced aging management technical evaluation of concrete structures in nuclear power plant. The information presented in this paper is the sole opinion of the authors and does not necessarily reflect the views of the sponsoring agencies.

7. References

Ideker, J. H., *et al.*, "The Importance of Outdoor Exposure Site Testing", *Proc. of 14th Int. Conf. on Alkali-Aggregate Reaction*, 051412-IDEK, 2012.

Kawabata, Y. and Yamada, K., "Evaluation of alkalinity of pore solution based on the phase composition of cement hydrates with supplementary cementitious materials and its relation to suppressing ASR expansion", *Journal of Advanced Concrete Technology*, Vol. 13, pp. 538-553, 2015.

Kawabata, Y., *et al.*, "Correlation between laboratory expansion and field expansion of concrete: Prediction based on modified concrete expansion test", *Proc. of 15th Int. Conf. on Alkali-Aggregate Reaction*, 15ICAAR2016_034, 2016.

Lindgard, J., *et al.*, "Alkali-silica reaction (ASR) – performance testing: Influence of specimen pre-treatment, exposure conditions and prism size on alkali leaching and prism expansion", *Cement and Concrete Research*, Vol. 42, pp. 223-243, 2013.

Yamada, K., *et al.*, "CPT as an evaluation method of concrete mixture for ASR expansion", *Construction and Building Materials*, Vol. 64, pp. 184-191, 2014.

Yamada, K., *et al.*, "Importance of alkali-wrapping in Concrete Prism Tests", *Proc. of 15th Int. Conf. on Alkali-Aggregate Reaction*, 15ICAAR2016_084, 2016a.

Yamada, K., *et al.*, "Exact effects of temperature increase and alkali boosting in concrete prism test with alkali wrapping", *Proc. of 15th Int. Conf. on Alkali-Aggregate Reaction*, 15ICAAR2016_203, 2016b.

Structures Damaged by ASR and DEF

Improving the Prognosis of Structures Damaged by Expansive Concrete with Physico-Chemical Modelling

STÉPHANE MULTON* — ALAIN SELLIER* — ÉTIENNE GRIMAL** — ÉRIC BOURDAROT**

** LMDC, Université de Toulouse*
INSA/UPS Génie Civil, 135 Avenue de Rangueil, 31077 Toulouse cedex 04, France
stephane.multon@insa-toulouse.fr, alain.sellier@insa-toulouse.fr

*** Centre d'Ingénierie Hydraulique*
Génie Civil – Structure, Savoie Technolac, 73373 Le Bourget-du-Lac, France
etienne.grimal@edf.fr, eric.bourdarot@edf.fr

ABSTRACT. *Expansions due to chemical reactions between components of concrete can cause severe and permanent damage to large concrete structures. Several methods have been proposed to reassess structures damaged by swelling. But, using expansions test on specimens to represent swelling of massive structures is still in question. Indeed, expansions on structures appear to be always longer and larger and the differences cannot be only explained by temperature and moisture effects. The first part describes the scale effects which disturb alkali-silica reaction (ASR) expansion tests performed in laboratory. The impacts of reactants (alkali, water) transfers on expansion measured on specimens are analysed. The physicochemical phenomenon of ASR-damaged concrete can be investigated with multi-scale modelling in order to take into account the moisture transfer, the consumption of alkali by ASR-gel production and the decrease of alkali concentration in pore solution due to external leaching. This approach can also be applied to structures damaged by delayed ettringite formation (DEF). The second part of the paper details the physicochemical mechanisms which should be taken into consideration to evaluate the behaviour of structures submitted to DEF. The final aim of multi-scale approach is to take benefit of ASR or DEF expansion tests carried out in laboratory to perform structural calculations with the same modelling and data. On purpose, the models should be enough simple to be usable for structural calculations but able to reproduce the scale effects due to water and alkali transfers. That is the reasons why they have to consider some physico-chemical details to predict expansion in different environmental and chemical conditions.*

KEYWORDS: *alkali-silica reaction (ASR), delayed ettringite formation (DEF), prognosis, scale effects, structures.*

1. Introduction

Different expansive phenomena can occur in concrete structures and cause irreversible damage. The swelling origin can be found in the reactive nature of concrete components (aggregate and/or cement) and in the environmental conditions (temperature, moisture). The consequences of the swelling are cracking and displacement in damaged-structures. It can lead to the dysfunction of the structures (checking of Serviceability Limit State, SLS) or to call into question the structural capacity (Ultimate Limit State, ULS). One of the most studied is the Alkali-Silica Reaction (ASR). ASR is responsible of numerous degradations of structures all over the world. Delayed Ettringite Formation (DEF) in structures can also cause similar irreversible damage. These two reactions can act alone or be combined. In the case of structures damaged by these mechanisms, the origin, and thus the potential of expansion, is linked to physicochemical aspects. It is thus important to propose methodology of prognosis at the structure scale with consideration of the mechanisms at the material scale. The reassessment of structures damaged by expansive concrete should be performed in the framework of multiscale analysis. Modelling should be enough simple to be usable for structural calculations and enough complete to reproduce the scale effects due to water, alkali and reaction products transfers in specimens.

2. Alkali-Silica Reaction

2.1. *Aggregate Scale*

Two stages have to be considered in modelling to be representative of usual observations, one for the ionic movements and one for the reactive mechanisms (Multon and Sellier 2016; Multon, Sellier, and Cyr 2009). According to the nature of the aggregate, transport and/or reactivity can drive the type of damage (Ponce and Batic 2006) and the kinetics of expansion due to ASR (Figure 1).

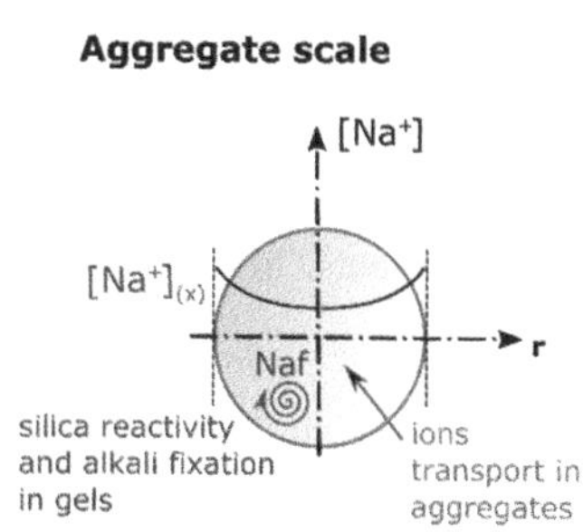

Figure 1. *Aggregate scale – Ions transport and silica reactivity (Na$_f$: alkali fixed in ASR-gels)*

The transport of ions is the main phenomena for very reactive aggregate (as opal or glass). It can be modelled by the following mass balance:

$$p_{agg}\frac{\partial(S_r \cdot [Na^+])}{\partial t} = -div\left(-D_{agg}\overrightarrow{grad}[Na^+]\right) + S_{Na} \tag{1}$$

p_{agg}, is the aggregate porosity, S_r, the saturation degree, $[Na^+]$, the alkali concentration, D_{agg}, the diffusion coefficient in the aggregate and S_{Na}, the rate of alkali bound in ASR-gels by unit of time and of aggregate volume.

For aggregate with slow reactivity, the main phenomena controlling ASR-kinetics are the reactive mechanisms: silica attack and formation of ASR-gels. In simplified modelling, the number of ASR-gels can be supposed to be proportional to alkali bound in gels (Multon *et al.* 2009). Therefore, the two reactive mechanisms are represented by only one equation:

$$S_{Na} = -\frac{dNa_f}{dt} = -\frac{\langle [Na^+]-[Na^+]^{thr}_{(Ca,T)}\rangle^+}{\tau_{ASR}} \tag{2}$$

t_{ASR} is the characteristic time of silica attack. It can be considered as a quantification of silica reactivity and is highly dependent of aggregate nature. $[Na^+]^{thr}_{(Ca,T)}$ is the alkali concentration threshold lower which the reaction products cause little expansion. It can be evaluated according to temperature and calcium concentration as proposed in (Multon and Sellier 2016).

For the most of reactive aggregates with intermediate reactivity, the two mechanisms are necessary to obtain relevant ASR-kinetics.

2.2. *Concrete Scale*

The expansive mechanisms are impacted by transfers occurring at concrete scale. Transfers are responsible of structural effects due to alkali and water gradients between core and skin even in small specimens used in laboratory (Figure 2). Such gradients can lead to skin cracking and impact measured expansion. As a consequence, it is very difficult to obtain data representative of the real material expansion without measurement artefacts. Multi-scale analysis can help to analyse such measurements.

In the case of the alkali diffusion, it is necessary to take into account the coupling between diffusion and alkali consumed by the reaction. The mass balance can be written:

$$p_c \frac{d(S_r \cdot [Na^+])}{dt} = -div\left(-D_c\ \overrightarrow{grad}[Na^+]_c\right) + S^c_{\Sigma Na_f} \tag{3}$$

p_c is the concrete porosity, S_r, the degree of saturation, $[Na^+]_c$, the alkali concentration in concrete solution, D_c, the diffusion coefficient in concrete, and $S^c_{\Sigma Na_f}$, the sink term of alkali bound in ASR-gels. The sink term can be evaluated from the alkali flow in aggregate (Figure 2):

$$S^c_{\Sigma Na_b} = -\sum_{agg} N_{agg} \cdot 4\pi R^2_{agg} \cdot \overrightarrow{\varphi^{agg}_{Na}}\left(r = R_{agg}\right) \tag{4}$$

N_{agg} is the number of reactive aggregates of each size per concrete volume, R_{agg}, the aggregate size and $\varphi^{agg}_{Na}\left(r = R_{agg}\right)$, the alkali flow in aggregate.

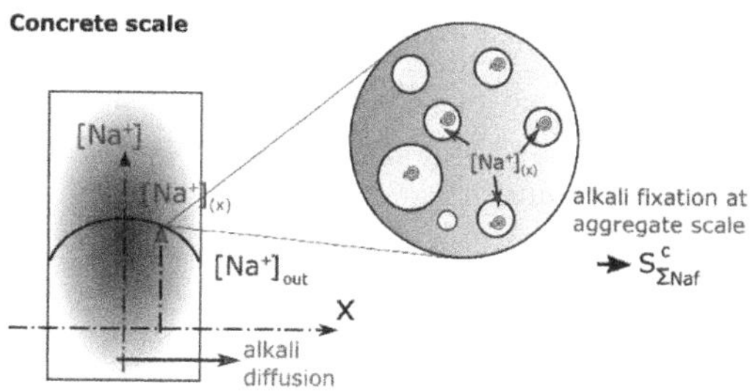

Figure 2. *Concrete scale – ions diffusion and alkali fixation*

2.3. *Impact of Alkali Leaching on Expansion*

Alkali leaching is an important cause of disturbance for expansion test in laboratory (Lindgård *et al.* 2013; Rivard *et al.* 2003). The impact of alkali leaching on ASR-expansion has been recently studied by the multi-scale approach presented in (Multon and Sellier 2016). Alkali content in concrete has to be evaluated through diffusion process in concrete and consumption in ASR-products. Even if the representation of the physico-chemical mechanisms is simplified to be usable for structural assessment, the multi-scale modelling is able to reproduce the scale effect of ASR-expansion due to alkali leaching. Such modelling can extrapolate results to evaluate expansion without leaching which can be more representative of expansion in large structures (Multon and Sellier 2016).

2.4. *Impact of Moisture Conditions on Expansion*

Moisture conditions can be another cause of disturbance when expansion test are used in structural prognosis. As ASR expansions are larger in conditions of high relative humidity, most of ASR-expansion tests in laboratory are carried out at 95%

RH while the moisture conditions of real structures are various. At the start, the degree of saturation of concrete is often about 85% due to cement hydration. External conditions act on moisture conditions on long time scale to modify the initial degree according to external water supply or to drying conditions. Therefore, expansion can take place in concrete with smaller degrees of saturation than in laboratory in large parts of massive structures. Physicochemical modelling have to take into account the dependence of ASR-expansion to degree of saturation. The dependence of alkali diffusion coefficient to degree of saturation has been assumed as an exponential function:

$$D_{agg} = D_{agg}^{0} \times e^{k_w(Sr-1)} \tag{5}$$

D_{agg}^{0} is the diffusion coefficient for the saturated media. k_w is a constant characteristic of the pore network. It can be calibrated on measurement. For concrete, most of data referenced in (Zhang and Zhang 2014) can be reproduced with k_w lying between 3 and 7. For aggregates, pore networks are probably different and dependant of aggregate nature.

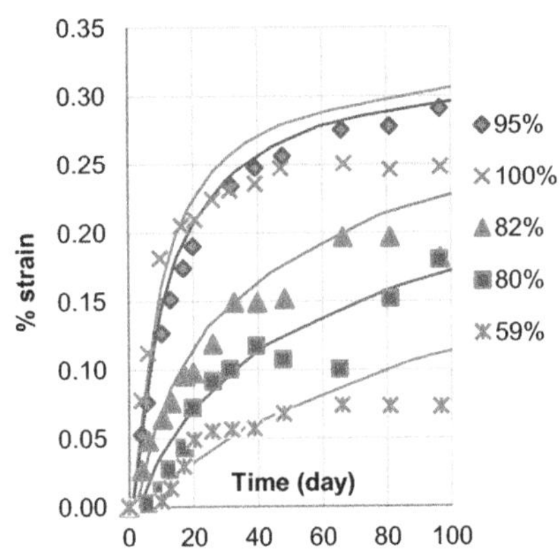

Figure 3. *Expansion according to moisture conditions in (Poyet et al. 2006) (points) and modelling*

In Figure 3, the experimental data drawn from (Poyet *et al.* 2006) have been reproduced with a coefficient k equal to 7. For the highest degree of saturation (100%), leaching can be important and explain why the expansion is smaller than for 95% RH (Poyet *et al.* 2006) which has not been taken into account in this calculation. For the smallest degree of saturation (59%), the kinetics is not perfectly reproduced. It can be due to the progressive drying of concrete. In this calculation, the degree of saturation has been taken constant and equal to 0.35. In reality, it decreases between 0.7 (equilibrium at the beginning of the measurement) and 0.35

(equilibrium reached for 59% RH) (Poyet 2003). Future works will consider leaching and moisture transfer to obtain more precise calibration.

3. Delayed Ettringite Formation

3.1. *Principle*

The main physicochemical mechanisms of the Delayed Ettringite Formation (DEF) lies on thermodynamics disequilibrium. The complete interactions between the different components of the cement paste and even with concrete aggregate according their mineralogy lead to highly nonlinear calculation (Salgues *et al.* 2014). Such modelling is difficult to use in structural context (Salgues 2013). Simplified equations are necessary to obtain modelling able to reproduce a part of laboratory experiments and to evaluate the capacity of DEF-damaged structures. First, it is necessary to identify the main parameters which can act on DEF kinetics and expansion level in structures to take into account their effects.

Structures can be damaged by DEF due to high temperature during hydration of cement. In such cases, the temperature reached in the structure depends on thermal boundary conditions and is variable according to the location: at core, concrete can experiment high temperature during long time while close to the skin, the temperature is lower. DEF-kinetics is also highly dependent on the alkali concentration of storage and the impact of alkali leaching is great for concrete (Famy *et al.* 2001). Therefore, simplified modelling should take into account the dependence of DEF-expansion level to thermal history of concrete and the dependence of kinetics to the concrete alkali concentration.

In order to obtain modelling able to reproduce these effects and to reassess damaged structures, the simplified modelling proposed in this paper is based on simplified equations (Sellier and Multon. 2016). First, a system of equations evaluates the content of the different aluminate phases in cement paste and the sulfate concentration during the period at high temperature. During the service life at ambient temperature, another system calculates the mass balances of aluminum and sulfate consumed by the delayed ettringite formation.

At the start of the structure life, monosulfate (AFm) and ettringite (AFt) contents are estimated from SO_3/Al_2O_3 ratio. If the ratio is lower than 1, all the sulfates are supposed to form monosulfate. If it is upper than 3, the ettringite content is assumed to be equal to initial aluminum content. For intermediate cases, the two phases are supposed to exist.

During the period at high temperature, modelling decreases the aluminum content fixed in AFm and AFt, to fix it progressively in hydrogarnet (Dilnesa *et al.* 2014). These last are supposed not to be available for DEF when concrete returns at ambient temperature. At the end of the high temperature period, the aluminum content available for DEF is:

$$\tilde{A} = A^c - (AFt + AFm + C_3AH_6) - (DEF + HG) \quad (6)$$

A^c is the initial aluminum content, C_3AH_6 is aluminate phases which can exist if the initial SO_3/Al_2O_3 ratio is lower than 1. In this case, all the aluminium ions are not consumed by the formation of AFm and can participate to other aluminate phases. The initial AFt and delayed DEF are distinguished since only DEF is supposed to cause damage to concrete.

The same assumptions lead to estimate the sulfate available for DEF:

$$\tilde{S} = S^c - (3\,AFt + AFm + G) - (3\,DEF) \quad (7)$$

S^c is the initial aluminum content, G is gypsum which remains if the initial SO_3/Al_2O_3 ratio is upper than 3. In this case, all the sulfate ions are not consumed by the formation of AFt and can participate to gypsum formation.

3.2. *Impact of the Duration of High Temperature Curing*

By convention, initial conditions on monosulfate and ettringite are equal to the amount formed in standard condition (High humidity, ambient temperature) (Sellier and Multon. 2016). Of course if temperature during hydration is high, AFm and AFt are rapidly dissolved (as if they would have never been formed). The temperature leading to the destabilization of primary sulfo-aluminates is θ^{th}. The presence of alkali in the concrete solution leads to the modification of thermodynamics equilibrium and impacts the value of this temperature (Salgues *et al.* 2014). This phenomena is quantified by an empirical law in Kchakech's PhD thesis (Kchakech 2015). It can be evaluated by:

$$\theta^{th} = \theta^{ref}.\begin{cases} Na^{-n} \; if \; Na > Na^{th} \\ Na^{th^{-n}} \; if \; Na \leq Na^{th} \end{cases} \quad (8)$$

In Kchakech's PhD thesis, $n \approx 0.18$, Na is the concentration in mol/l of water in porosity and $\theta^{ref} \approx 64°C$. In the present modelling, the threshold Na^{th} is added to avoid obtaining very high (and unrealistic) destabilisation temperature in case of very low alkali content. The value $Na^{th} \approx 0.8\,mol/l$ is fixed to limit the destabilization temperature at a maximum temperature compatible with literature results (Brunetaud 2005).

When the temperature is upper than θ^{th}, the dissolution rate of AFm and AFt is supposed to be proportional to residual content of these phases:

$$\frac{\partial(AFt+AFm)}{\partial t} = -\frac{AFt+AFm}{\tau^d} \quad (9)$$

τ^d is the characteristic time of dissolution, which is supposed to be the same for all the phases by sake of simplicity. It depends on the diffusion rate of ions in the cement paste surrounding reactive sites and to the real kinetics of dissolution.

During high temperature, hydrogarnet (HG) is formed and consumes aluminums ions. The HG formation rate is supposed to be proportional to the aluminum content available for the formation of aluminate phases:

$$\frac{\partial HG}{\partial t} = \frac{\tilde{A}}{\tau^f} \quad (10)$$

τ^f is the characteristic time of aluminum fixation in HG. For concrete containing lime addition or limestone aggregate, the formation of carboaluminate can occur. This phase could also be responsible of irreversible fixation of aluminum ions and thus modified the conditions of formation of secondary AFt. For the moment, all the phases fixing irreversibly aluminum ions are grouped in the same term HG. But future works could study the interest to separate HG and carboaluminate.

As a consequence, the rate of the aluminum content available for the formation of aluminate phases can be evaluated by (11) which models the evolution of aluminates available for DEF versus time. As this equation leads to an evolution of $\tilde{A}$ with a maximum, it explains the pessimum effect of heating duration on DEF illustrated in Figure 4.

$$\frac{\partial \tilde{A}}{\partial t} = -\frac{\partial(AFt+AFm)}{\partial t} - \frac{\partial HG}{\partial t} \quad (11)$$

During the high temperature period, the increase of sulfate concentration is supposed to be due to the dissolution of sulfoaluminate phases formed at the end of curing period at ambient temperature. The proportionality coefficient is the ratio between the sulfate contained in phases which can be dissolved in high temperature (S_0^1) and the corresponding aluminum content (A_0^1):

$$\frac{\partial \bar{S}}{\partial t} = -\frac{S_0^1}{A_0^1}\frac{\partial A^1}{\partial t} \quad (12)$$

with:

$$S_0^1 = AFm_0 + 3\,AFt_0 \tag{13}$$

and

$$A_0^1 = AFm_0 + \,AFt_0 \tag{14}$$

C-S-H bind more sulfate ions for high alkali concentration (Divet and Randriambololona 1998). As a consequence, the AFt and AFm dissolution rate are faster for high alkali concentration. From literature results (Brunetaud 2005; Kchakech 2015; Martin 2010), an empirical law is proposed to take into account this dependence (Sellier and Multon. 2016):

$$\frac{\tau^d}{\tau^{d^{ref}}} = e^{\left(-\frac{Na}{Na^{th}}\right)} \tag{15}$$

Na^{th} is a threshold in alkali concentration calibrated on literature results. The decrease of the characteristic time of dissolution corresponds to the acceleration of the corresponding mechanisms at high alkali concentration.

The analysis of literature results (Brunetaud 2005) shows that the hydrogarnet formation could be slowed down by alkali concentration. Perhaps, the presence of sulfate ions bound in C-S-H could prevent their formation. To reproduce literature results, it is necessary to propose an empirical law to describe this dependence (Sellier and Multon. 2016):

$$\frac{\tau^f}{\tau^{f^{ref}}} = 1 - e^{\left(-\frac{Na}{Na^{th}}\right)} \tag{16}$$

Na^{th} is the same threshold in alkali concentration than for the previous equations and calibrated on literature results.

Finally, the delayed ettringite formation can occur in ambient temperature. When the temperature is lower than θ^{th}, the precipitation rate of DEF is supposed to be proportional to the difference between the final DEF content (DEF_∞) and the content at the time of calculation:

$$\frac{\partial DEF}{\partial t} = \frac{1}{\tau^p}\left(DEF_\infty - DEF\right) \tag{17}$$

DEF_∞ is supposed to be equal to the lowest value between the available aluminum ($\tilde{A}$) and the third of available sulfate ($\tilde{S}$). τ^p is the characteristic time of DEF precipitation. It depends on the diffusion of ions in the cement and to the real kinetics of precipitation.

During this period, aluminum and sulfate available for DEF have to be reassess according to their consumption in DEF:

$$\frac{\partial \tilde{A}}{\partial t} = -\frac{\partial DEF}{\partial t} \tag{18}$$

and

$$\frac{\partial \tilde{S}}{\partial t} = -3\frac{\partial DEF}{\partial t} \tag{19}$$

Temperature acts on thermodynamics equilibrium but also on the properties of ionic transport and as a results on the characteristic times of dissolution, fixation and precipitation. Based on experimental works, it is possible to propose simplified relationships inspired by Vant'Hoff and Arhenius laws to quantify the effect of temperature on the three kinetics (Kchakech 2015; Martin 2010; Salgues 2013).

The two characteristic times of the high temperature period, τ^d and τ^f, have been calibrated for several mortars and concrete from the literature works (Brunetaud 2005; Famy *et al.* 2001; Kchakech 2015; Martin 2010). A unique couple can be used for all these mix-designs (Sellier and Multon. 2016) and allows the reproduction for a same set of parameters of the pessimum effect of high temperature curing duration on DEF-expansion (Figure 4) as observed in (Brunetaud, Divet, and Damidot 2008).

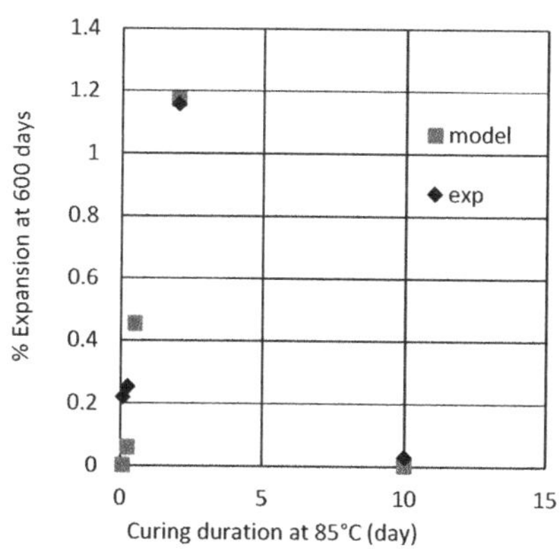

Figure 4. *Pessimum of expansion according to curing duration at 85°C, experimental points from (Brunetaud* et al. *2008)*

For small durations, AFt and AFm are not destabilized (τ^d is too large compared to the curing duration) and little expansions are obtained for durations of 2 and 6 hours (Figure 4). Modelling does not reproduced perfectly the experimental results for 2 and 6 hours. It can be seen that the start of expansion is highly nonlinear since expansion greater than 0.4% is predicted for duration of 12 hours for the same concrete. It explains why it is difficult to obtain a better prediction for small curing duration with this simplified modelling. For very large curing duration, no expansion occurred since aluminium ions are supposed to be bound by hydrogarnet formation (τ^f is smaller than the curing duration). Maximal expansion is reached for intermediate durations (Figure 4).

3.3. *Impact of Alkali Leaching*

Alkali acts on C-S-H binding capacity (Famy *et al.* 2001) and on calcium concentration in cement paste. AFt and AFm formations are thus highly dependent of alkali concentration (Famy *et al.* 2001). Modelling developed by (Salgues *et al.* 2014) gives a quantification based on thermodynamics equilibrium. However, it leads to highly nonlinear calculations replaced herein by an empirical relationship based on experimental results (Famy *et al.* 2001):

$$\frac{\tau^{pref}}{\tau^{p}} = \begin{cases} \left(1 - \frac{Na}{Na^{bl}}\right)^{m} \; if \; Na < Na^{bl} \\ 0 \; if \; Na \geq Na^{bl} \end{cases} \tag{20}$$

Na is the alkali concentration in concrete porosity and Na^{bl} is the alkali concentration which prevents DEF. m is the exponent necessary to reproduce (Famy *et al.* 2001) results. Na^{bl} is supposed not to be an absolute value but to depend on concrete composition and is taken to be equal to two (2) times the initial concentration of concrete to reproduce (Famy *et al.* 2001) results (Figure 5).

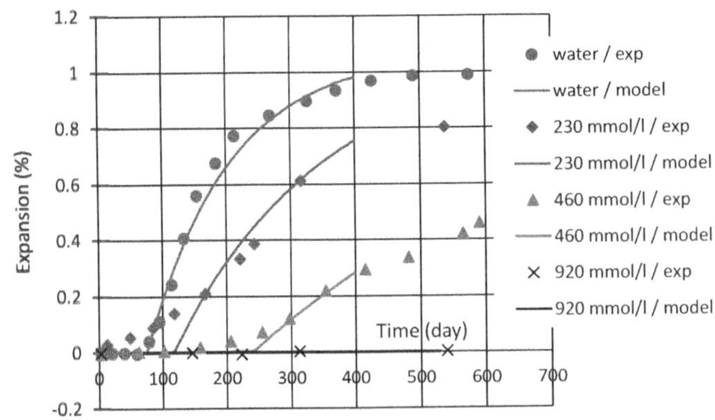

Figure 5. *DEF-expansion according to alkali leaching, experimental points from (Famy* et al.*, 2001)*

This relative reference is necessary to reproduce expansion for concrete with small initial alkali concentration as concrete studied in (Brunetaud *et al.* 2008). It considers the dependence of thermodynamics equilibrium reached during hydration on the initial alkali content.

4. Conclusion

In order to obtain relevant assessment of structures damaged by ASR or DEF, prognosis should be based on objective expansion tests in laboratory. However, several mechanisms influence these tests (alkali and moisture transfer, reaction products permeation) and expansion measured on specimens with usual tests are difficult to transpose to structural calculations without a minimal consideration of implied chemical phenomena. This paper proposed modelling based on physicochemical mechanisms with simplified assumptions to be able to analyze specimens' expansion and assess structural behavior with the same parameters. For ASR, multi-scale modelling based on two main phenomena (diffusion in aggregate and silica reactivity) is able to reproduce the impact of alkali leaching and moisture transfer in laboratory. For DEF, a simplified modelling based on the mass balances of aluminum and sulfate ions has been developed to reproduce the impact of the duration of the high temperature curing and of alkali leaching on DEF

5. References

Brunetaud, X., 2005. "Etude de l'influence de différents paramètres et de leurs interactions sur la cinétique et l'amplitude de la réaction sulfatique interne au béton." Ph.D. Thesis, Laboratoire Central Des Ponts et Chaussées.

Brunetaud, X., Divet, L., and Damidot, D., 2008. "Impact of Unrestrained Delayed Ettringite Formation-Induced Expansion on Concrete Mechanical Properties." *Cement and Concrete Research* 38:1343–48.

Dilnesa, B.Z., Lothenbach, B., Renaudin, G., Wichser, A., and Kulik, D., 2014. "Synthesis and Characterization of Hydrogarnet Ca3(AlxFe1−x)2(SiO4)y(OH)4(3−y)." *Cement and Concrete Research* 59:96–111.

Divet, L., and Randriambololona R., 1998. "Delayed Ettringite Formation: The Effect of Temperature and Basicity on the Interaction of Sulphate and C-S-H Phase." *Cement and Concrete Research* 28(3):357–63.

Famy, C., Scrivener K.L., Atkinson A., and Brough A.R., 2001. "Influence of the Storage Conditions on the Dimensional Changes of Heat-Cured Mortars." *Cement and Concrete Research* 31:795–803.

Kchakech, B., 2015. "Etude de l'influence de l'échauffement subi par un béton sur le risque d'expansions associées à la réaction sulfatique interne." Ifsttar, Université Paris-Est.

Lindgård, J., *et al.* 2013. "Alkali-Silica Reaction (ASR) - Performance Testing: Influence of Specimen Pre-Treatment, Exposure Conditions and Prism Size on Alkali Leaching and Prism Expansion." *Cement and Concrete Research* 53:68–90.

Martin, R-P. 2010. "Analyse sur structures modèles des effets mécaniques de la réaction réaction sulfatique interne du béton." Ph.D. Thesis, LCPC Laboratoire Central Des Ponts et Chaussées Paris.

Multon, S., and Alain S. 2016. "Multi-Scale Analysis of Alkali-Silica Reaction (ASR): Impact of Alkali Leaching on Scale Effects Affecting Expansion Tests." *Cement and Concrete Research* 81:122–33.

Multon, S., Alain S., and Martin C. 2009. "Chemo-Mechanical Modeling for Prediction of Alkali Silica Reaction (ASR) Expansion." *Cement and Concrete Research* 39:490-500.

Ponce, J.M., and Batic O.R. 2006. "Different Manifestations of the Alkali-Silica Reaction in Concrete according to the Reaction Kinetics of the Reactive Aggregate." *Cement and Concrete Research* 36:1148–56.

Poyet, S. *et al.* 2006. "Influence of Water on Alkali-Silica Reaction: Experimental Study and Numerical Simulations." *Journal of Materials in Civil Engineering* 18(August):588–96.

Poyet, S. 2003. "Etude de la dégradation des ouvrages en béton atteints par la réaction alcali-silice : approche expérimentale et modélisation numérique." PhD Thesis, University Marne La Vallée.

Rivard, P., Bérubé, M.A., Ollivier, J.-P., and Ballivy G. 2003. "Alkali Mass Balance during the Accelerated Concrete Prism Test for Alkali-Aggregate Reactivity." *Cement and Concrete Research* 33:1147–53.

Salgues, M. 2013. "Modélisations Des effets structuraux des réactions sulfatiques internes et alcali-granulats : application aux barrages en béton." PhD Thesis, University of Toulouse.

Salgues, Marie, Alain Sellier, Stéphane Multon, Eric Bourdarot, and Etienne Grimal. 2014. "DEF Modelling Based on Thermodynamic Equilibria and Ionic Transfers for Structural Analysis." *European Journal of Environmental and Civil Engineering* 18(4):377–402.

Sellier, A., and Multon, S.. 2016. *Modélisation Simplifiée de La Réaction Sulfatique Rapport Interne LMDC /2016-2.* Technical Report, LMDC/INSA/UPS, Toulouse.

Zhang, Y., and Zhang, M. 2014. "Transport Properties in Unsaturated Cement-based Materials - A Review." *Construction and Building Materials* 72:367–79.

Can Certain Alkali Minerals Explain the Slow Reactivity of Granitic Aggregates in Dams?

ANTONIO SANTOS SILVA* — ISABEL FERNANDES** — ANA RITA FERRAZ** — DORA SOARES*

** National Laboratory for Civil Engineering*
Department of Materials, Av. Brasil 101, 1700-066, Lisbon, Portugal
ssilva@lnec.pt, dsoares@lnec.pt

*** University of Lisbon, Faculty of Sciences*
Department of Geology, Campo Grande, 1749-016, Lisbon, Portugal
mifernandes@fc.ul.pt, Ana.rita.e.f@gmail.com

ABSTRACT. *Swelling processes due to alkali-silica reactions are a major limitation to durability of concrete dams and hydraulic structures. These reactions occur in high humidity environments between the aggregates and the alkaline cement interstitial fluids, resulting in the formation of expansive alkali-silica gels. The speed, or even the occurrence, of these reactions is quite difficult to predict because it is not related simply to the mineralogical nature of the aggregates. A mineral can even prove to be stable or unstable in different rocks and the relative reactivity of various minerals is also different from aggregate to aggregate. So, there is controversy about which are the factors (chemical composition, structural type, micro-deformation, ...) that effectively control the alkali reactivity observed in some aggregates. Besides, there is no consensus about what kind of alkaline minerals are more likely to release alkalis to the interstitial cement solution, and also on the factors that accelerate that solubilisation. This paper aims to contribute to the ongoing discussion of this topic, and so different granitic aggregates, with sound and altered factions, were subjected to alkaline solubility tests in order to measure the alkalis released and to identify the mineral phases that contribute to alkali media raise. The results obtained confirm that granitic aggregates release alkalis in alkaline environment, being this release dependent on the degree of alteration of these aggregates. Natural altered aggregates have less capacity to release alkalis than the non-altered aggregates. In the tested granite samples we observe a higher release of potassium, which is attributed to the higher natural alteration of plagioclase in comparison to K-feldspars.*

KEYWORDS: *ASR, alkalis release, granites, alteration, quantitative evaluation.*

1. Introduction

Concrete swelling is one of the main degradation mechanisms that affects the hydro concrete structures nowadays. The alkali-silica reactions (ASR) are probably the ones that deserve more concern worldwide. These internal reactions involve an alkaline attack to the siliceous aggregates, being their development rate very dependent on water or humidity supply, which is permanently available in hydraulic structures. The reaction products formed, referred as alkali-silica gels, are able to adsorb water from the surrounding cement paste or from the environment, inducing enough expansive pressure to damage concrete that may compromise the operation and safety of these structures (Soares et al., 2015).

Several dams in Portugal show signs of deterioration due to ASR, but most of these dams started to exhibit signs of degradation only several decades after construction. This situation is not exclusive of Portugal, since other countries presents a similar situation (Curtis, 2000).

To develop ASR in a concrete structure the following prerequisites must be simultaneously fulfilled: presence of alkali reactive minerals in the aggregates, high content of alkalis in concrete and sufficient moisture supply (at least 80%). Prevention of ASR is therefore intended to eliminate at least one of the three aforementioned requisites. However, in the case of concrete dams some of these conditions may be difficult or impracticable to implement (e.g. reduce the access of moisture or the availability of local non-reactive aggregates).

Controlling the alkalinity of the concrete is the preventive measure that is in general recommended to avoid or suppress ASR. Pozzolans or "supplementary cementitious materials" (SCMs), such as low-lime fly ash, silica fume or metakaolin, have demonstrated to be effective in this respect (Thomas, 2011). However, some of these SMCs can supply significant amounts of alkalis to the concrete pore solution therefore modifying the alkali control level of ASR.

Besides the alkalis present in the cementitious materials and cement, some aggregates are supposed to contribute to ASR by releasing alkalis from minerals such as alkali feldspars and mica, despite the fact that alkalis are in stable chemical bounds within the minerals. This subject has been studied by different authors (e.g. Bérubé et al., 2002; Constantiner and Diamond, 2003; Lu et al., 2006) and methods of analysis were suggested in order to determine the content of alkalis that can be released when the aggregates are reacting with strong alkaline solutions, such as sodium and potassium hydroxide (Bérubé and Fournier, 2004; Soares et al., 2015, 2016). Constantiner and Diamond (2003) examined mortars containing low-alkali cement and aggregates with alkaline feldspars and concluded that these minerals can contribute with alkalis and therefore participate in ASR. Granitic aggregates have also been identified as alkalis suppliers being this release dependent on the mineralogy and particle size of the aggregate (Soares et al., 2015).

Granites contain feldspars, quartz and mica as main minerals and are classified as slow reactive as structures with these aggregates can present manifestations of ASR decades after being built. Table 1 presents the features that can be identified by petrographic analysis and which are supposed to contribute to the potential alkali reactivity of these rocks. The batches of granitic aggregates sold by quarries to concrete manufacturers usually contain rocks with variable degrees of alteration due to the heterogeneities of the rock masses. The objective of the present work is to determine the content of alkalis released by some known Portuguese granites and to identify the mineral phases that contribute to alkali media raise.

Type	**Type of feldspar**	**Perthite**	**Presence of micas**	**Argillization**	**Myrmekite**
Input	Type of alkalis release	Alkalis release	Alkalis release	Alkalis release	Alkalis and silica release
Photo	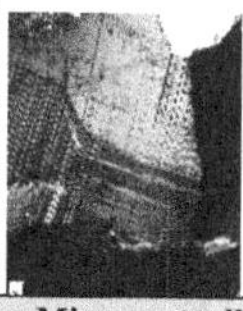	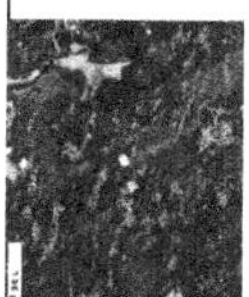	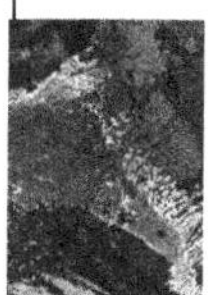	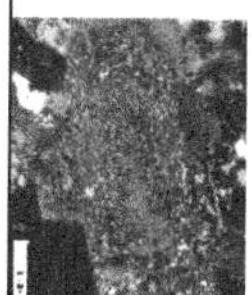	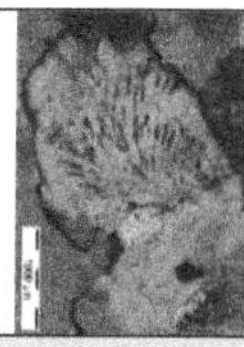
Type	**Microcrystalline quartz**	**Undulation extinction**	**ub-granulation**	**Fracture**	
Input	Silica release	Silica release	Silica release	Penetration of the interstitial fluids into the aggregate particle	
Photo	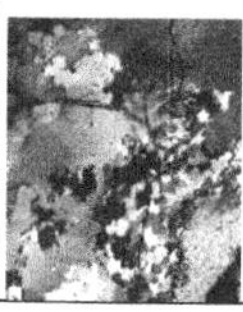	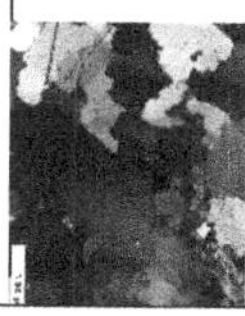	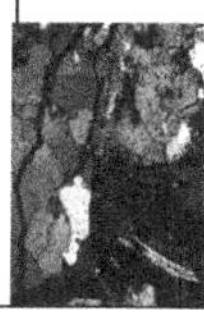	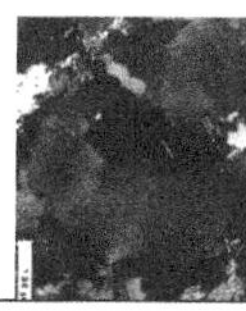	

Table 1. *Characteristics observed under petrographic microscope of granitic rocks which may contribute to ASR*

2. Materials and Methods

2.1. *Granitic Aggregates and Samples Preparation*

Four granitic aggregates identified as GR2, GR23, GR24 and GR29 were selected. For each of the samples selected, two different fractions were tested. One set of samples is composed of sound, not altered rock (N), the other set corresponds to altered granites (A), although with variable degree of alteration. The alteration is identified in these rocks by the dull appearance of the feldspars due to the replacement by clay minerals and by the chloritization of biotite.

These aggregates were previously tested by the accelerated RILEM AAR-4.1 concrete expansion test (RILEM AAR-4.1, 2016), which confirmed their potential reactivity. However, the expansion values obtained are different, namely GR23 and GR29 samples have expansion values of 0.03% at 15 weeks, whilst GR2 and GR29 have expansions of 0.06% at same age.

These granites were also characterized by chemical and petrographic analysis. The chemical analysis was performed on a representative sample of aggregate, crushed and grinded to pass a 106 μm sieve. For petrographic analysis, hand samples of different granitic aggregates were selected for production of 30 μm thickness thin-sections.

Finally, to evaluate the alkalis released by aggregates, the granitic samples were crushed and grinded to pass a 150 μm sieve.

2.2. *Chemical Analysis*

The chemical analyses of main oxides were performed by $LiBO_2/Li_2B_4O_7$ fusion at inductively coupled plasma (ICP).

2.3. *Petrographic Analysis*

Petrographic characterization was performed by analysis of thin-sections under polarizing microscope, coupled with automatic point-counter. This method enables to quantify the relative volume of potentially reactive components to alkalis, as well as the mineral content which may provide the alkalis for the system (e.g. plagioclase, K-feldspars, muscovite and biotite).

The assessment of petrographic alkali reactivity was based on local experience with granitic aggregates, namely on the dimensions of the quartz crystals (<100 μm = microcrystalline quartz) and according to the classes established by RILEM AAR-1.1 (2016) and by the Portuguese specification LNEC E 461 (2007). Myrmekitic quartz was included in the microcrystalline quartz group and considered as a potentially reactive form of silica (Ramos et al., 2016).

2.4. *Alkalis Release*

The method used to evaluate the alkali content released by the aggregates aims the extraction of Na^+ e K^+ ions in alkaline solutions simulating the interstitial pore solution of concrete (RILEM AAR-0, 2016). Therefore, an amount of representative aggregate sample (100 g) was immersed in a specific volume (400 mL) of alkaline solution at a ratio of 1:4 aggregate/solution. All sample materials were placed into hermetically sealed polyethylene bottles, and conditioned in a chamber at 38°C and RH> 95% (test conditions of RILEM AAR-3). Twice a week, the test containers were gently rolled back for about 10 seconds, and in predefined time intervals (e.g.: 2, 7, 13 and 26 weeks) a sample was taken from the test solution (10 mL). After filtration and acidification, the determination of alkali released content (Na^+ and K^+) of the extraction solutions was performed by atomic absorption spectroscopy (AAS).

Two alkaline solutions were employed: NaOH 0.7M (for the evaluation of K supply) and KOH 0.7M (for the Na supply).

3. Results and Discussion

3.1. *Chemical Analysis*

The results obtained for chemical bulk analysis (Table 2) show quite similar contents in SiO_2 and Al_2O_3 and low contents in Fe_2O_3 and CaO. Regarding the alkalis, all the samples are richer in K_2O than in Na_2O showing a ratio K_2O/Na_2O varying from 1.54 to 1.70.

	GR2 (Ramos, 2013)	GR23	GR24	GR29
SiO_2	72.76	72.66	72.31	72.71
Al_2O_3	14.13	14.04	14.55	14.06
Fe_2O_3	1.58	2.13	1.65	1.95
MnO	0.03	0.03	0.02	0.04
MgO	0.48	0.48	0.38	0.29
CaO	0.57	0.85	0.64	0.74
Na_2O	2.81	3.00	2.99	3.26
K_2O	4.43	4.96	5.19	5.08
$Na_2O_{eq.}$	5.72	6.26	6.41	6.60
TiO_2	0.21	0.26	0.26	0.17
P_2O_5	0.33	0.3	0.44	0.23
LOI	1.63	1.05	1.21	0.55
Total	98.98	99.77	99.64	99.09

Table 2. *Chemical composition of the granites (%)*

Considering the sodium equivalent content of the granites tested, with the individual values ranging from 5.72% (GR2) to 6.60% (GR29), it can be said that the potential for alkalis release should be higher in GR29, assuming that higher total alkalis content was related to a higher alkalis release.

3.2. *Petrographic Analysis*

The aggregates analysed are mainly fine- to medium-grained two mica granites, composed of quartz, K-feldspar, plagioclase (essentially albite, An_{0-10}, and oligoclase, An_{10-30}), muscovite and biotite, with variable contents of chlorite and other accessory minerals. Aggregate GR2 is a coarse-grained granite exhibiting hypidiomorphic porphyritic texture with quartz, K-feldspar and plagioclase as main components, biotite and muscovite is much lower content. The rock is intensely deformed with stretched crystals, ribbons, sub-graining and fractures. In GR2-A, K-feldspar, mainly microcline, usually shows signs of alteration. The plagioclase (oligoclase) is more altered than K-feldspar and the argillization is stronger in the core of the crystals (Figure 1). Biotite lamellar plates are partially to totally replaced by chlorite and muscovite larger crystals are deformed, showing kink-folds and symplectic fringes. There are some crystals of small dimensions of muscovite between the crystals of the main components of the rock, as well as microcrystalline quartz. In the altered fraction these characteristics are more intense, and biotite is almost totally altered to chlorite.

Figure 1. *Aggregate GR2: deformation of the rock reflected in sub-graining of the quartz. Plagioclase (pl) is altered, mainly in the core of the zoned crystal, and microcline (mc) does not show alteration (Crossed Polarized Light – CPL)*

GR23 is fine- to medium-grained granite exhibiting hypidiomorphic porphyritic texture. K-feldspar is composed of microcline and orthoclase in megacrysts containing inclusions of plagioclase. These minerals show alteration, which is more intense in the cores of the plagioclase crystals in the altered fraction of the aggregate (GR23-A). The larger crystals of plagioclase are zoned, with albite in the outer rims and Ca-richer fractions in the cores which show sericitization and crystals of muscovite developed along the cleavage planes (Figure 2). In some crystals also calcite is formed due to the alteration. Myrmekites are quite common. Muscovite larger plates show symplectic fringes. Biotite is altered to chlorite mainly along the cleavage planes.

Figure 2. *Aggregate GR23-A: crystals of microcline (mc) and plagioclase (pl). Plagioclase is altered to sericite and calcite (cc), mainly in the cores of the crystals (CPL)*

GR24 is a fine to medium-grained granite in which plagioclase forms crystals of variable sizes. The more abundant have the size of the crystals of the matrix and correspond to albite, the smaller are mainly oligoclase and occur as inclusions in the K-feldspar crystals. Fraction GR24-A exhibits argillization of the feldspars, which is stronger in the core of the crystals of plagioclase. K-feldspar, mainly microcline and some orthoclase, both perthitic, is also altered although in a lower intensity than plagioclase. Muscovite often shows symplectic fringes, mainly in the larger tabular crystals, but it can also occur as clusters of small crystals disposed in radial layout. Biotite lamellae are partly altered to chlorite. The main characteristic related to deformation of this rock is the presence of strain corridors and a fracturing network often lined by fine-grained quartz (Figure 3). In GR24-A also microcrystalline muscovite is present lining the fractures.

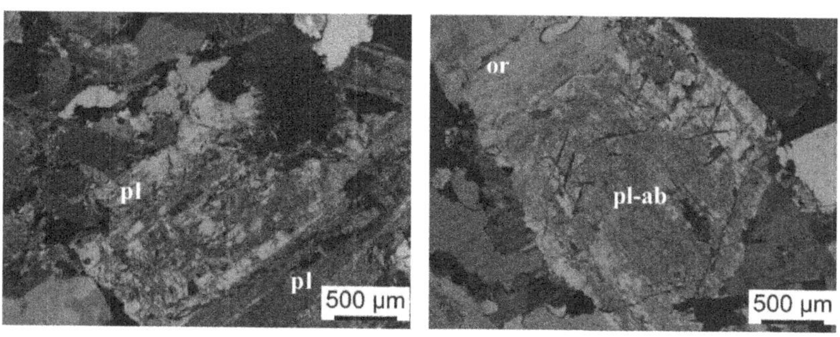

Figure 3. *Aggregate GR24-A: crystal of zoned plagioclase (pl). Albitization of K-feldspar (orthoclase - or) crystal and alteration of albite (pl-ab) to clay minerals (CPL)*

Aggregate GR29 is medium-grained granite. Perthitic orthoclase, which is the dominant K-feldspar, and microcline form the largest crystals in the rock and show some argillization, mainly in the altered fraction of the rock (GR29-A). There are two types of plagioclase, one occurring as small inclusions in the K-feldspar and mostly composed of oligoclase, and the largest crystals corresponding essentially to albite. These minerals show signs of seritization, which is more intense in the cores of the crystals, although it may affect, with less intensity, the whole crystals (Figure 4). In the altered fraction, crystals of muscovite occur along the cleavage planes of the plagioclase. Muscovite is the most common mica in the aggregate and has irregular boundaries. Biotite, less common, is frequently altered to chlorite.

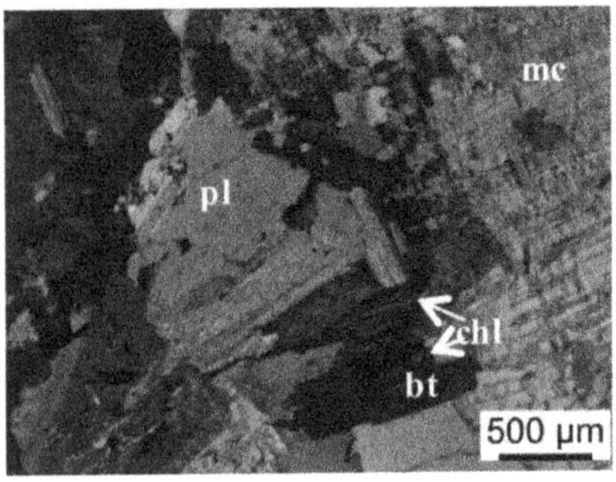

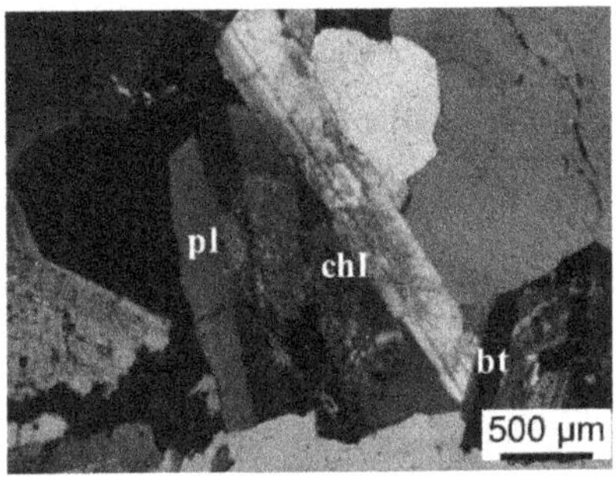

Figure 4. *Aggregate GR29: perthitic microcline (mc) and plagioclase (pl) crystals. Biotite (bt) lamellae are partially replaced by chlorite (chl) (CPL)*

Table 3 presents the mineral composition obtained by point-counting.

	GR2 (Ramos, 2013)	GR23	GR24	GR29
K-feldspar	26.3	29.9	31.4	32.9
Plagioclase	26.1	19.5	20.3	24.6
Quartz	34.2	34.4	33.7	32.7
Muscovite	9.7	7.5	10.0	4.0
Biotite + chlorite	3.1	8.1	4.1	5.7
Accessory minerals	0.6	0.6	0.6	0.3

Table 3. *Mineral composition of the granites obtained by point counting*

Quartz is the main component of three of the samples (exception is GR29) and K-feldspar (26.3 to 32.9%) is more abundant than plagioclase (19.5 to 26.1%). The content of muscovite is higher than biotite + chlorite in GR2 and GR24 with a ratio of 3.1 and 2.4, respectively. Biotite + chlorite are slightly more abundant in GR23 and GR29.

3.3. *Alkalis Release*

Figure 5 presents the results obtained for the Na_2O and K_2O contents released (mass%) during 26 weeks by the tested granites.

The analysis of these charts shows that for all the samples the altered fractions (A) release less alkalis than the non-altered fractions (N), although in the altered fractions there are more cracks and voids allowing the access of the pore fluids to the interior of the aggregate particles. In the altered fractions, feldspars have lost K^+ and Na^+ due to the alteration to clay minerals, which means that in these fractions there is lower alkalis content left to be leached during the tests. From these results, it can be concluded that the natural mechanisms of alteration have a stronger role in the releasable alkalis than the fact of the facilitated access of the fluids and the larger specific area available to the alkaline attack due to cracking.

It can also be verified that the content of K_2O released is higher than Na_2O in all the samples. This fact can be explained by the higher content in potassium in all the samples but, mainly, due to the cause above explained. In fact, the petrographic analysis allows to identify the alteration of the feldspars, which is in all the cases stronger for the plagioclase crystals than for the K-feldspars. Alteration in the cores of plagioclase crystals is visible even in the non-altered fractions. Therefore, there will be less Na available than K. Also, the results of alkalis released show a good correlation with the K-feldspar content obtained by petrographic point counting. The granites GR24 and GR29 are the ones with higher $Na_2O_{eq.}$ released (Figure 6), and the same that have the higher content in K-feldspar (Table 3).

Regarding the content in mica minerals obtained by petrographic analysis, it seems that the content on these minerals is not affecting much the results of alkalis released. For example GR23 has 7.5% of muscovite, against 4.0% for GR29, although the last one presented a higher release of alkalis.

Taking into consideration these facts, it can be said that the granites behaviour in the accelerated expansion tests should not be attributed only to the reactive silica constituents but also to the degree of alteration of the feldspars. The natural alteration that is present in some alkaline minerals of the granitic rocks, in this particular the plagioclase and the K-feldspars, can perhaps explain the slow reactivity of these rocks in hydro structures. Besides, when considering the degree of alkalis release it seems to depend on the alteration state of the rock, and can justify also the slow alkali-reactive behaviour of the granites.

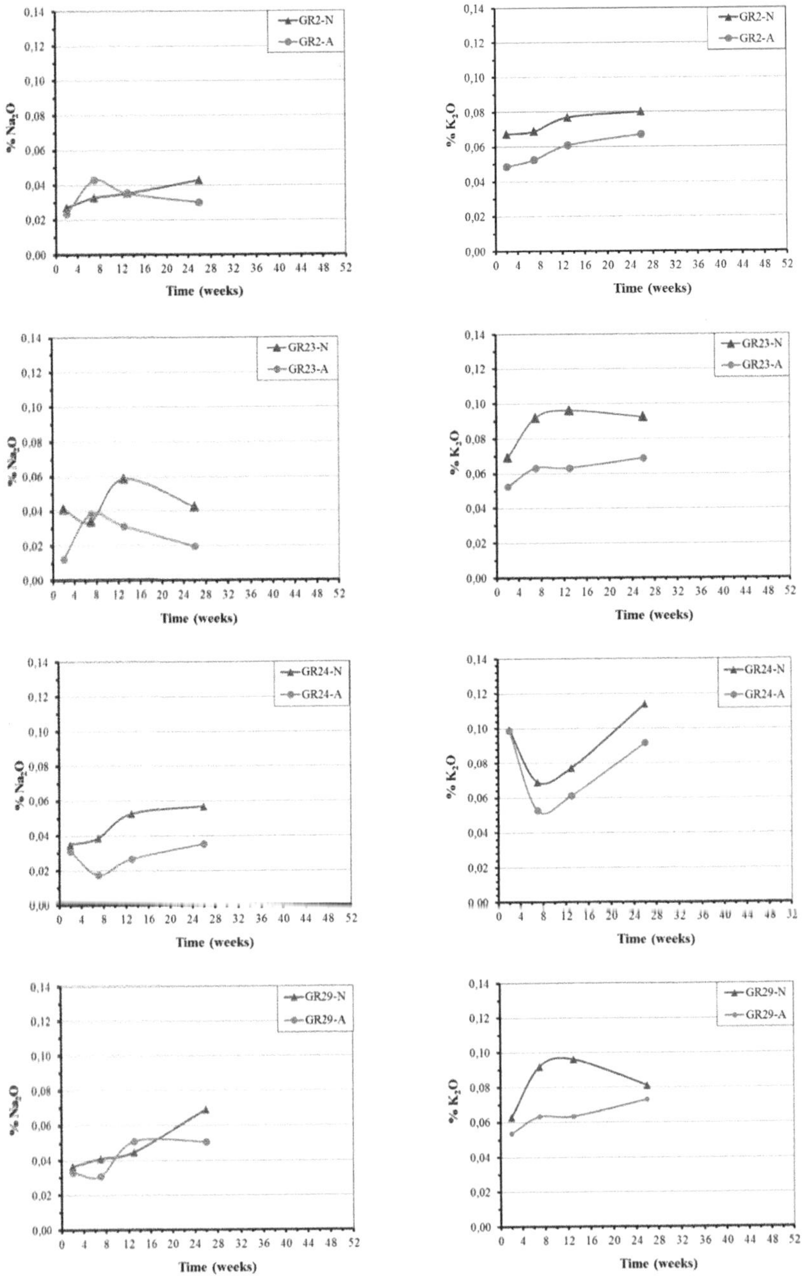

Figure 5. *Na_2O and K_2O (%) contents released with time by granitic aggregates in alkaline solutions (Notation: N – not altered rock; A – altered rock)*

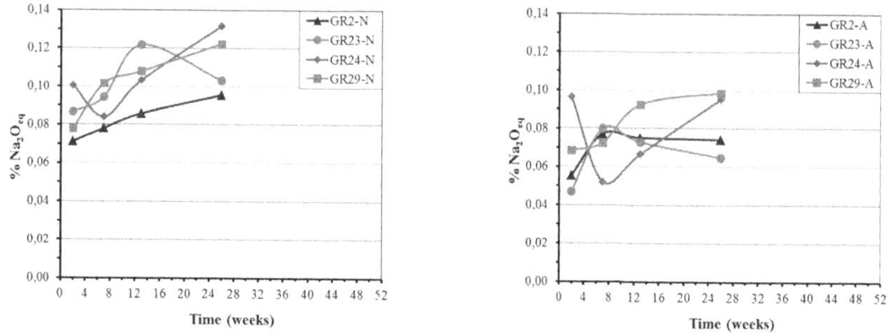

Figure 6. *$Na_2O_{eq.}$ (%) content released with time by granitic aggregates in alkaline solutions (Notation: N – not altered rock; A – altered rock)*

4. Conclusions

The work carried out is ongoing but confirms that granitic aggregates containing alkaline minerals in their composition are susceptible to release alkalis in alkaline environment. This release is dependent on the degree of alteration of these rocks. It has been found that natural altered aggregates, namely in terms of the dull appearance of the feldspars due to the replacement by clay minerals and by the chloritization of biotite, have less capacity to release alkalis than the non-altered aggregates. Also, for the four granites tested higher K extraction has been found, which is attributed to the higher natural alteration of plagioclase in comparison to K-feldspars.

Future studies should be conducted to confirm these findings in tests with different aggregates, and at longer extraction ages.

5. Acknowledgements

The authors wish to acknowledge LNEC for the financial support under project RE-IMPROVE – Expansive reactions in concrete – prevention and mitigation of their effects.

6. References

Bérubé, M.A., Fournier, B., "Alkalis release by aggregates in concrete – significance and test methods", In: Tang, M and Deng, M (eds). *Proceedings of the 12th International Conference on Alkali-Aggregate Reaction*, Beijing, China: 17-30, 2004.

Bérubé, M.A., Duchesne, J., Dorion, J.F., Rivest, M., "Laboratory assessment of alkali contribution by aggregates to concrete and application to concrete structures affected by alkali-silica reactivity". *Cement and Concrete Research* (32): 1215-1227, 2002.

Constantiner, D., Diamond, S., "Alkali release from feldspars into pore solutions", *Cement and Concrete Research* (33): 549-554, 2003.

Curtis, D.D., "A Review and Analysis of AAR-Effects in Arch Dam" In: M.A. Bérubé, B. Fournier, B. Durand (eds). *Proceedings of the 11th International Conference on Alkali-Aggregate Reaction*, Centre de Recherche Interuniversitaire sur le Béton, Université Laval, Quebec, Canada: 1273-1282, 2000.

LNEC E 461, "Betão. Metodologias para prevenir reacções expansivas internas", *Especificação LNEC*, Lisboa, Portugal: pp 6, 2007.

Lu, D., Zhou, X., Xu, Z.. Lan, X., Tang, M., Fournier, B., "Evaluation of laboratory test method for determining the potential alkali contribution from aggregate and the ASR safety of the Three-Gorges dam concrete". *Cement and Concrete Research* (36): 1157-1165, 2006.

Ramos, V., "Characterization of the potential reactivity to alkalis of Portuguese aggregates for concrete", PhD Thesis on Geology, University of Porto, Porto, Portugal: pp 417, 2013.

Ramos, V., Fernandes, I., Noronha, F., Santos Silva, A., Soares, D., Leal, S., Fournier, B., "Assessment of the potential reactivity of granitic rocks. Petrography and expansion tests". *Cement and Concrete Research*, (86): 63–77, 2016.

RILEM Recommended Test Method: "AAR-0 - Outline Guide to the Use of RILEM Methods in the Assessment of the Alkali-Reactivity Potential of Aggregates", RILEM Recommendations for the Prevention of Damage by Alkali-Aggregate Reactions in New Concrete Structures. State-of-the-Art Report of the RILEM Technical Committee 219-ACS. In: Philip J. Nixon and I. Sims (Eds). Springer Dordrecht Heidelberg New York London: 5-34, 2016.

RILEM Recommended Test Method: "AAR-1.1 - Detection of Potential Alkali-Reactivity - Part 1: Petrographic Examination Method", RILEM Recommendations for the Prevention of Damage by Alkali-Aggregate Reactions in New Concrete Structures. State-of-the-Art Report of the RILEM Technical Committee 219-ACS. In: Philip J. Nixon and I. Sims (Eds). Springer Dordrecht Heidelberg New York London: 35-60, 2016.

RILEM Recommended Test Method: "AAR-3 - Detection of Potential Alkali-Reactivity – 38°C Test Method for Aggregate Combinations Using Concrete Prisms", RILEM Recommendations for the Prevention of Damage by Alkali-Aggregate Reactions in New Concrete Structures. State-of-the-Art Report of the RILEM Technical Committee 219-ACS. In: Philip J. Nixon and I. Sims (Eds). Springer Dordrecht Heidelberg New York London: 79-97, 2016.

RILEM Recommended Test Method: "AAR-4.1 - Detection of Potential Alkali-Reactivity – 60°C Test Method for Aggregate Combinations Using Concrete Prisms", RILEM Recommendations for the Prevention of Damage by Alkali-Aggregate Reactions in New Concrete Structures. State-of-the-Art Report of the RILEM Technical Committee 219-ACS. In: Philip J. Nixon and I. Sims (Eds). Springer Dordrecht Heidelberg New York London: 99-116, 2016.

Soares, D., Santos Silva, A., Mirão, J., Ramos, V., Fernandes, I., Menéndez, E., "Assessment of alkalis released by aggregates. Contribution to the alkalinity increase and AAR development in concrete", In: Pina, C., Portela, E., Caldeira, L., Batista, A., Dias, I., Santos, R. (Eds): *Proceedings of the 2nd International Dam World Conference*, April 21-24, Lisbon, Portugal: 249-258, 2015.

Soares, D., Santos Silva, A., Mirão, J., Fernandes, I., Menéndez, E., "Study on the factors affecting alkalis release from aggregates into ASR", *Proceedings of the 15th International Conference on Alkali Aggregate Reaction*, São Paulo, Brazil: pp 9, 2016.

Thomas, M.D.A., "The effect of supplementary cementing materials on alkali–silica reaction: a review", *Cement and Concrete Research* (41): 1224–1231, 2011.

Experimental Study on Effects of Aggregates Mineralogical Composition and Preservation Conditions on DEF in Concrete

MARIE MALBOIS*,** — LOIC DIVET** — STÉPHANE LAVAUD** — JEAN-MICHEL TORRENTI**

** LMT – Cachan*
ENS Cachan, 61, avenue du Président Wilson, 94235 Cachan Cedex, France
malbois@lmt.ens-cachan.fr

*** IFSTTAR*
14-20, Boulevard Newton, Cité Descartes, Champs-sur-Marne, 77447 Marne la Vallée Cedex 2, France
marie.malbois@ifsttar.fr, loic.divet@ifsttar.fr, stephane.lavaud@ifsttar.fr, jean-michel.torrenti@ifsttar.fr

ABSTRACT. *DEF in concrete is likely to develop in massive civil engineering structures with major securities issues. DEF could lead to swelling and cracking which may impact in a significant manner mass transfer and mechanical properties. The on-going parametrical experimental campaign presented in this paper aims at quantifying the impact of aggregates chemical composition and of preservation conditions on DEF. Two concrete compositions are made, composed with calcareous or siliceous aggregates. The samples are then continuously preserved in water, or subjected to immersion and drying cycles. All compositions are monitored and tested at several deadlines to determine the evolution of mechanical (compressive and flexural tests). First results highlight the significant influence of aggregates composition on the formation kinetics, and also reveal that exposing samples to drying tends to slow DEF.*

KEYWORDS: *DEF, aggregates composition, drying cycles.*

1. Introduction

Delayed ettringite formation in concrete can have a significant impact on the long term behaviour of massive concrete structure. Ettringite is a crystal characterised in 1970 by Moore and Taylor (Moore *et al.*, 1970) and its formation controls the concrete hardening but also the material workability at young age. Under a particular set of conditions, ettringite formation could be delayed and could lead degradations of the material. The swelling induced by the development of delayed ettringite generates internal and external cracking and so affect the durability of the concrete structures. Finally, the economic impact of DEF is significant due to the maintenance and repairing, or even destruction fees.

It is unanimously admitted that DEF develops only under peculiar conditions:

– Heat treatment (>65°C) during the concrete hardening, due to the cement hydration process or to thermic treatments undergone by pre-casted concrete parts. (Kchakech, 2012) took an interest in defining the impact of the length and the temperature of the thermic treatment on the DEF.

– Humid environment: DEF only develops in concrete in contact with water (Al Shamaa, 2012) (Heinz *et al.*, 1989) (Odler *et al.*, 1995) due to the departure of alkalines and the pH drop.

– Cement composition: (Pavoine, 2012) studied the impact of sulphates, aluminates and alkalis proportions and also the cement fineness on DEF.

– Aggregates mineralogical nature: (Monteiro *et al.*, 1986) (Grattan-Bellew *et al.*, 1998) (Brunetaud, 2005) demonstrated that sand or aggregates petrographic nature can impact the DEF, in terms of kinetics but also amplitude.

The presented study takes an interest in quantifying the effects of aggregates nature and also of the environmental conditions by submitting samples to soaking and drying cycles.

2. Experimental Procedure

2.1. *Materials*

The concrete formulation used in this study is detailed Table 1.

The Portland cement CEMII/A L with 6% of calcareous additions (42.5 MPa) is used with a 0.57 water-cement ratio. The aggregates are either calcareous (Boulonnais aggregates) or siliceous (Palvadeau aggregates).

Materials	Composition
Cement	350 kg
Water	201 kg
Sand (0-5 mm)	858 kg
Aggregates (5-12.5mm)	945 kg
Viscosity modifying admixture	5,6 kg
W/C	0,57
A/S	1,10
Density	2354 kg/m^3

Table 1. *Concrete composition*

2.2. *Procedure*

2.2.1. *Thermic Treatment*

In order to simulate the heating induced by cement hydration in massive concrete structures, samples undergo a thermic treatment detailed in Figure 1, according to the LCPC N°66 method, with a 98% relative humidity. After treatment, samples are immersed in water (20°C ±3°C). Samples from the first study will be continuously kept under water, whereas samples from the second study will undergo soaking and drying cycles.

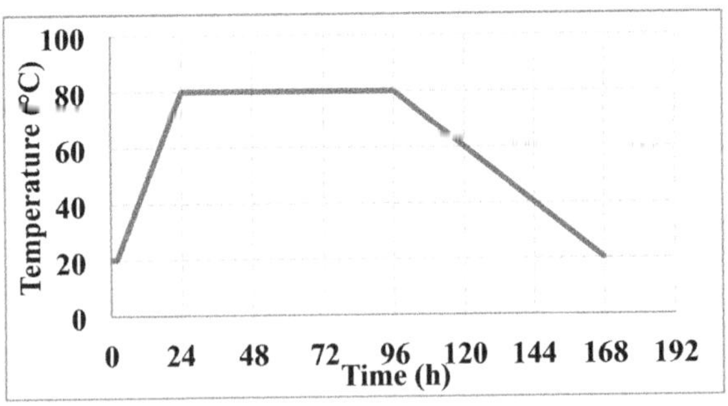

Figure 1. *Heating treatment*

2.2.2. *Protocols*

Two types of testings are performed in this study. The first one is the monitoring throughout the procedure of three 11*22 cm cylindrical samples for:

– mass variation;

– swelling, following the n°66 LCPC method;

– Dynamic Young Modulus, following the method detailed in (Brunetaud, 05) and (Al Shamaa, 12) with the “FDR” prototype.

The second type is mechanical testing at several deadlines, chosen to fit singular points of the expansion curves:

– 7 days, characterisation of healthy materials;

– a few months to a few years, reorientation of the expansion curves;

– from one to several years, stabilisation the expansion curves.

The tests performed are:

– compressive tests: three 11*22 cm cylindrical samples tested following the French norm (NF EN 12390, 2003), assessing the resistance strength and Young Modulus;

– three-points bending tests: two 6*10*40 cm notched prismatic samples, with a 3cm notch depth, tested following the (RILEM 95 method) to determine the flexural strength and fracture energy.

3. Effect of Aggregates Mineralogical Composition on DEF

3.1. *Effects on Expansion*

In his thesis, (Brunetaud, 05) identify a four phases degrading phenomenon of samples undergoing DEF: initiation phase, latent phase (swelling under 0.1%), acceleration phase (swelling between 0.1 and 2%) and a stabilisation phase.

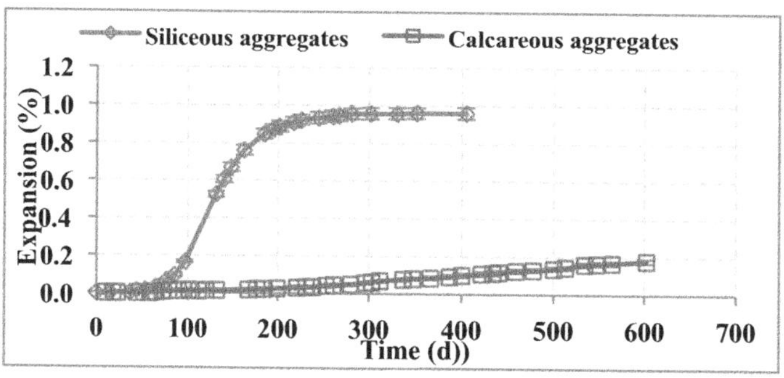

Figure 2. *Swelling of concrete samples with siliceous or calcareous aggregates*

The evolution of mean expansions is presented Figure 2. The curve for the samples made with siliceous aggregates displays the complete profile of DEF induced expansion and is stabilised after one year with a maximum relative swelling of 0.959%; while the curve for the samples made with calcareous aggregates only just left the latent phase and is in the beginning of the swelling phase. The use of siliceous aggregates induces an important increase of the kinetic of delayed ettringite formation in this concrete.

3.2. *Effects on Mechanical Properties*

3.2.1. *Effect of the Thermic Treatment on Mechanical Properties*

Both formulations are tested seven days after casting, on treated and not treated samples to determine the impact of the thermic treatment on the material. Results of compressive and three points bending tests are gathered in Table 2.

First, when taking an interest in the Young Modulus and compressive strength, the two formulations display a decrease, but it is more significant in the case of siliceous aggregates. Then, it can be noticed that the formulations have an opposite behaviour on the cracking energy results. Indeed, the "siliceous" samples undergo a cracking energy decrease of almost 15% after heating, whereas the "calcareous" samples cracking energy increase of more than 30%. These observations could be due to the nature of the interfacial transition zones and their evolution under heating.

		No treatment		**With treatment**		
		Value	*Ecart type*	*Valeur*	*Ecart type*	
Rc (MPa)	**Siliceous**	33.3	*0.9*	23.5	*0.9*	**- 29.5%**
	Calcareous	36.9	*1.1*	29.5	*0.5*	**- 20.1%**
E (GPa)	**Siliceous**	36.9	*0,13*	31.4	*0.75*	**- 14.8%**
	Calcareous	36.0	*0.14*	33.2	*0.30*	**- 7.7%**
Gf (N/m)	**Siliceous**	133.3	*9.6*	113.9	*6.5*	**- 14.7%**
	Calcareous	73.4	*2.5*	96.9	*7.4*	**+ 32%**

Table 2. *Effects of thermic treatment on mechanical properties*

3.2.2. *Effects of the Aggregate Composition on Dynamic Modulus Evolution*

As the expansion, Dynamic Modulus has several evolution phases (Brunetaud, 05): a latent phase; a significant decrease of the Modulus, corresponding to the material stiffness loss due to the initialisation of cracking induced by the formation of delayed ettringite and an increase of the Modulus, corresponding to stiffness recovering due to filling of the voids by the ettringite.

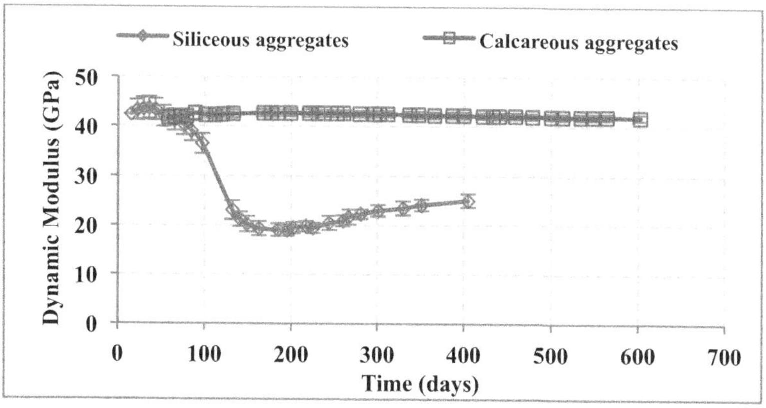

Figure 3. *Dynamic Young Modulus evolution of concrete samples with siliceous or calcareous aggregates*

On Figure 3, it can be observed that after one year, the samples with siliceous aggregates are at an advanced stage in the last phase, whereas the samples with calcareous aggregates are still in the dormant phase.

3.2.3. *Discussion*

The mineralogical nature of aggregates has a significant impact on the development of the pathology. The phenomenon is certainly induced by the different interaction between cement paste and aggregates. As a matter of fact, the roughness of siliceous or calcareous aggregates is different, leading to different adsorbed water on their surfaces, which could influence the transition zone properties and the quality of the interface. But also, siliceous aggregates are chemically inert during cement hydration, whereas calcareous aggregates have a better chemical affinity, and they react with the cement paste and form a hydrate which may strengthen the ITZ and decrease its porosity.

Moreover, the decrease of mechanical properties after the heating cure could perhaps reveal that in case of siliceous aggregates, that the ITZ is fragile and the cure induces a first damage on the samples. In that case, the delayed ettringite could freely crystallise in the ITZ porosity and the cracks induced by the heating treatment. The samples with calcareous aggregates seem to offer less favourable conditions to the development of delayed ettringite, corresponding with a delayed appearance of DEF, and a slower evolution.

We will now take an interest in the evolution of the mechanical properties with the formation of delayed ettringite in samples made with siliceous inclusions.

4. Impact of DEF on Mechanical Properties

4.1. *Evolution of Mechanical Properties*

The formation of delayed ettringite in concrete structure leads to superficial an internal cracking, which could jeopardise the durability of structure and so its stability and the users security due to the alteration of mechanical properties. We will now see in which extend the DEF in samples composed with siliceous aggregates modifies the mechanical behaviours.

Results from compressive and bending tests at the three deadlines mentioned previously are gathered in Table 3.

	7 days		**5 months**		**1 year**	
	Value		*Value*		*Value*	
Rc (MPa)	23,5	*0,9*	12,3	*0,2*	12,5	*0,1*
E (GPa)	31,4	*0,75*	7,8	*0,18*	10,2	*0,28*
Gf (N/m)	113,9	*13,0*	100,5	*14,0*	97,8	*3,3*

Table 3. *Impact of DEF on mechanical properties*

These results highlight the decrease of performance as soon as the expansion of samples begins for all indicators. However the decrease rates are the same for each factor:

- Compressive strength: -50%.
- Static Young Modulus: -70%.
- Dynamic Young Modulus: -45%.
- Cracking Energy: -15%.

The recovering of the Young Modulus or compressive strength noticed at in the final testing; but also leads to a more limited decrease; is certainly due to phenomenon described before: the delayed ettringite fills the voids of the concrete and brings stiffness to the material.

4.2. *Viewing of the Ettringite*

To verify the causes of the damages and confirm the development of DEF, samples were controlled by the use of X-ray micro-tomography at every deadline and observed with a scanning electron microscope. The images in Figure 4 are slides from the micro-tomography scans made at the mechanics laboratory of Lille (LML).

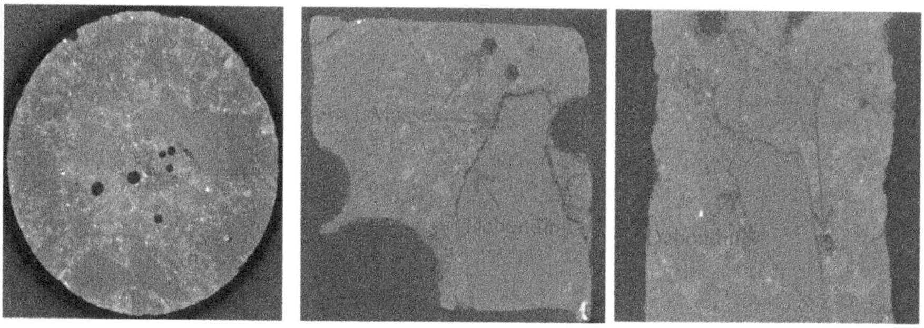

a b c

Figure 4. *Tomographic monitoring of samples with siliceous aggregates. Mean diameter: 6mm, voxel size: 4µm. a.7 days, b. 5 months, c. 1 year*

Image 4a is the reference image, scanned at the beginning of the study. The sample scanned after five months (Image 4b) shows an important porosity in which it can be noticed the formation of masses and also an important debonding phenomenon at the cement paste-aggregates interfaces. The cracks and porosity is progressively filled by the masses, as observed in the scan after one year (Image 4c).

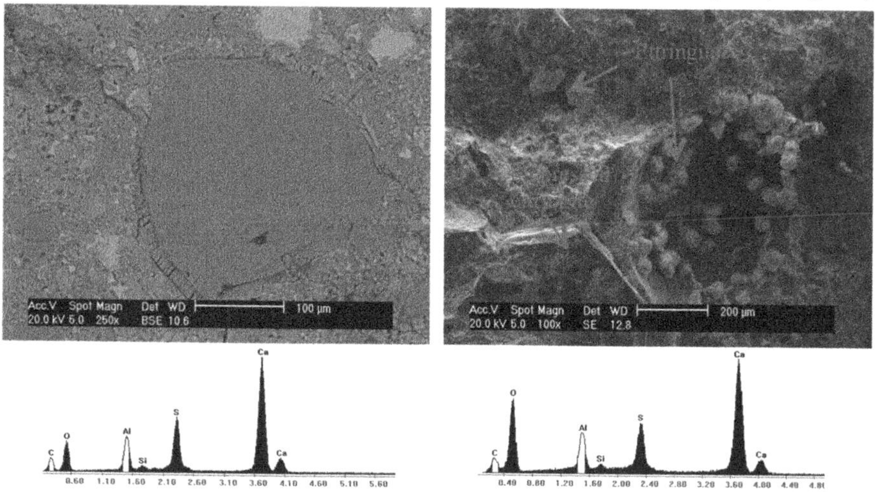

a b

Figure 5. *Observations of samples with a scanning electron microscop. a. Vein of compressed ettringite at the aggregate-cement paste interface, b. Void with free ettringite formation and aggregate covered with compressed ettringite*

In order to determine the nature of these masses, the sample scanned after a year was observed with a scanning electron microscope in the laboratory IFSTTAR. Figure 5 displays the observations. Image 5a; obtained on a polish surface; indicate the presence of typical delayed ettringite formation as the form of a vein of compressed ettringite in an ITZ. Figure 5b, obtained on a fresh surface, presents two vacuoles with ettringite masses, and also an aggregate covered with compressed ettringite. The spectrums obtained with the Energy Dispersive X-Ray Spectroscope of the microscope confirm that the observed species is ettringite.

5. Effect of Soaking – Drying Cycles on DEF

A second set of samples with calcareous aggregates was casted and submitted to thermic treatment, and is submitted to drying (20 ± 3°C – 50%RH during 14 days) and soaking (20 ± 3°C during 7 days) cycles. These preservation conditions represent a more realistic environment of some structures which are not in constant contact with water. The swelling monitoring of this new study is compared Figure 6 to the one of the composition previously studied and kept continuously in water. It seems that the cycles restrict the delayed ettringite formation in the material.

The curves presented Figure 7 highlight that the cycles slow down the formation, and do not prevent it. In other words, in the long term, the expansion will reach the same range than the one of immersed samples.

This delay could be due to the fact that thirty-two water molecules are needed to the crystallisation of ettringite (3CaO.Al2O3.3CaSO4.32H2O) and during drying periods, the lack of water prevents its formation. Also, it could reveal the necessity of water as a reaction medium for DEF.

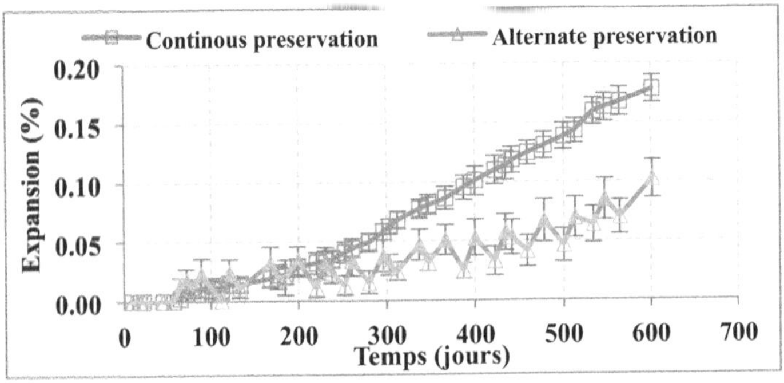

Figure 6. *Impact of drying cycles on the swelling of concrete samples*

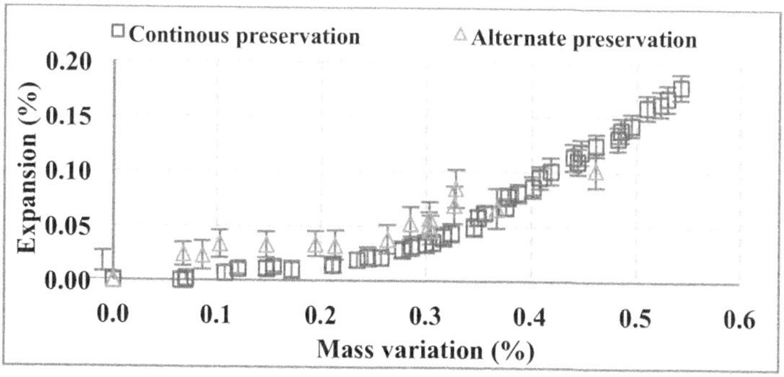

Figure 7. *Impact of drying cycles on DEF amplitude*

6. Conclusion

The use of siliceous aggregates in heat-treated concrete leads to rapid and significant damaging due to DEF of the material; whereas the use calcareous aggregates in concrete leads to slow manifestation of DEF damaging. We make the hypothesis that the main reason of these different behaviours could lay in the nature of aggregates-cement paste interfaces. The porosity and the mechanical behaviour of the ITZ are linked to the adsorbed water on the aggregate surfaces which is very dependant of aggregates nature. Also, calcareous aggregates have a better affinity with cement paste and leads to the formation of a hydrate which strengthen the ITZ. Moreover the decrease of mechanical properties of concrete made with siliceous indicates a pre-damage of the material, which could offer additional voids in which the delayed ettringite can easily crystallise.

Exposure to drying cycles tends to slow down the formation of delayed ettringite, indicating that water act as a reagent in the DEF but also as a reaction medium.

7. Acknowledgements

This research was achieved in the framework of the French research Project ANR MOSAIC. The authors would like to extend their appreciation and gratitude for the financial support provided by the French National Research Agency.

8. References

Al Shamaa M., "Etude sur le risque de développement d'une réaction sulfatique interne et de ses conséquences dans les bétons de structure des ouvrages nucléaires", Thesis, Université Paris-Est, France, 2012 (in French).

Brunetaud X., " Étude de l'influence de différents paramètres et de leurs interactions sur la cinétique et l'amplitude de la réaction sulfatique interne au béton", Thesis, École Centrale des Arts et Manufactures, France, 2005 (in French).

Grattan-Bellew P.-E., Beaudoin J.-J., Vallee V.-G., "Effect of aggregate particle size and composition on expansion of mortar bars due to delayed ettringite formation". *Cement and Concrete Research*, Vol. 28, n°8, 1998, p. 1147-1156.

Heinz D., Ludwig U., Rüdiger I. (1989), Delayed ettringite formation in heat treated mortars and concretes, *Concrete Precasting Plant and Technology*, Vol. 11, 1989, p. 56-61.

Kchakech B. (2015), "Etude de l'influence de l'échauffement subi par un béton sur le risque d'expansions associées à la Réaction Sulfatique Interne", thèse de doctorat, Université Paris – Est, France, 2015 (In French).

Méthode d'essai des LPC n°66, "Réactivité d'un béton vis-à-vis d'une réaction sulfatique interne. Essai de performance, techniques et méthodes des laboratoires des ponts et chaussées", LCPC, 2007.

Monteiro P.J.M. & Mehta P.K., "The transition zone between aggregate and type K expansive cement", *Cement and Concrete Research*, Vol. 16, 1986, p. 111-114.

NF EN 12390-3, "Essai pour béton durci – résistance à la compression des éprouvettes", 2003.

Moore A.E., Taylor H.F.W. (1970), Crystal Structure of Ettringite, *Acta Crystallographica*, 26(4), 1970, p. 386- 393.

Odler I., Chen Y., "Effect of cement composition on the expansion of heat-cured cement pastes", *Cement and Concrete Research*, 25(4), 1995, p. 853-862.

Pavoine A., Brunetaud X., Divet L., "The impact of ciment parameters on Delayed Ettringite Formation", *Cement and Concrete Composites*, 34, 2012, p. 521-528.

RILEM 95, "Determination of the fracture energy of mortar and concrete by means of three-point bend tests on notched beams".

The Identification, Extent and Prognosis of Alkali-Aggregate Reaction Related to Existing Dams in Switzerland

RUSSEL MICHAEL GUNN* — KAREN SCRIVENER** — ANDREAS LEEMANN***

** Swiss Federal Office of Energy (SFOE)*
Mühlestrasse 4, 3063 Ittigen, CH-3003 Bern, Switzerland
russell.gunn@bg-21.com

*** EPFL, École polytechnique fédérale de Lausanne*
Laboratory of Construction Materials
EPFL-STI-IMX-LMC, Station 12, CH-1015, Lausanne, Switzerland
karen.scrivener@epfl.ch

**** Empa Materials Science and Technology*
Überlandstrasse 129, CH-8600, Dübendorf, Switzerland
andreas.leemann@empa.ch

ABSTRACT. *Based on visual observations, instrumentation monitoring and subsequent behavioural evaluations of the structure, a campaign of in-situ testing with specific short, medium and long-term (perhaps with remedial measures) goals may be undertaken. The main objective of this document is to present appropriate test procedures and methods that provide the necessary quality information needed to assess AAR within dams and how this may affect the safety of the structure over its operating life span. The practical and theoretical developments, advantages, disadvantages, cost and time implications related to these tests are the key elements of the presentation. The extent, prognosis and numerical modelling of the AAR phenomenon specifically within the context of mass (unreinforced) concrete dams in Switzerland are also developed. A practical decision making tool for the diagnosis of alkali-aggregate reaction in dams is also exposed and a novel approach using data measurements and numerical methods is proposed to be able to select core samples from the dam for subsequent AAR testing that produces representative and reliable results of the existing structure.*

KEYWORDS: *alkali-aggregate reaction, alkali-silicate reaction, dams, diagnosis, prognosis, testing (and all words in the text that are written in italics).*

1. Introduction

Between 1945 and 1970, Swiss hydropower and the construction of many concrete dams went through the boom period and accounted for almost 90% of the domestic energy production. By 1985, with the commissioning of five nuclear power stations, this figure fell to 56%, still representing an important reliance on hydropower as a source of renewable energy. Today, 69 gravity, 51 arch, 2 multiple-arch and 2 buttress concrete dams and 27 river weirs comprise the portfolio of hydropower schemes in Switzerland. Four of these large dams are greater than 200 m and 25 are greater than 100 m. Concrete dams therefore represent an important source of energy supply and revenue in Switzerland.

In *Switzerland*, mass concrete dams and their associated normally *reinforced* appurtenant structures (spillways, water intakes, etc.) are ageing and some have already been rehabilitated following the swelling effects of AAR (alkali-aggregate reaction) (SwissCOD 2015). These rehabilitation works and related studies were performed over a period of more than 20 years running in parallel to the development of new AAR research projects that have greatly increased the understanding of the AAR phenomenon (Ben Haha 2006, Dunant 2009, Giorla 2013, Cuba Ramos 2017).

In the early stages of AAR identification many conflicting test results were obtained about the potential or existence of AAR in dams. In one particular case (a gravity dam), for which in-situ irreversible displacements and visual observations intimated the presence of AAR, subsequent laboratory testing indicated that on a petrographic level the aggregates were *potentially reactive*, however on a core sample level (residual expansion tests) the mass concrete was far below AAR limits and therefore *potentially non-reactive*. Shortly afterwards, modified laboratory testing procedures and a critical reinterpretation of the initial expansion phase due to water absorption for the same dam core samples indicated the contrary of previous results and that the entire structure was comprised of potentially reactive materials. Rehabilitation of the structure then consisted of releasing AAR induced stresses by saw cutting selected vertical joints in the dam body.

Experience gained from such projects highlights many important issues related to the way that AAR is and can be realistically identified within Swiss dams. This article capitalises on past studies that have been performed under partial or full mandate of the Swiss Federal Office of Energy such as Empa (Leeman *et al.* 2013) and the EPFL (Ben Haha 2006, Dunant 2009, Giorla 2013, Cuba Ramos 2017) and provides a return of experience within the context of dam surveillance and safety. In short, a procedural tool for the identification, extent and prognosis of AAR is proposed that is relevant to existing Swiss dams in terms of their geographic location/altitude, geological/hydrogeological foundation properties (SwissCOD

2006), construction materials (SwissCOD 2000), seasonal/environmental boundary conditions, operating regimes (normal and pumped storage production), elementary and combined static and seismic loading, the resulting displacements and stresses within the structure and finally, surveillance and dam safety (SwissCOD 2003).

2. AAR in the Context of Dam Engineering and Safety

The majority of large Swiss concrete dams are built on gneiss, granite limestone and slate foundations. Typically, the aggregates in dam concrete mixes use locally sourced materials. The problematic minerals (mainly strained quartz) in some of these aggregates may slowly (20 to 30 years post-construction) react to different degrees with the alkaline pore solution (pH > 13) of concrete which in a moist environment (> 80%) may lead to concrete swelling. AAR encompasses reactions with silica (SiO_2), silicates (e.g. feldspar, mica, etc.) and carbonates, ASR, ASSR (slower rates than ASR) and ACR respectively and herein we focus mainly on ASR.

2.1. *Geological Context and Supplementary Cementitious Materials*

The geological context within which the dam is located, hence the reason why we specifically refer to dams in Switzerland herein, needs to be well understood to identify and understand the ASR phenomenon. ASR in highly reactive aggregates such as Opal is identified by dark rims around the aggregate, whereas for granite and gneiss, the ASR reaction process starts within the aggregate. Moreover, for the latter the reaction process is slower (Ben Haha 2006).

Investigations on microbars/mortar-bars, MBT, (SIA Cahier technique 2042) have shown that from 79 quarries distributed over Switzerland, approximately 85% of the rock sources can be classified as potentially reactive (expansion > 0.11%) and that the *slow reaction process starts within the aggregate* (SwissCOD 2015, Leemann *et al.* 2013). This information paints a broad picture of what is to be expected from first stage AAR petrographic (SN 670 115 & 116) and microbar (SIA Cahier technique 2042) testing.

It should also be mentioned that since the first concerns of AAR were published (Stanton 1940), mass concrete dams typically require a two-staged process of thermal control: pre-and post-cooling (Stucky & Derron 1957). The pre-cooling phase addresses the heat of hydration which can be controlled by the addition of pozzanlic materials in the mix design, supplementary cement materials, SCM, which are also a counter-measure against the effects of ASR. Hence, to control the temperature rise in mass concrete, most dams contain SCM, which are known to mitigate ASR.

In conclusion, it is vital to clearly identify both the aggregate types and SCMs used in mass concrete mix designs, of which there may be many within the dam body, so as to be able to identify the signs of ASR both in terms of the *spatial distribution* and *extent per pocket* of ASR gel.

2.2. *Behaviour of Dams Undergoing Expansion Due to AAR*

ICOLD (Bulletin 79 1991) provides a useful insight into the swelling of concrete dams due to AAR in terms of dam safety whereby it is has been stated that "*there are no potentially catastrophic or rapid failure safety issues raised by AAR in dams*". However, the extent of the phenomenon world-wide, the impact on the operation of hydraulic equipment governing the control of reservoir storage volumes and in the general safety evaluation of the structure to fulfil its design purpose, means that AAR is an important issue in the context of dam engineering and safety.

Generally, in Switzerland the expansion of dams between 100 and 250 m in height due to AAR may vary between 2 to 15 mm and 5 to 40 mm *per year* respectively (ca. 0.002% and 0.016% *per year*). AAR expansions may be restricted to regions of the dam, for example, crest, mid-section and/or base or to all of the structure and may generate compressive, tensile or mixed mode stresses depending on the boundary conditions and hyperstatic nature of the dam. Comparing AAR expansion values with those typically generated in the dam due to self-weight and full hydrostatic pressure, it becomes apparent that the magnitude of AAR irreversible displacements is quite important. Typically, vertical "uplift" displacements and an upstream horizontal drift, especially at the crest elevation, are associated with cracking in other areas of the structure which may be classified as structurally stable or unstable (Amberg 2013). Micro and meso scale cracking also leads to a reduction of tensile, compressive and elastic modulus material properties in the order of 50 to 70%, 25 to 60% and 60 to 70% respectively (SwissCOD, 2015)

2.3. *Factors Influencing the Identification of AAR in Existing Dams*

Typically, we start AAR investigations with the observed global behaviour of the structure based on visual inspection, dam instrumentation and/or geodetic readings that show a progressive increase with time in irreversible deformations that *could be* attributed to AAR and that cannot be attributed to other behavioural mechanisms such as plasticity and/or creep.

Experience has shown that the identification and the measure of extent of AAR in dams is highly dependent on the *selection of material samples*, whether it be from aggregate quarries used during the construction period, noting that quarries are difficult to access many years after construction for technical and economic reasons, or external crest, upstream, downstream faces of the dam or within the structure such as cores extracted from control and access galleries.

The investigation process is often hampered by the fact that the visual signs of AAR such as *gel exudations, surface map cracks lined with dark rims and expansion drifts* are not always visible (*low strains*) (Godart *et al.* 2013) making it difficult to select *non-affected reference specimens,* and *affected* areas of the structure from which core samples should be extracted and tested in the laboratory.

Clearly, *all the actors* in AAR studies such as geologists, material scientists, engineers and analysts need to know the factors that influence and impact on *how, where and to what extent* AAR is identified in the structure.

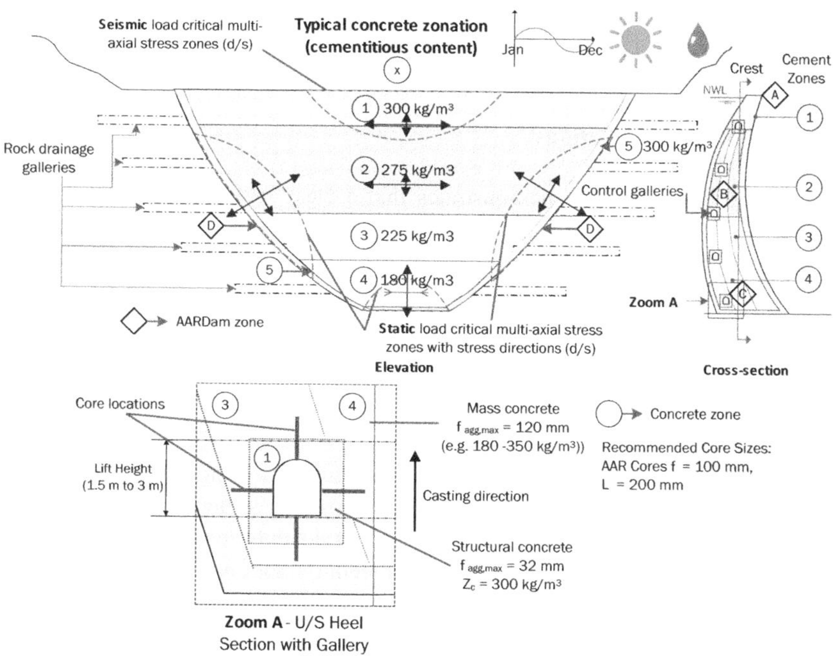

Figure 1. *Schematic view of elements affecting the selection of AAR testing core samples*

Amongst the *constituent parts* of mass concrete used in dams, we highlight:

– aggregate *types* noting also that mass concrete unlike structural concrete may have *maximum aggregate diameters* of 120 mm and *shapes* that can be angular (extracted from rock quarries) or rounded (rivers beds);

– cement or cementitious material *types* that are typically selected to have a low heat of hydration and *quantities* per cubic metre ranging from 150 to 400 kg/m^3 of concrete (Figure 1);

– external boundary conditions:

- temperature, humidity and multi-axial principal stress conditions in terms of magnitude and direction (Giorla A. B., 2013);

– core samples:

- size of extracted cores, how they are *handled*, *prepared* and *tested* are important for short- and long-term AAR assessments (Merz et al. 2013). The recommended diameter and length are 100 and 200 mm respectively.

In general, the characteristics and behaviour of the structure as well as the material properties at the micro, meso and macro scales are needed to benefit from current state-of-the-art field and laboratory investigation methods as described below.

3. AAR Material Investigations

Experience in Switzerland demonstrates that a clear *strategy for AAR material investigations* is needed. In addition to the points mentioned previously, it is important to define the goals at the very onset of the studies. Hence, one needs to distinguish between those tests related to *identification*, *extent* and *prognosis* of AAR to be able to define for example, rehabilitation AAR counter measures.

3.1. *AAR Prevention for New Dams*

This paper targets AAR investigations for existing dams, however many of the test procedures and methods described herein for existing structures can be applied to new dams. For economic reasons, dams are constructed of local materials that can be sourced and processed with crushing, screening and batching plants. Rock quarries are typically opened and closed following the completion of the structure. The main counter-measure against AAR is the use of natural or processed (i.e. fly ash) pozzolan.

Swiss Technical Note SIA 2042, 2015 unifies and updates existing AAR testing procedures and methods mostly for new structures and formulates a three-level procedure (P_1, P_2, & P_3) to prevent AAR. Dams in Switzerland are mostly classified as P3 and would therefore be subjected to all three levels of AAR testing.

3.2. *AAR Investigations for Existing Dams*

The following strategy is proposed for the investigation of AAR in existing dams:

– Phase 1: *AAR stage tests*, material source(s), e.g. quarries, and visual inspection and core sampling should cover all stages of the AAR process in the *short, medium and long term* of the structures service life including information needed for numerical modelling (constitutive material law calibration tests).

The stage or extent of AAR dictates the *type of AAR tests* that are to be performed.

– Phase 2: *AAR monitoring and data analysis*. Using statistical, deterministic (finite element modelling) and hybrid data analysis techniques to capture the potential zones of AAR swelling (visible and non-visible) and hence, perform analyses to assess the behaviour and safety of the dam.

– Phase 3: *AAR follow-up tests* should be systematically performed *every 5 years* (expert review period for dams under Swiss *direct surveillance guidelines*) from the onset of AAR. Field, laboratory and desktop studies are proposed herein to define the scope of testing.

It is vital to capture the historical development of the swelling process to better understand AAR and how it affects the structure during its service life.

3.3. *Phase 1: AAR Stage Tests – Look-Up Tables*

The objective of this section is to present *standard* and *supplementary/validating* AAR tests in easy to refer to *look-up tables* that allow the AAR project team to rapidly understand, identify and select current *state-of-the-art* AAR test methods for their particular project. By "*standard test*", we refer to tests that have passed uniform test procedures such as *round-robin* tests providing results that should be reproducible in different testing facilities. "*Validating or supplementary tests*" refer to tests that have yet to be validated as a standard and therefore, the results may be test facility/laboratory dependent.

To achieve this goal, we split the AAR test *look-up tables* into *three distinctive parts*:

– **Classification** (Table 1) of AAR tests by *category*, *location* (field, laboratory), *phase* (design, build, service life), *type* (chemical, mechanical, numerical, visual, crack index, optical and non-destructive), *sample type* (structure, core, polished slices and constituent materials such as aggregates), *classification* (qualitative, quantitative, identification, severity, prognosis), reliability (well-established,

established, rarely used, research domain, test still in its infancy), *quality of information* (excellent, good, satisfactory, poor, unsatisfactory) and *sample preparation* (difficult, standard, easy or none).

– **Influencing Factors** (Table 2 and Table 4): specimen size, quantity of cement, additives, relative humidity, temperature (°C), time (days) and cost (in Swiss Francs)

– **Results and Conclusions** (Table 3 and Table 5): *Standard limits* above which AAR is deemed to be potentially deleterious or otherwise, *laboratory and field observations* with an AAR classification index from levels 1 to 4 indicating low to high amounts of AAR respectively and finally, a summary of the advantages and disadvantages of the tests.

Column 2 of Table 1 to Table 5 provides *literature references* that may be referred to in order to ascertain *detailed information* about the specificities of test objectives, methods, results and conclusions.

To assist in the AAR test selection process, the authors provide their choice of recommended tests denoted by * — ** and ***.

3.3.1. *AAR Tests for New Dams*

AAR material testing for *new builds* and to a certain extent for existing dams have mainly focused on *three tests* (refer to references in brackets for details):

– *Petrographic* tests on aggregates (ASTM C-294, SN 670'115/116),

– *Microbars*, MBT on crushed aggregates (ASTM C-1260, AFNOR XP 18-594) and;

– *Concrete Performance* testing, CPT (ASTM C-1293, AFNOR P18-454). CPT normally requires the mix design with the highest cement content.

These tests are predominantly performed on *sourced aggregate samples* that are *readily available* (used for regular industrial applications such as *structural concrete*) which may not be the case for some existing dam locations whereby quarries have been closed and/or rendered difficult to access.

These tests are interrelated in the process of AAR identification and in general not used as standalone tests. They have also been further developed over time to provide quicker (*rapid, ultra-rapid*) and more reliable results (avoiding *leaching/lixiviation – alkali control*), but basically, they have a common thread in terms of method with the variables being those of *specimen size, chemical environment*, for example the use of sodium hydroxide/oxide, *temperature* and *humidity*.

3.3.2. *AAR Tests for Existing Dams*

Table 1 presents a list of *standard* and *supplementary* AAR tests for dams indicating recommended tests (*denoted by* * — ** *and* ***) as based on the authors' experience and engineering judgement. In general, standard tests are those described in National Standards/Directives implying a minimum level of evaluation/checking that is to be satisfied for new or existing dams. Supplementary tests validate the existence of AAR in dams and provide further *qualitative* and *quantitative* information pointing to the *extent* of AAR at a given moment in time (Leemann *et al.* 2013).

	Test Method	Ref.	N°	A	B	C	D	E	F	G	H	I	J
Standardised Testing Procedures	**Petrographic***	ASTM C-295 SN 670'115	1	1	1,2	1,3	1	4, 6	2,3,4	1,2,3,4	1	3	2
	Chemical	ASTM C-289	2	1	2	1,3	1	1	4	1,2,3	1	4	2
	Mortar-Bar	ASTM C-227	3	1	2	1,3	2	2	4	1,2,3	1	4	1
	Mortar-Bar (Modified)	ASTM C-1260	4	1	2	1,3	2	2	4	1,2,3	1	3	1
	Mortar-Bar - MBT* (Aggregates-Ultra-Rapid)	SN 2042	5	1	2	1,3	2	2	4	1,2,3	1	3	1
	Mortar-Bar (Accelerated)	ASTM C-1567	6	1	2	1,3	4	2	4	1,2,3	1	3	1
	Concrete Prism	ASTM C-1293	7	1	2	1,3	3	2	4	1,2,3	2	2	2
	Concrete Prism (Accelerated)	ASTM C-1293	8	1	2	1,2,3	3	2	4	1,2,3	2, 4	3	2
	Concrete performance Test - CPT*	SN 2042	9	1	2	1,2,3	3	2	2	1,2,3	1	3	2
	Petrographic tests on Hardened Concrete	ASTM C-856	10	1	2	1,2,3	3	4,6	2,3	1	2	3	2
	Mechanical Tests Compression, Traction (Brazilian), E modulus	SN 206-1	11	1	2	1,2,3	3	2	2	1,2	1	5	2
Validating/Supplementary Testing Procedures	**Field Surveillance**** Structural inspection (dam faces, galleries, etc.) & monitoring data evaluations (DAMBASE)	OFEN	12	2	1	3	3	4	1	1,2,3	2	2	3
	Field Test - CI** Crack Index (France) (LCPC)	LCPC	13	2	1	3	4	4-7	1,2,3	1,4	2	2	2
	Field Test - DRI Damage Rate Index - USA / ICOLD	ICOLD	14	2	1	2,3	3	4-7	1,2,3	1,4	2	2	2
	Field Test - NCM Norwegian Crack Method	NCM	15	2	1	2,3	3	4-7	1,2,3	1,4	2	2	2
	Optical microscopy** (Polarised light)	EMPA	16	2	2	1-3	4	4, 5, 6	3	1,2,3,4	1	2	1
	Microscopy** Scanning Electron Microscopy (SEM)	EMPA	17	2	2	1-3	3	1,5,6	3	1, 2, 3, 4	2	2	1
	Microscopy Energy Dispersive X-Rays (EDX)	EMPA	18	2	2		3	1,5,6	3	1, 2, 3, 4	2	2	1
	Microscopy Electron-Probe Micro Analyser (EPMA)	EMPA	19	2	2		3	1,5,6	3	1, 2, 3, 4	2	2	1
	Non-destructive (NDE) UPV (linear) NEWS (non-linear) X-Ray Tomography (3D) - XTM	EMPA	20	2	1,2	3	3	7	1,2	1	4,5	3	1
	Prognosis Testing** Residual Expansion concrete prisms	EMPA (CPT SN-2042)	21	2	2	1-3	3	2	2	5	1	2	2
	Prognosis Testing*** Numerical Modelling (Constitutive Laws)	EMPA EPFL	22	2	3	1-3	4	-	1-4	5	4	3	4

Category A	Location B	Phase C	Specimen D	Type E	Sample F	Classification G	Reliability H	Info Quality I	Preparation J
1 - Standard	1 - Field	1 - Design	1 - Aggregates	1 - Chemical	1 - Structure	1 - Qualitative	1 - Well established	1 - Excellent	1 - Difficult
2 - Suppl.	2 - Lab.	2 - Build	2 - Mortar	2 - Mechanical	2 - Core	2 - Quantitative	2 - Established	2 - Good	2 - Standard
3 - Other	3 - Other	3 - Service	3 - Concrete	3 - Numerical	3 - Slices (Polished)	3 - Identification	3 - Rarely used	3 - Satisfactory	3 - Easy
			4 - Other	4 - Visual/Eye	4 - Constituents	4 - Severity	4 - Research domain	4 - Poor	4 - None
				5 - Crack Index	5 - Others	5 - Prognosis	5 - Infancy	5 - Unsatisfactory	
				6 - Optical					
				7 - Non-destructive					

* *Recommended Standard Tests for AAR in Dams;* ** *Recommended Validation Tests for AAR in Dams;* *** *Recommended Prognosis Test*

Table 1. *Classified (letters A to J) Standard & Validating/Supplementary AAR Identification, Extent and Prognosis Tests*

Columns G to J of Table 1 are clearly *subjective* in nature and attempt to provide the investigator with a rapid overview of the information acquired from the test in terms of *classification, reliability, quality* and *level of difficulty* in the preparation.

Table 2 gives an overview of the factors influencing the various *standard* test methods with their corresponding limits as defined in the test procedural statements (refer to the references in column 2). The relative humidity applied in tests is in general above 85% and temperatures while in storage can vary depending on the duration of testing, with higher temperatures indicating a shorter test duration. The external alkali environment surrounding the samples is in general adjusted to limit the effects of leaching and the test duration ranges between *1 day* in the case of the ultra-accelerated test and more than *1 year*.

	Test Method	Ref.	N°	Sizes	Cement (kg/m3)	Adds	RH (%)	Temp (°C)	Time (days)	Cost-kCHF min.	max.
Standardised Testing Procedures	**Petrographic***	ASTM C-295 SN 670'115	1	Φ NX53-100 mm (Undeveloped quarries cores) 45 kg (selected quarries -sand, gravel)	-	-	-	38	2-4	1	2
	Chemical	ASTM C-289	2	3-25g crushed, Sieved: Passed (300-μm), Retained (150-μm)	-	1M NaOH	-	80	<1	1	2
	Mortar-Bar	ASTM C-227	3	25x25x285 mm, grading (crushed), 4.75mm-300-μm, (>4 specimens)	Variable	-	>50	38	14-365 then @+180	1	2
	Mortar-Bar (Modified)	ASTM C-1260	4	25x25x285 mm grading (crushed), 4.75mm-300-μm (>3 specimens)	Variable	1M NaOH	>50	80	16	1	2
	Mortar-Bar - MBT* (Aggregates-Ultra-Rapid)	SN 2042	5	10x10x40 mm grading (crushed) 16μm-63μm (4 specimens, 3 mixes: C/A 2, 5,10)	CEM I 42.5 480, 600	1.5% Na_2O_{Eq}	>95%	**100 (4h) 150 (6h)**	<1	1	2
	Mortar-Bar (Accelerated)	ASTM C-1567	6	25x25x285 mm, grading (crushed), 4.75mm-300-μm, (>3 specimens for each cementitious and aggregate combination)	Variable	1M NaOH	>50	80	16	1	2
	Concrete Prism	ASTM C-1293	7	75x75x285mm, (3 per mix), 19mm-4.75mm	420	Na_2O_e 1.25%	100%	38	7,28,56,90,180,270, 365 then @+180	1	2
	Concrete Prism (Accelerated)	ASTM C-1293	8	75x75x285mm, (3 per mix), 19mm-4.75mm	420	Na_2O_e 1.25%	100%	**60**	7,28,56,90,180,270, 365 > @+180	2	3
	Concrete performance Test - CPT*	SN 2042	9	70x70x282mm, (3 per mix), max 22.4mm	Variable	Na_2O_e Formula	100%	**60**	7,28,56,70,84 then @+28	2	3
	Petrographic tests on Hardened Concrete	ASTM C-856	10	Φ150X300mm (min), 200X400mm for better results	Variable	-	-	38	2-4	1	2
	Mechanical Tests Compression, Traction (Brazilian), E modulus	SN 206-1	11	Φ150X300mm (min), Φ = 2-3Dmaxagg, H = 2Φ	Variable	-	-	-	1	2.5	2.5

Table 2. *Influencing Factors – Standard AAR Identification, Extent and Prognosis Tests*

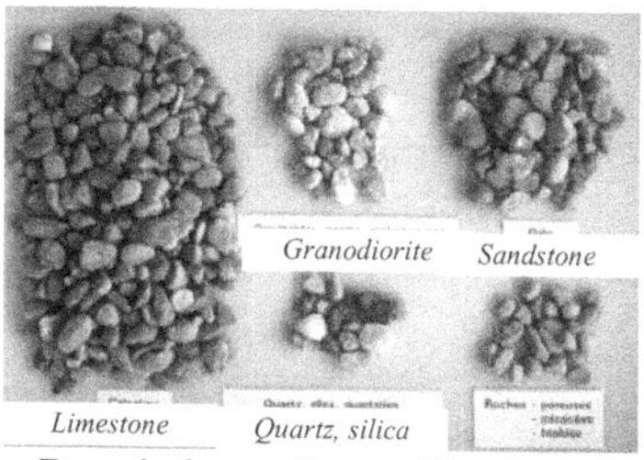

Description and quantification of rock types with an estimation of the relative concentration of potentially reactive aggregates

Petrographic (P)

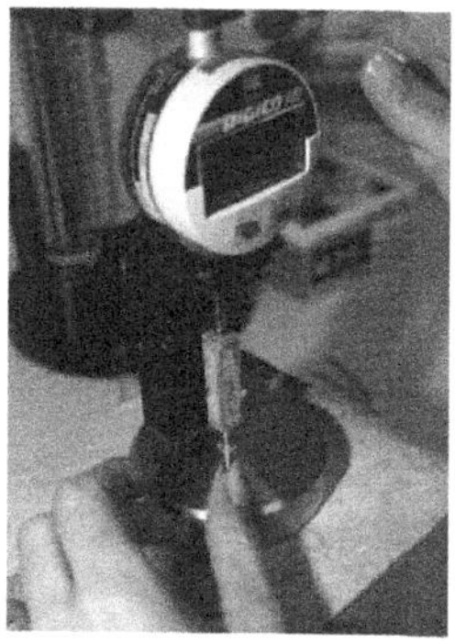

Microbar (MBT)

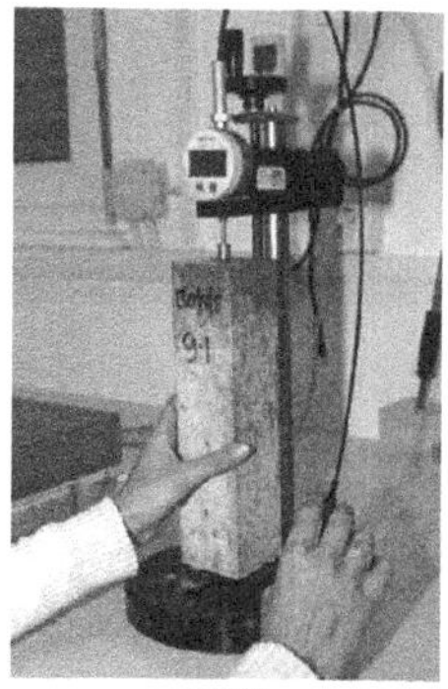

CPT
Concrete Performance

Figure 2. *Recommended (*) and other Standard AAR Tests (Hammerschlag et al. 2005)*

Figure 2 illustrates the test sample constituents, sizes and test apparatus for the recommended petrographic, MBT and CPT tests.

Table 3 provides a brief summary of AAR expansion limits based on: Standards, observations (laboratory, field and level – 1 high AAR, 3 low AAR), advantages and disadvantages of methods supporting the need to perform *supplementary/validation* tests to substantiate the identification and extent of AAR. Tests have been described in detail by Leemann *et al.* 2013 and a summary of the decisive elements for the selection of AAR tests, including recommended sample sizes, is provided in Table 4.

Field surveillance and monitoring is a vital source of information in the diagnostic and prognostic evaluation of dams. The measure of AAR damage at a given time (index T) and stress direction (index N) using the *crack index* at both field (C_{ITN}) and laboratory (CS_{ITN}) scales allows *displacements* (from dam instrumentation monitoring), *crack opening* (from field LCPC and laboratory optical microscopy methods), and AAR expansions (*strain,* measured at the laboratory scale) to be compared and coupled in a systematic way.

This way of coupling displacements, crack openings and strains at both the laboratory and structural levels is vital in structural mechanics (scale-effects, upscaling methods for constitutive laws).

	Test Method	Ref.	N°	Limits (max) Standards	ASR Observations Laboratory	Field	Level	Advantages	Disadvantages
Standardised Testing Procedures	**Petrographic***	ASTM C-295 SN 670'115	1	N/A	- Mineralogy, particle conditions, etc.	5	1	- Uranyl acetate facilitates ASR detection (microscopy)	- Not a standalone method for ASR detection, - Experienced petrographic expert & geologist required
	Chemical	ASTM C-289	2	Add result to chart: 1. Innocuous, 2. potentially deleterious, 3. deleterious	- Dissolved silica, reduction in alkalinity	-	4	- Low cost	- Poor reliability, severe test, - fails to identify slow reacting aggregates (low expansion for high silica content)
	Mortar-Bar	ASTM C-227	3	- 0.05%(3mo), 0.10% (6mo)	- Expansion	-	4		- Problems with leaching - no detection of slow AAR reactions - expansion not always related to AAR
	Mortar-Bar (Modified)	ASTM C-1260	4	@16 days, Δ_L> 0.10% (more tests), Δ_L > 0.20%, Potential deleterious, 0.10%<DL<0.20%, innocuous/delete.	- Expansion	-	4	- Fast and easy to perform	- many aggregates fail the (severe) test - suppl. tests needed for mid-range expansions (0.10% < Δ_l < 0.20%)
	Mortar-Bar - MBT* (Aggregates-Ultra-Rapid)	SN 2042	5	Non-reactive: @ < 1 day if, Δ_L< 0.110% and conforms to petrographic tests for all C/A mixes, If Δ_L > 0.110% for 1 of 3 C/A mixes - aggregate is reactive (more tests needed for that aggregate), C/A = 5	- Fabrication, initial measurement, steam cure 4h, alkaline cure 6h in autoclave at 150 °C, lower temperature to 20°C in containers, final measurement: Expansion, no information related to the degree of reactivity.	-	4	- Fast and easy to perform	- many aggregates fail the (severe) test, - supplementary tests needed for expansions in the order of 0.10% < Δ_l < 0.20%, - no correlation between MBT and CPT according to SN 2024 - Aggregate source may not be available for some existing dam locations where quarries have been closed and rendered difficult to access.
	Mortar-Bar (Accelerated)	ASTM C-1567	6	@16 days, Δ_L> 0.10%, Potential deleterious	- Expansion	-	4	- fast and easy to perform - Useful for cementitious materials typically used for dams (heat control)	- Perform also ASTM-1260 without SCM (pozzolan, etc.) - may overestimate the reactivity of some aggregate types used with SCM
	Concrete Prism	ASTM C-1293	7	Potential deleterious: @365 >= 0.04% or @730 for SCM >=0.04%	- requirement: w/c ratio 0.42-0.45, concrete alkali content = 5.25 kg/m3	-		- Considered to be the most representative test of field observations	- Meaningful results only after long time lapse (1 year)
	Concrete Prism (Accelerated)	ASTM C-1293	8	Potential deleterious: @365 >= 0.04% or @730 for SCM >=0.04%		-		- Faster test with possible reliable results	- Leaching leads to reduction in pH, expansion limit = 0.10%
	Concrete performance Test - CPT*	SN 2042	9	<u>Concrete non-reactive if 1 of 3 criteria is satisfied:</u> Criterion [1] @140 Δ_{Lmean}< 0.02% Δ_{Lmax}<0.025%, @90,120,150 2 of 3 values <0.0025, sum of 3 values < 0.01. Criterion [2] @150 deferred expanison, Δ_{Lmean}< 0.01%, Δ_{Lmax}<0.015%. Criterion [3] @365 conditions 1 & 2 not respected, Δ_{Lmean} < 0.03% Δ_{Lmax}<0.035%	- Only longitudinal deformation considered			- Modified AFNOR P18-454:2004 + Swiss updates - Standard provides a methodology for AAR testing related to Swiss Standard 206-1	- Validation tests are needed to compare CPT with cores from structure (SEM, EDX, etc.) - no correlation between MBT and CPT according to SN 2024 - Aggregate source may not be available for some existing dam locations where quarries have been closed and rendered difficult to access.
	Petrographic tests on Hardened Concrete	ASTM C-856	10	-	1-6: reactive aggregates, crack orientation, ASR gel formation	-	4	- Well established test	- Not a standalone method for ASR detection, - Experienced petrographic expert required for ASR evaluation
	Mechanical Tests Compression, Traction (Brazilian), E modulus	SN 206-1	11	-	- AAR damage may lead to problems in the test and the evaluation of the results	-	3	- semi-quantitative results	- large scatter - low sensitivity to AAR - no independent reference values

ASR Observations		
Laboratory	**Field**	**Level**
Signs	**Movements**	**ASR**
1 - Rims	1 - Raised Crest	1 - Low
2 - Pockets	2 - Horizontal "drift"	2 - Med.
3 - Exudation	3 - Crack > 2.5mm/m	3 - High
4 - Surface	4 - Other	4 -All
5 - Map cracks	5 - N/A	-
6 - Expansion	-	

** Recommended Standard Tests for AAR in Dams*

Table 3. *Limits, Observations, Advantages and Disadvantages of Recommended (*) and Standard AAR Identification, Extent and Prognosis Tests*

	Test Method	Ref.	N°	Sizes	Cement (kg/m3)	Adds	RH (%)	Temp (°C)	Time (days)	Cost-kCHF min.	max.
Validating/Supplementary Testing Procedures	**Field Surveillance**** Structural inspection & monitoring data evaluations (DAMBASE)	OFEN	12	-	Variable	-	-	Field	3	4	6
	Field Test - CI** Crack Index (France) (LCPC)	LCPS	13		Variable	-	Site	Site	7	2	4
	Field Test - DRI Damage Rate Index - USA / ICOLD	ICOLD	14	Structure	Variable	-	Site	Site	7	2	4
	Field Test - NCM Norwegian Crack Method	NCM	15	Structure	Variable	-	Site	Site	7	2	4
	Optical microscopy** (Polarised light)	EMPA	16	- Thin sections, 20-30 µm, (35 x 50, 90 x 50 mm²)	-	-	-	Field	60	6	6
	Microscopy** Scanning Electron Microscopy (SEM)	EMPA	17	- Thin sections, 20-30 µm, (35 x 50, 90 x 50 mm²), - MBT, CPT sections	Variable	-	-	-	60	3	3
	Microscopy Energy Dispersive X-Rays (EDX)	EMPA	18	- Thin sections, 20-30 µm, (35 x 50, 90 x 50 mm²), - MBT, CPT sections	Variable	-	-	-		3	3
	Microscopy Electron-Probe Micro Analyser (EPMA)	EMPA	19	- Thin sections, 20-30 µm, (35 x 50, 90 x 50 mm²), - MBT, CPT sections	Variable	-	-	-		3	3
	Non-destructive (NDE) UPV (linear), NEWS (non-linear) X-Ray Tomography (3D) - XTM	EMPA	20	-	Variable	-	-	-	30	8	8
	Prognosis Testing** Residual Expansion concrete prisms	EMPA (see also CPT SN-2042)	21	Recommended: Φ100X200mm or Φ150X300mm	Variable	Vary: Air, Water NaOH	>95%	38	170	7	7
	Prognosis Testing*** Numerical Modelling (Constitutive Laws)	EMPA EPFL	22	-	Variable	-	-	-	-	-	-

** *Recommended Validation Tests for AAR in Dams,* *** *Recommended Prognosis Test*

Table 4. *Influencing Factors for Validating/Supplementary AAR Identification, Extent and Prognosis Tests*

Table 5 highlights the observations, advantages and disadvantages of *validation/supplementary* AAR tests and proposes recommended tests (** — ***) that are deemed necessary for dam AAR investigations.

4. Proposed Method for Dam AAR Identification, Extent and Prognosis Evaluations

Having described *standard* and *supplementary* AAR tests and methods in chapter 3, the next step is to define a procedure that clearly defines which tests are to be performed at which stage of AAR investigations and how they are implicitly inter-related.

	Test Method	Ref.	N°	ASR Observations Laboratory	Field	Level	Advantages	Disadvantages	Remarks
Validating/Supplementary Testing Procedures	**Field Surveillance**** Structural inspection & monitoring data evaluations (DAMBASE)	OFEN	12	1,4,5	1,2,3,4	2,3	- fast and easy to perform	- only surface evaluation - accessibility could be an issue	
	Field Test - CI** Crack Index (France) (LCPC)	LCPS	13	5, 6	5, 6	2,3	- AAR status at given time (mm/m), - Simple method to validate/calibrate other AAR test methods	- Test operator dependent - Cracking not only due to AAR	- Operator dependent
	Field Test - DRI Damage Rate Index - USA / ICOLD	ICOLD	14	5, 6	5, 6	2,3	- Test is dam orientated (ICOLD)	- Test operator dependent - Essential to ensure that the concrete is frost resistant, - DRI dependent on the weighting factors assumed (subjective)	- Cracks continuing from the aggregates into the cement paste are not considered.
	Field Test - NCM Norwegian Crack Method	NCM	15	5, 6	5, 6		- AAR status at given time (mm/m) - Simple method to validate/calibrate other AAR test methods	- Test operator dependent - Cracking not only due to AAR	- Operator dependent - Reportedly a more reliable method compared to DRI
	Optical microscopy** (Polarised light)	EMPA	16	- Difficult to distinguish between AAR products and silicate rock fragments.		1-3	- Large surface areas evaluated. - Rapid first large scale evaluation method	- No chemical information retrieved. - Difficult to distinguish between AAR products and silicate rock fragments.	
	Microscopy** Scanning Electron Microscopy (SEM)	EMPA	17	1,2,3,4,5,6	1,2,3	1-3	Good resolution, Chemical composition with EDX, Good representativeness of dam.	Established test, but widely spread industry. Small surface area covered by tests.	
	Microscopy Energy Dispersive X-Rays (EDX)	EMPA	18	1,2,3,4,5,6	1,2,3	1-3	Good resolution, Chemical composition with EDX, Good representativeness of dam.	Established test, but widely spread industry. Small surface area covered by tests.	- Cores are taken based on suspect AAR locations derived from field observations, assumed stresses
	Microscopy Electron-Probe Micro Analyser (EPMA)	EMPA	19	1,2,3,4,5,6	1,2,3	1-3	Good resolution, Chemical composition with EDX, Good representativeness of dam.	- Established test, but widely spread industry. - small surface area covered by tests.	
	Non-destructive (NDE) UPV (linear), NEWS (non-linear) X-Ray Tomography (3D) - XTM	EMPA	20	XTM method is complementary to optical microscopy. Size to resolution choice means the size must be small for higher resolutions		2,3	- large volume of the structure analysed.	- not much experience of the tests (research domain). - test still under development. - interpretation of results difficult to confirm AAR.	- UPV can be a proxy for an actual mechanical parameter - XTM provides a 3D image with lower resolution than SEMs
	Prognosis Testing** Residual Expansion concrete prisms	EMPA (see also CPT SN-2042)	21	Phase 1 - Capillary action Phase 2 - Non-linear, then linear expansion Phase 3 - Irreversible expansion	5	1-3	- information on kinetics and magnitude obtained. - good correlation between CPT (end of test) and residual expansion tests for crack index, irreversible expansion and expansion	- leaching is a problem that renders the results unclear. - link with test and numerical model not clear.	- Core diameter (< 100 m), agg. diam., temperature, humidity & stress state can influence the results
	Prognosis Testing*** Numerical Modelling (Constitutive Laws)	EMPA EPFL	22	-	-	1-3	- numerical models may cover the entire structure as well as samples	- link to concrete properties not given. - often a black box.	- Research is still needed to validate constitutive models against lab and field tests.

ASR Observations		
Laboratory	**Field**	**Level**
Signs	**Movements**	**ASR**
1 - Rims	1 - Raised Crest	1 - Low
2 - Pockets	2 - Horizontal "drift"	2 - Med.
3 - Exudation	3 - Crack > 2.5mm/m	3 - High
4 - Surface	4 - Other	4 - All
5 - Map cracks	5 - N/A	-
6 - Expansion	-	

*** Recommended Validation Tests for AAR in Dams, *** Recommended Prognosis Test*

Table 5. *Limits, Observations, Advantages and Disadvantages for Validating/Supplementary AAR Identification, Extent and Prognosis Tests*

The proposed procedural flowchart for AAR testing in dams is given in Figure 3 whereby references denoted by [x] therein refer to column 3 ("N°") of Table 1. Hence, AAR test methods and procedures are unified in approach.

The main objectives of the method as presented in Figure 3 are as follows:

– *Where to select cores:* To make the Engineer aware of the importance of required strength zoning (cement content per m^3 of concrete), at the dam faces, core section and dam-foundation interface and of principal stress orientations, geographic orientation of the structure (south and north facing) and ambient temperature conditions, *five AAR investigation zones* are proposed (Figure 1 and Figure 3, zones 1, 2, 3, 4, 5).

– Core samples should try to intersection *several concrete zones* and it is recommended to discard the first 10 to 15 cm of core drilling that may have suffered damage from frost, freezing and thawing. In addition, care should be taken to avoid horizontal lift and vertical contraction joints since they may have different material properties compared to the main dam concrete zones.

– *Surveillance and monitoring*: Dams do not necessarily swell uniformly and AAR may not always be visible to the eye. Therefore, the application of statistical and deterministic methods is proposed for *four regions* (A, B, C and D) to target swelling within the structure. This information helps to *select core locations*, *monitor the long-term* behaviour of the structure and calculate a fictitious expansion/crack (micro) opening (CF_{ITN}) *corroborating* field and laboratory measurements.

4.1. *Phase 2: AAR Monitoring and Data Analysis*

The process of AAR identification starts with dam surveillance and monitoring data. AAR may be fully or partially visible (*Direct Approach*) or non-visible (*Indirect Approach*) to the eye. For the former case, the *Crack Index* C_{ITN} (mm/m. year) in accordance with the LCPC method (LCPC 1997) can be evaluated and core samples taken accordingly. For the latter, *four AAR zones* (*crest, crown base, mid-height section and abutments*) are prioritised and AAR displacements are estimated using statistical and/or deterministic (finite element) methods. The main objective at this phase of the studies is to define the location of AAR within the dam and compute a *fictitious AAR crack index* CF_{ITN} (mm/m. year) as a function of section height.

4.1.1. *Application of Statistical Data Monitoring Methods to AAR*

Statistical methods (*regression analyses*) uncouple the *causes and effects* and enable reversible & irreversible displacements (*drifts*) such as those due to AAR to be estimated (Gerber *et al.*, 2015). The procedure involves data regression analyses of the *control (comparison)* and *prediction* data periods based on the selection of mathematical models for water level fluctuations, temperatures (concrete, air, reservoir) and an exponential term for inelastic behaviour (AAR, damage, etc.) (Weber 2002).

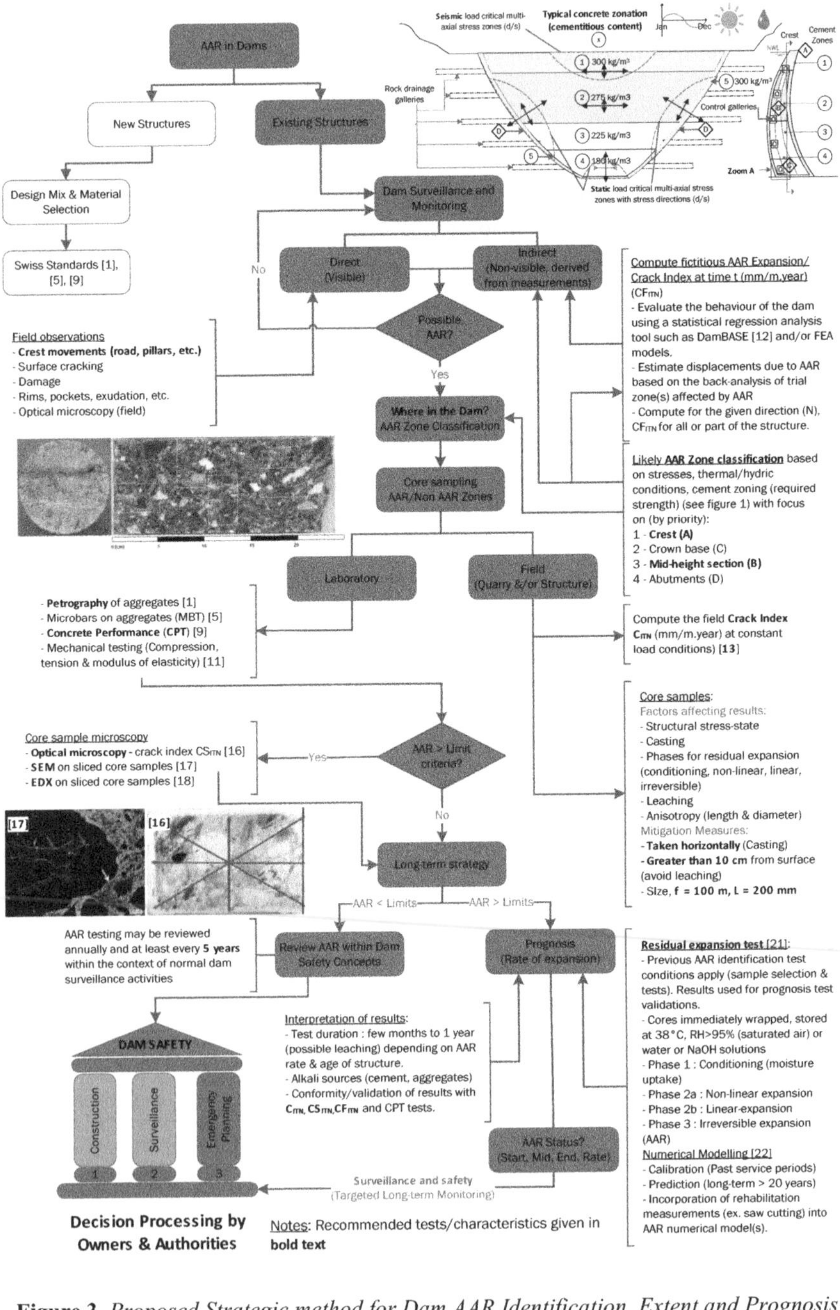

Figure 3. *Proposed Strategic method for Dam AAR Identification, Extent and Prognosis Evaluations (Squared-bracketed numbers refer to the references provided in table 1 to 5)*

The mathematical models for inelastic behaviour require a clear understanding of the potential phenomena and engineering judgement. The method and software developed by Weber and updated by Gerber *et al.* uses the *t-test* (importance of variables), *variance inflation factor* (to avoid matrix ill-conditioning) and a flexible software tool allowing model variables to be *switched on and off*.

Statistical analyses can be applied over *part or the entire life-span* of the structure and hence *before* and *after* the onset of AAR which normally develops some years after the first impounding of the reservoir (SwissCOD 2003). Hence, it seems possible to make a first estimation of the AAR displacements based on statistical methods, for example using the software tool developed in Switzerland called DamBASE (Gerber *et al.* 2015) and/or deterministic methods as described in chapter 5.

Having defined/validated the location of AAR within the dam and computed the *fictitious AAR crack index* CF_{ITN} (mm/m. year) as a function of section height, the process of AAR identification and testing may start in the laboratory and field for selected core samples and zones of the dam.

4.1.2. *Proposed AAR Identification and Extent Procedure*

The first round of laboratory tests which are mainly used for the prevention of AAR in new structures, (petrographic, MBT, CPT and mechanical) shall define if the AAR limits have been exceeded and if so, as a first priority, *optical microscopy* shall be performed in a global manner over the *entire AAR investigated area* of the structure followed by *image microscopy* (SEM, EDX, etc.) on *selected parts of the investigated areas* of the structure which consist of thin polished core slices impregnated with uranyl acetate so that AAR becomes more visible under UV light. Scanning electron microscope, SEM and energy dispersive X-ray, EDX tests are performed to identify AAR and the chemical composition of the sample respectively.

Optical microscopy on core samples defines a third measure of crack index, *crack sample index*, CS_{ITN} (mm/m. year) allowing the extent of AAR to be assessed for different parts of the structure at a given time and principal stress direction within the structure.

This method provides the Engineer with *three measures of AAR extent* which have been found to correlate with MBT and CPT tests (Leemann *et al.*, 2013) thus clearly facilitating the *identification, extent and prognosis* of AAR in dams.

4.2. *Phase 3: AAR Follow-Up Tests*

Figure 3 presents the long-term strategy for AAR investigations which is split into two parts: *Prognosis* and *Periodic Review* (*every 5 years*) for the cases when AAR limits are and are not exceeded respectively. The former implies residual expansion tests on core samples (Leemann *et al.*, 2013) and numerical modelling and the latter, continued monitoring.

Finally, the decision processing is performed by Dam Owners in collaboration with the authorities.

5. Deterministic Approach to AAR Identification, Extent and Prognosis

The objectives of the approach presented herein are to identify *areas or zones of AAR* in the structure (upstream and downstream faces) and fictitious AAR expansion/crack displacements and openings (CF_{ITN}) for part of or the entire dam. This information is useful for the selection of core samples when AAR is not visible to the eye, confirms visible AAR zones, provides a measure of AAR *extent* and gives an indication of the *long-term effects* of AAR.

The proposed idea is essentially a derivative of the deterministic approach described in reference SwissCOD 2003. This approach can also be used with statistical methods such as DamBASE and other closed form solutions for thermal loading for example, Stucky & Derron, 1957.

Deterministic models have already been used in the evaluation of AAR. In their complex form, constitutive material laws may be used to predict the effects of AAR at the *micro, meso* and *macro* scales (Dunant 2009, Cuba Ramos 2017). The formulation, implementation, calibration against material tests and the ability to reliably model the physical phenomenon in the medium to long term still remains a challenge (Saouma *et al.*, 2013).

In Switzerland, AAR research is progressing and already much headway has been made over the years and the results are looking very promising both from the experimental and numerical points of view (Ben Haha 2006, Dunant 2009, Giorla 2013, Cuba Ramos 2017).

5.1. *Formulation of the Deterministic Approach to AAR*

The basic idea proposed herein is to firstly calibrate a finite element model of the structure (dam and foundation) subjected to normal operation loads for a period of time *without* the effects of AAR (say the first 5 to 10 years of operation). Hence, the mechanical material properties (*moduli, Poisson's ratio and coefficient of thermal expansion*) are defined. It is noted that *creep effects* normally predominate at the early life cycle of the structure and therefore are implicitly included in the calibrated moduli values of rock and concrete (Gunn *et al.*, 1996).

The finite element mesh is overlaid with a *grid* of intersecting lines consisting of *vertical* instrumentation (plumblines) and *horizontal* bands of unit temperature control sections (ranging from 1°C to 0°C) used for both external and internal fictitious AAR temperature loadings. Hence, *AAR swelling is simulated by an increase/decrease in temperature at intersecting control points.*

The FEA displacement response of the structure (dam and foundation) for annual variable hydrostatic and unit temperatures defines *influence lines* for which *control point* coefficients can be stored in a spreadsheet. Hence, the response of the structure can be obtained from the spreadsheet database and *no further FE analyses* are required.

Using the spreadsheet, the known reservoir level and external ambient temperatures at a given time provide *control point* (instrumentation reading points) displacements. The difference between the computed and measured values can be attributed to AAR and temperatures fixed for AAR horizontal bands at different elevations until a match is obtained between the computed and measured values using a spreadsheet solver. *The resulting AAR fictitious temperatures define AAR core sampling zones.*

The principle of superposition for a linear system is used to sum the effects of mechanical loads (hydrostatic pressure and thermal loading) and hence, the global displacement at point M is given by:

$$u(M) = U^T(M) + U^M(M) + U^{AAR}(M) \qquad [1]$$

where U^T is the displacement field of the dam due to the thermal loading and U^M the displacement field due to mechanical loading (i.e. hydrostatic pressure). The mechanical displacement field is written as:

$$U^M(M) = \sum_{k=1}^{K} \alpha_k \, u^M_{WL_k}(M) \qquad [2]$$

where $u^M_{WL_k}(M)$ is the displacement at point M due to the hydrostatic pressure for a water level of WL_k. Activation coefficient α_j is defined by $\alpha_k = 1$ if WL_k is close enough to the considered water level, otherwise zero. Hence, only one α_j coefficient is equal to 1, the others are 0. Different mechanical loadings are given by the K water levels WL_k(base to the crest elevation in steps of say 5 or 10 m) where K is the number of hydrostatic loads. Elementary loadings are applied within the model to evaluate and store the displacement fields at instrumentation locations (e.g. plumblines).

Now the thermal displacement field is written in accordance with the *theory of influence lines* given by elementary thermal loads:

$$U^T(M) = \sum_{i=1}^{N} \Delta T_i \, u_i^T(M) + \delta T_i v_i^T(M) \qquad [3]$$

where $u_i^T(M)$ is the displacement at point M due to the elementary thermal load case i defined by the temperature field $f_i^T(M)$. The *constant* relative temperature $\Delta T_i = T_i - T_i^{ref}$ and $v_i^T(M)$ is the displacement at point M due to the elementary temperature *gradient* δT_i through the dam at level i. Finally, N is the number of elementary thermal load cases and T_i^{ref} is the reference temperature at level i. The constant and gradient temperature components define the temperature variation across the thickness of the dam.

These terms can be computed from the thermocouples in the dam section or from external water and air temperatures used in the Stucky-Derron method to provide *equivalent constant* and *gradient* temperatures at a given location and time. The elementary load cases have to verify the partition of unity, such as:

$$1 = \sum_{i=1}^{N} f_i^T(M) \qquad [4]$$

The elementary temperature fields $f_i^T(M)$ are defined below noting that a linear variation applies for intermediate values:

$$f_i^T(M \; at \; level \, j = i) = 1 \text{ and } f_i^T(M \; at \; level \, j \neq i) = 0 \qquad [5]$$

To demonstrate, the case for six temperature fields is given below.

The difference between computed and measured displacements is attributed to AAR which is then simulated *only* by an increase in *constant* temperature for each level i.

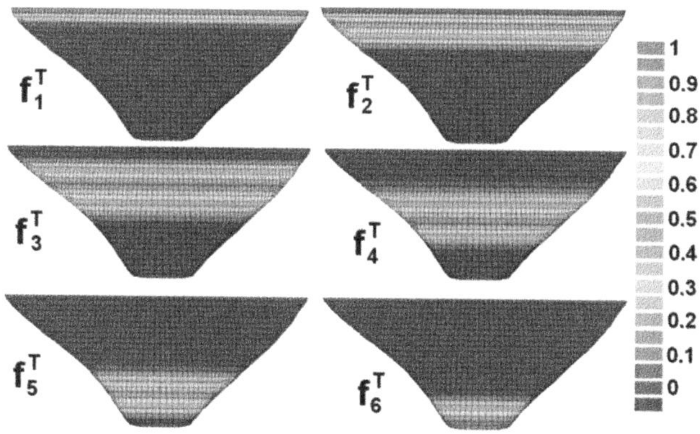

Figure 4. *Temperature field influence lines*

The method is summarised below:

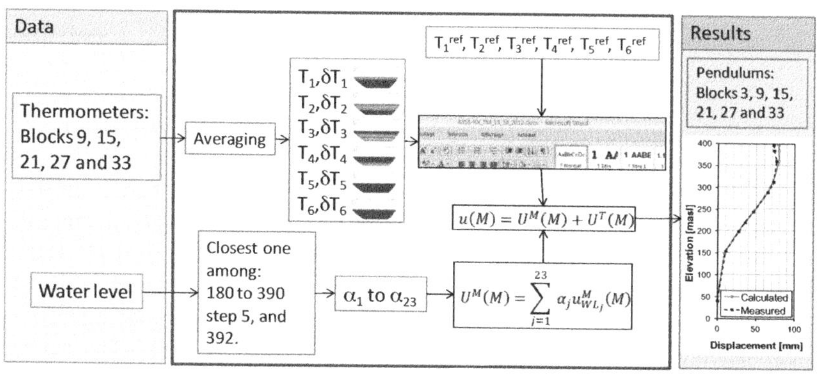

Figure 5. *Influence line Method for AAR Core Sample Zoning*

The resulting AAR zones can be concluded on the basis of fictitious temperature zoning as shown in the figure below whereby both the intensity and location of possible AAR reactions is now exposed or made visible.

5.2. *Schematic Results of the Deterministic Approach Applied to AAR Investigations*

Applying the concepts presented previously should lead to the results given in Figure 6. Clearly the schematic presentation of AAR zones in Figure 6 indicates where core samples should be taken and also provides the analyst with a good idea about how AAR swelling displacements could be predicted by an *equivalent temperature distribution* over the entire structure. Such an equivalent temperature distribution over the height and width of the dam (left to right banks) to simulate both non-uniform horizontal and vertical expansions has also been presented within the context of AAR investigations of a dam in Switzerland (SwissCOD 2015).

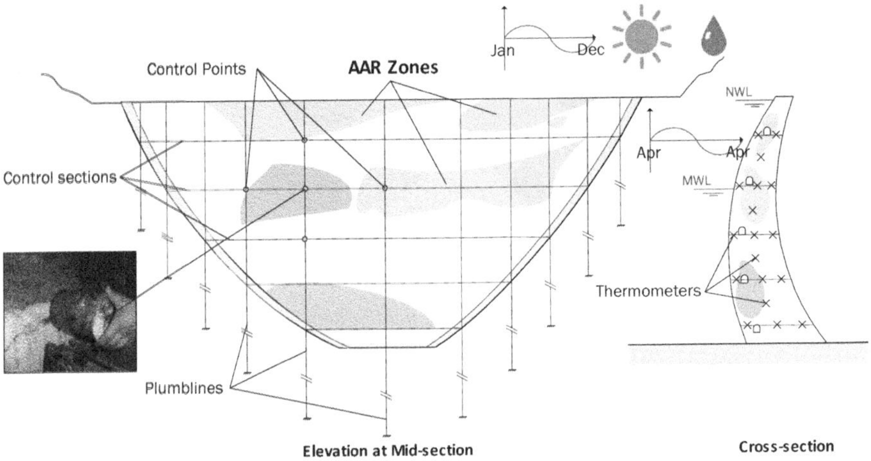

Figure 6. *AAR Zoning for Core Sample Selections (colours indicate different levels of affected and non-affected AAR regions)*

Contrary to previous investigations, the approach described herein unifies the notion of an *equivalent temperature increase or decrease* at specific regions to obtain the swelling displacements of the structure with the physical phenomenon rather than an artefact (temperature rise/lowering) to obtain a desired result (AAR swelling displacements). Hence, the AAR equivalent temperature loading also has a physical meaning that the dam engineer can associate with the real behaviour of the structure improving the knowledge base decision process related to risk and safety.

It should be noted that the above-mentioned method provides only an *approximation* of the effects of AAR on the structure because differential displacements (difference between measured and computed values) can also be attributed to other irreversible mechanisms such as *plasticity, creep and damage* (Gunn 2001, Gunn 2001). However, with good field measurements, appropriate statistical/deterministic data evaluations and engineering judgement, it should be possible to not only define the potential areas of AAR, but also to get a much better insight into the behavioural mechanisms of the swelling structure which of course shall be important when coming to the conception, design and analysis of rehabilitation works.

6. AAR Prognosis Testing

Table 1, Table 4 and Table 5 recommend two types of prognosis tests. The *first* consists of 170-day residual expansion tests performed on core samples (recommended size: 100 mm diameter and 200 mm in length) taken from both affected and non-affected AAR areas in the existing structure for comparison purposes (*uncoupling other effects such as creep*). These tests are described in the reference (Merz *et al.*, 2013) for which the results have been validated & correlated against CPT and optical microscopy test results (expansion and crack opening respectively). These findings are important since they establish several important links related to AAR tests performed on concrete for new and existing structures.

The residual expansion test overcomes the problem of using aggregate sources (quarries from the construction period) that may be encountered for MBT and CPT tests and moreover, it provides additional information related to the status of irreversible AAR swelling (*expansion at a given time, extent and rate*).

The proposed method does not consider the initial water swelling/creep effects at the initial phases of the test and therefore targets *true* AAR. To better understand the results and conclusions of this work, the measured longitudinal expansion on core samples (*f* = 100 mm, L = 200 mm), the swelling phases and test results are schematically outlined below.

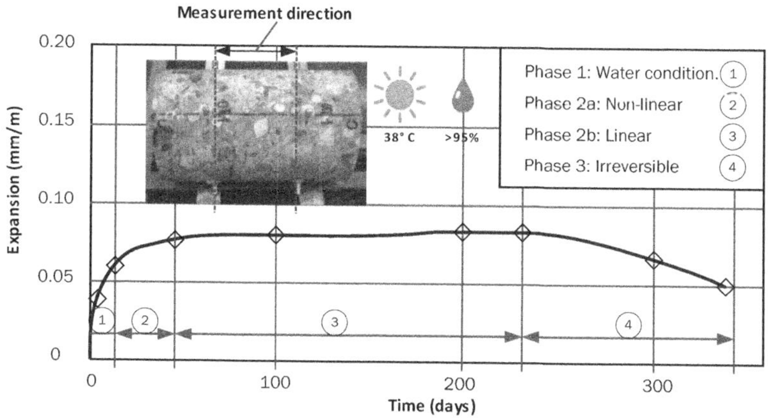

Figure 7. *Different Phases of AAR Residual Expansion Prognosis Tests based on Longitudinal Direction Expansion Measurements on Core Samples (Merz et al., 2013)*

The *second* prognosis test is based on numerical studies whereby the constitutive models are calibrated against laboratory test results and swelling behaviour predicted with time. These advanced studies form part of the next phase of AAR investigations currently being performed at the EPFL in Lausanne.

7. Conclusions and Recommendations for Further Studies

The identification, extent and prognosis of AAR remain a subject of great importance in dam engineering and safety evaluations. This contribution presents a methodical and detailed approach to AAR investigations for existing dams in Switzerland that remains to be validated against case studies.

An overview of AAR within the context of *Dam Engineering and Safety* with specific emphasis on the geological properties of aggregates, the use of supplementary cementitious materials, cement zoning within the dam, principal stress orientations, environmental conditions (temperature and humidity), the behaviour of dams undergoing expansion due to AAR with typical expansion values in *mm per year* and the determining factors that influence the identification of AAR within existing dams is given. This background information is deemed to be critical in AAR investigations because it targets *how, where and to what extent AAR* is identified in the structure.

A detailed review of tests and investigations allowing the *identification, extent and prognosis* of AAR to be evaluated for both new and existing dams is provided. Three phases for the investigation of AAR are suggested. Phase 1, *AAR Stage tests*, proposes "Look-up tables" for AAR Standard and Supplementary Tests that firstly classify AAR tests in terms of *scope, time, cost and quality* of information and then, the essential factors that may *influence test results* and conclusions are given. This provides the AAR investigative project team with a *state-of-the-art* overview of AAR tests.

Phase 2 of AAR investigations looks at the process by which AAR *monitoring and data analysis* may be logically used. A strategic flow-chart approach to identification, extent and prognosis evaluations is given. This approach captures all of the essential elements of AAR within the context of dam safety evaluations and proposes that *Phase 3 AAR follow-up tests* are performed *every 5 years* in the event that AAR has been initially detected in the structure. This proposal is deemed to be essential to further validate the expansion in the structure and the prognosis tests presented in chapter 6.

Finally, a *new deterministic approach* is proposed for the identification of AAR within structures for which no visual signs of AAR, apart from expansion measured by *dam instrumentation,* is apparent. This approach defines three expansion/crack displacement and opening indexes that schematically illustrate AAR affected zones of different magnitude and importance throughout the entire dam body. Knowledge of these AAR zones with their respective extents allows test core samples to be selectively targeted in the dam body. Moreover, the method defines *fictitious temperatures* that allow correlation tests to be performed against laboratory residual expansion tests enabling the long-term effects of AAR and hence appropriate rehabilitation works to be estimated.

8. Acknowledgements

The authors would like to thank Dr. Johannes Maier of SFOE, Supervision of Dams Section for reviewing the text and providing his valuable comments and to Manuel Alvarez and Daniel Lüthy of the Swiss Federal Office of Transport for sharing their AAR knowledge and experience related to structures other than dams.

9. Disclaimer

The work presented in this paper has been largely performed at SFOE, however the statements and opinions herein are those of the authors and do not necessarily reflect a stance position of SFOE.

10. References

Amberg, F., "Performance of Dams affected by Expanding Concrete", *Dams and Reservoirs under Changing Challenges*, edited by Robert M. Boes, CRC Press, pp. 115–122, 2011.

ASTM C-294, "Standard descriptive Nomenclature for Constituents of Concrete Aggregates".

ASTM C-1260, "Standard Test Method for Potential Alkali Reactivity of Aggregates (Mortar-Bar Method)".

ASTM C-1293, "Standard test method for concrete aggregates by determination of length change of concrete due to alkali-silica reaction".

Ben Haha, M., "Mechanical Effects of Alkali Silica Reaction in concrete studied by SEM-image analysis", PhD Thesis N° 3516, École Polytechnique Fédérale de Lausanne, May 2006.

Cuba Ramos, A.I., "Multi-Scale Modeling of the Alkali-Silica Reaction in Concrete", PhD Thesis, École Polytechnique Fédérale de Lausanne, Spring 2017.

Dunant, C., "Experimental and Modelling Study of the Alkali-Silica–Reaction in Concrete", PhD Thesis N° 4510, École Polytechnique Fédérale de Lausanne, November 2009.

Gerber, M., Bühlmann, M., "DamBASE Version 1.0 May 2015", Versuchsanstalt für Wasserbau Hydrologie und Glaziologie, VAW, 2015.

Giorla, A.B., "Modelling of Alkali-Silica Reaction under Multi-Axial Load", PhD Thesis N° 5982, École Polytechnique Fédérale de Lausanne, October 2013.

Godart, B., de Rooij, M., Wood, J.G.M., "Guide to Diagnosis and Appraisal of AAR Damage to Concrete in Structures: Part 1 Diagnosis (AAR 6.1)", April 2013.

Grimal, E., "Caractérisation des effets du gonflement provoqué par la réaction alcali-silice sur le comportement mécanique d'une structure en béton", Université Paul Sabatier-Toulouse III, February 2007 (in French).

Gunn, R.M., Bossoney, C., "Creep analysis of mass concrete dams", *Hydropower and Dams* Volume 3, Issue 6, 1996.

Gunn, R.M., "Non-linear design and safety analysis of arch dams using damage mechanics, Part 1: Formulation", *Hydropower and Dams*, Volume 1, Issue 1, 2001.

Gunn, R.M., "Non-linear design and safety analysis of arch dams using damage mechanics, Part 2: Applications", *Hydropower and Dams*, Volume 1, Issue 3, 2001.

Hammerschlag, J.G., "Analyse des réactions alcali-granulat – RAG, Méthodes d'essai et application au projet cemsuisse", Holcim Presentation, November 2005 (in French).

ICOLD Bulletin 79, "Alkali-Aggregate Reaction in Concrete Dams", 1991.

Larive, C., "Apports combines de l'expérimentation et de la modélisation à la compréhension de l'alcali-réaction et de ses effets mécaniques", École Nationale des Ponts et Chaussées Spécialité: Structures et Matériaux, June 1997 (in French).

LCPC, "Détermination de l'indice de fissuration d'un parement de béton", Méthode d'essai LPC 47, 1997 (in French).

Leemann, A., Griffa, M., "Diagnosis of alkali-aggregate reaction in dams", SFOE Publication N° 290934, Project N° 500863, Empa, Laboratory for Concrete/Construction Chemistry, Überlandstrasse 129, CH-8600 Dübendorf. http://www.bfe.admin.ch/themen, Dec. 2013.

Leemann, A., Merz, C., "An attempt to validate the ultra-accelerated microbar and the concrete performance test with the degree of AAR-induced damage observed in concrete structures", *Cement and Concrete Research* 49, pp. 29-37, 2013.

Merz, C., Leemann, A., "Assessment of the residual expansion potential of concrete form structures damaged by AAR", *Cement and Concrete Research* 52, pp. 182-189, 2013.

Saouma, V.E. "Numerical Modelling of AAR", CRC Press, 2014.

SIA Cahier technique 2042, "Prévention des désordres dus à la réaction alcalis-granulats (RAG) dans les ouvrages en béton", 2012, Correctif C1 2012 and Correctif C2 2015.

SN 670 115, "Minéralogie et pétrographie qualitative et quantitative des granulats", 2015.

SN 670 116, "Fillers – Minéralogie, pétrographie et minéraux argileux gonflants", 2015.

SN EN 206-1, "Béton – Partie 1: Spécification, performances, production et conformité", 2015.

Stanton, T.E., "Expansion of concrete through reaction between cement and aggregate", *Proceedings, American Society of Engineers*, Vol. 66, New York, pp. 1781-1811, 1940.

Stucky, A., Derron, M-H., "Problèmes thermiques posés par la construction des barrages-réservoirs", *Science & Technique*, Paul Feissly, libraire-éditeur, Lausanne, 1957.

SwissCOD (Swiss Committee on Dams), "Alkali-Aggregate Reaction (AAR) in Swiss Dams", Workshop held in Bern, Switzerland, 13th May 2015.

SwissCOD (Swiss Committee on Dams), "Concrete of Swiss Dams: Experiences and Synthesis", 2000, published in *Wasser Energie Luft (WEL) 92*. Jahrgang, 2000, Heft 7/8, CH-5401, Baden, Switzerland.

SwissCOD (Swiss Committee on Dams), "Die Felsfundationen Schweizerischer Betonsperren, Übersicht über die Fundationsverhältnisse", Herbst 2006 (in German).

SwissCOD (Swiss Committee on Dams), "Methods of analysis for the prediction and the verification of dam behaviour", April 2003, published in *Wasser Energie Luft (WEL)* Jahrgang, 2003, CH-5401, Baden, Switzerland.

Weber, B., "Vorhersage des Verhaltens von Talsperren mit Hilfe des Soll-Ist-Vergleichs - Statistischer Teil", April 2002 (in German).

Experimental Evidence for the Link Between Aggregate Degradation and Expansion and the Formulation of the Microstructural Model

CYRIL DUNANT* — KAREN SCRIVENER**

** University of Cambridge, department of Engineering, Use Less group*
Trumpington street, Cambridge, UK
cfd30@cam.ac.uk

*** EPFL, École polytechnique fédérale de Lausanne*
Laboratory of Construction Materials
EPFL-STI-IMX-LMC, Station 12, CH-1015, Lausanne, Switzerland
karen.scrivener@epfl.ch

ABSTRACT. *Establishing a prognosis for a structure affected by the ASR is difficult because the ASR changes the material properties of the concrete as it develops, as well as inducing strain. As the reaction develops over the long term, extrapolation of the structural conditions over decades is desirable. Therefore, empirical models are generally unsuitable as they offer no guarantee of correctness. Developing material models for the ASR based on the experimental evidence required the development of a number of numerical tools, novel in their own right, to integrate all the physical knowledge about the reaction which has been accumulated over the years. In turn, these models have allowed us to discard some hypotheses about the morphology of the damage in the concrete and the interaction of the reaction and the loads on the structure. This paper offers an overview of the considerations taken into account in the model developed at LMC and presents some results of simulations.*

KEYWORDS: *alkali-aggregate reaction, alkali-silicate reaction, dams, diagnosis, prognosis, testing (and all words in the text that are written in italics).*

1. Introduction

Modelling the ASR for the purpose of monitoring structures defines both the reliability constraints and the outputs of the model. ASR is first degradation at the microstructural level: cracks in aggregates and paste, gel pockets. Second, ASR induces degradation of mechanical properties at the macro-scale. The link between these two aspects is found in the phenomenological links between expansion and reaction and expansion and loss of stiffness. Confidence in a model comes from being able to verify both the practical outputs: degradation and expansion, as well as the underlying physical causes: the initiation and propagation of cracks in the microstructure.

The cracks induced by the ASR initiate *inside* the aggregates, from gel pockets. They then form a network connecting the gel pockets. As the reaction develops, the cracks percolate to the paste, and eventually join the aggregates. The relative amount of cracks in the paste and aggregates is critical for the resulting mechanical properties. First, the stiffness is affected differently by paste being weakened or the aggregates being weakened. But the relative amount of cracking in aggregate in paste is mechanically important also in terms of changing the creeping properties of the concrete. There is no easy analytical solution to that problem, however, and numerical modelling is necessary.

Concrete and mortars are frequently represented as two distinct, separable scales: simply representing the larger aggregates is possible, provided a model for the paste and aggregate mixture of the mortar. However, there is no clear separation of scale between concrete and mortar. Therefore, the model representing the microstructure must explicitly represent the aggregates, down to the smallest size possible. Further developments during the Giorla thesis introduced a creep model to the paste. A key aspect of ASR is the growth of gel pockets. The gel pockets are represented using extended finite elements in the LMC code. Such finite elements are used to represent the effects of interfaces or discontinuities without explicit meshing.

Microstructural modelling of ASR is a non-linear problem. The initiation and propagation of damage in the aggregates and paste must be computed. Due to the very high density of cracking in ASR, this is numerically difficult. Solving this problem required the invention of a new damage algorithm. This new damage algorithm is always stable. This allows the simulation to reach very high damage levels. Using this algorithm, it is possible to distinguish between numerical limits to the model (simulations become unstable because of the approximation) and physical limits (simulations fail because there exist no valid solution).

An energetically correct approximation of damage requires either careful calibration to the mesh, or a non-local approximation. The non-local damage method was introduces to represent the fracture process zone in metals. It can also be used to represent a degradation process occurring in a material which has an underlying, unrepresented microstructure.

Modelling creeping materials is commonly done using iterative finite difference schemes. Such schemes work well is the damage and creep problems can be decoupled. A simultaneous solving process is required here, however, because both creep and damage relax stresses. Therefore, as the gel grows, the stress is increased in the microstructure, but not as fast as in the elastic case. The effective rate of stress increase depends on the rate of gel growth and on the visco-elastic properties of the cement paste. Nevertheless, a critical stress may be reached at any point during this process which will initiate damage. The damage algorithm can be easily adapted provided the time at which this critical stress is known.

2. ASR Model Results

The model was run to help understand a number of experiments during the Dunant thesis. The LMC model was originally conceived as a tool to validate or invalidate our understanding of the physics and micromechanics of ASR. The model was not intended to be run to be able to make predictions on real structures. Rather, it was designed to be able to provide physics-based inputs to empirical models applicable at a larger scale.

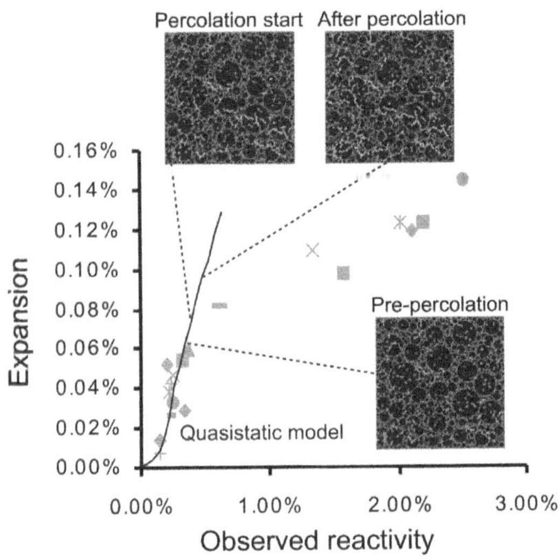

Figure 1. *Damage patterns as well as comparison between the predicted expansion and reaction and the experimental observations. The experimental observations overstate the damage in the aggregates due to the sample preparation.*

There are few free parameters in the model: the mechanical properties of the gel described in the literature are becoming better defined, although it seems that there is no single gel, but rather a range of possible compositions. Therefore the gel stiffness is assumed to be in the range of values reported in the literature, but can be adjusted depending on the aggregate. The PSD of the concrete is usually known. The mechanical properties of the paste and aggregates can be easily measured.

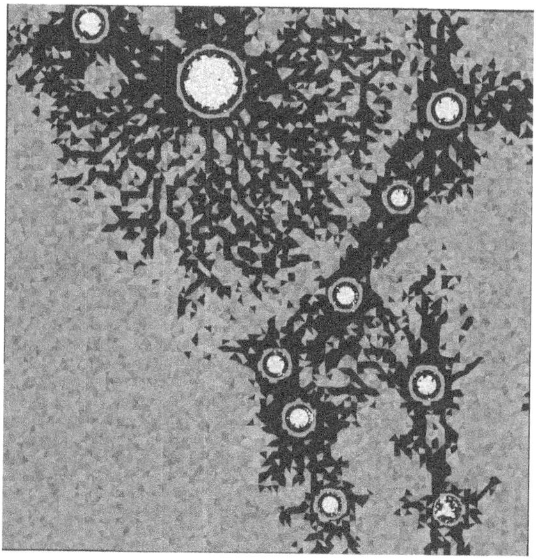

Figure 2. *The damage patterns predicted by a reaction occurring on the surface of the aggregates are very different from the experimental observations. Subsequent theoretical developments by Charpin corroborated this result*

The first set of simulations tried to reproduce the free expansion experiments normally conducted in the lab. Free expansion experiments are used in many case as a starting point to calibrate models. In the case of the LMC ASR model, they are used to validate the model as the parameters are all obtained from separate experiments or the literature. The only fit which needs to be performed is to measure the reaction rate of the gel using image analysis. As the model is quasistatic, it does not depend on kinetics. Rather, it relates the amount of gel produced to degradation, expansion and load.

A number of descriptions of the reaction assume that the gel forms on the surface of the aggregates and causes cracking in the paste. This hypothesis could be tested using the model. It was found that material degradation and expansion was vastly overestimated in this case. Further, the damage patterns which would result from this mode of reaction do not correspond to the microscopic observations.

The model, using the default parameters works very well in predicting the correlation between expansion and reaction as measured by image analysis up to approximately 0.6% of reaction. This corresponds to a very advanced state in practice, at which structures should not be considered safe. The quasistatic model stops following the experimental trend around 0.6% of reaction for two reasons. The first reason is that the measure of reaction by image analysis becomes less precise at this point due to the difficulty of keeping the aggregates intact: large holes in the aggregates, interpreted as reaction, are typically artefacts of sample preparation. The second reason is that this is also the point at which damage in the paste starts to percolate. Percolation of cracks in 2D has different consequences on the apparent strength of the sample than the percolation of cracks in 3D would.

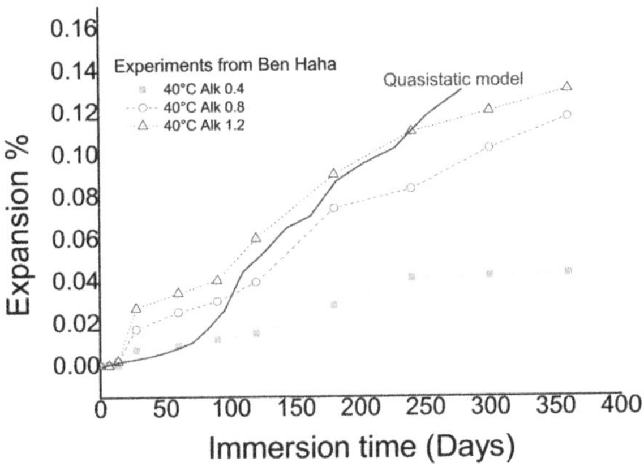

Figure 3. *Predicted and observed expansions. The experimental samples probably underwent leaching, which explains why he predicted expansion most closely matched the higher alkali samples*

The usefulness of the model was highlighted when it suggested that the expansion observed by Ben Haha levelled off too early. Indeed in the subsequent experiments by Dunant, using a simulated pore solution, the expansions were continued longer. This effect is visible when comparing the model to the expansions of samples with different amounts of alkalis in the mix: the model is the closest to the expansion using the highest concentration. To predict the long-term behaviour of dams affected by ASR, the evolution of the stiffness of concrete as well as the free expansion need to be considered. These two aspects are only loosely connected, as anisotropic loading of the concrete can affect the development of the damage.

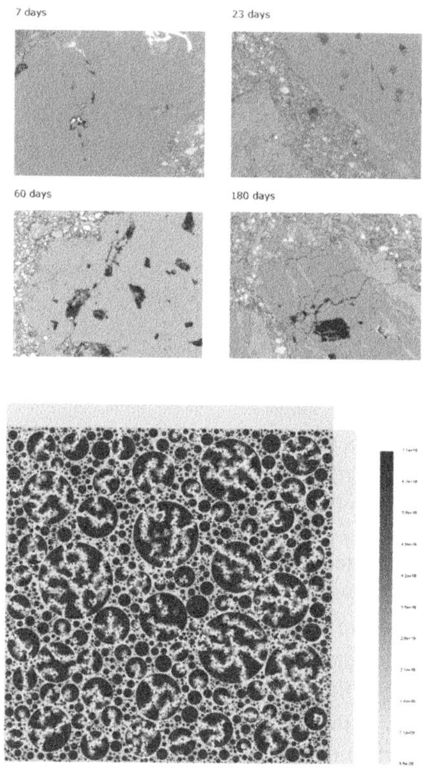

Figure 4. *Degradation of the concrete as predicted by the model. The morphology of the damage matches closely the microscopic observations*

The model predicts quite well the loss of stiffness before the cracks percolate, and therefore gives useful predictions in the range most relevant to massive concrete structures. Nonetheless, one of the important questions still open at the end of the Dunant thesis was why the damage predicted in the paste was so much larger than the loss of stiffness observed experimentally. As mentioned in the introduction, the damage in the paste has much larger consequences than the damage in aggregates. Therefore, the prediction from the LMC model are already more conservative than predictions of other physics-based micro-mechanical models where damage initiates in the paste rather than the aggregates. However, the damage being still too high, a hypothesis was put forward to explain the discrepancy: creep dissipated the stress from ASR.

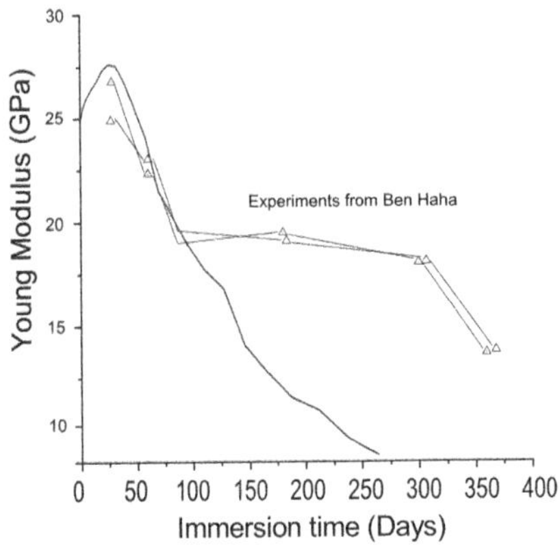

Figure 5. *The degradation predicted by the model follows well the experimental observation in the early stages of the reaction. In this instance, as creep is not taken into account, the predicted loss of stiffness eventually overestimates the experiment*

Indeed, expansions predicted as a function of the damage in the aggregates are closer to the experimental observations. The visco-elastic simulations showed that damage in the paste depends on the rate of expansion. Therefore experiments at higher temperature should exhibit steeper losses of properties for the same amount of reaction than field concrete. This illustrates how physics-based models help interpret observations. This finding is unique in the currently published ASR models, as the others are based on experiments subject to leaching, meaning this regime is never reached before the reaction is exhausted. In real dams, there are no reasons to believe exhaustion happens until long after the planned useful life of the structure.

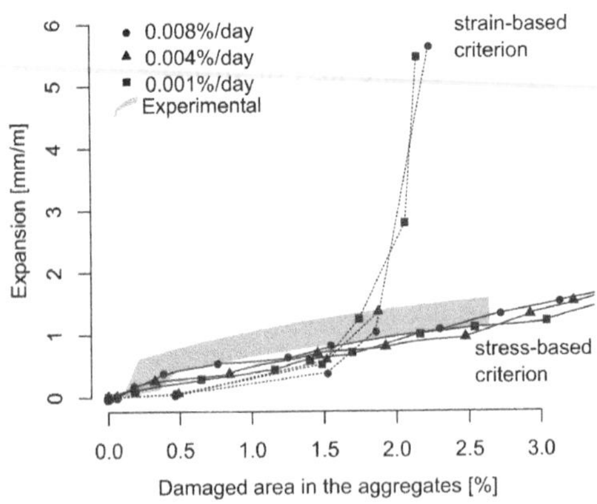

Figure 6. *Damage simulations using a visco-elastic paste model require distinguishing between strain and stress based fracture criteria, unlike quasistatic simulations for which they are equivalent. Using a stress-based criterion, the loads on the paste are relaxed*

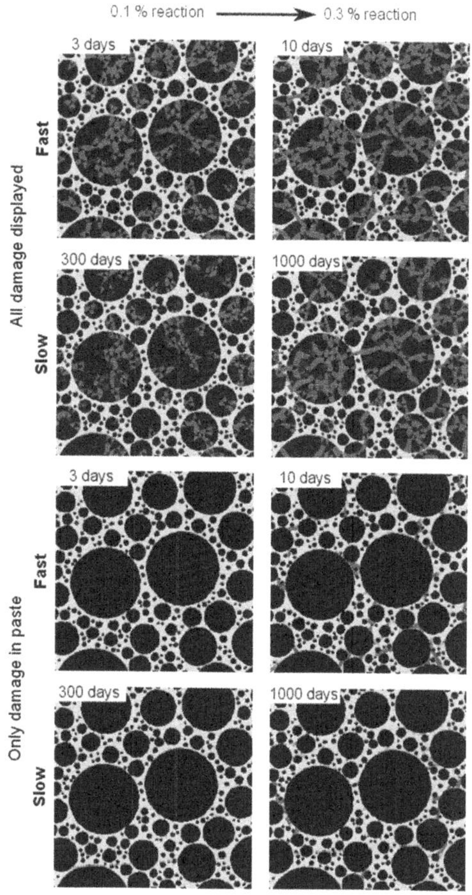

Figure 7. *Predicted damage in the paste and aggregates for two rates of loading. The difference in the damage pattern is clear when comparing to the simulations using a quasistatic model as presented above*

3. Conclusions

Developing numerical models which can simulate the ASR at the microstructure level proved to be enormously useful as a guide to interpreting experimental observations as well as provide the ground works for formulating structure-scale models which will allow the managers of structures to make informed decision on the safety and usefulness of their charges.

The key findings were that the morphology of the reaction, rings versus pockets or veins should strongly affect the expansion and degradation of the material, that the creep of the cement paste is key to mitigating the damage in the long term: damage in real structures occurring over long periods is always going to be lower than that predicted by accelerated experiments in the lab, and finally that ASR should not be understood as a phenomenon which induces stresses in structure, but rather as a phenomenon which weakens the material as it develops. This latter point is crucial in understanding the behaviour of affected structures.

4. References

Ben Haha M., "Mechanical Effects of Alkali Silica Reaction in concrete studied by SEM-image analysis", PhD Thesis N° 3516, École Polytechnique Fédérale de Lausanne, May 2006.

Cuba Ramos, A., I., "Multi-Scale Modeling of the Alkali-Silica Reaction in Concrete", PhD Thesis, École Polytechnique Fédérale de Lausanne, Spring 2017.

Dunant C., "Experimental and Modelling Study of the Alkali-Silica–Reaction in Concrete", PhD Thesis N° 4510, École Polytechnique Fédérale de Lausanne, November 2009.

Giorla A. B., "Modelling of Alkali-Silica Reaction under Multi-Axial Load", PhD Thesis N° 5982, École Polytechnique Fédérale de Lausanne, October 2013.

Charpin, Laurent, and Alain Ehrlacher. "A computational linear elastic fracture mechanics-based model for alkali–silica reaction." *Cement and Concrete Research* 42.4 (2012): 613-625.

Dunant, C. F., and E. C. Bentz. "Algorithmically imposed thermodynamic compliance for material models in mechanical simulations using the AIM method." *International Journal for Numerical Methods in Engineering* 104.10 (2015): 963-982.

Giorla, Alain B., Cyrille F. Dunant, and Karen L. Scrivener. "Role of creep on the microstructural damage induced by alkali-silica reaction. " *Newsletter* (2016)

Giorla, A. B., K. L. Scrivener, and C. F. Dunant. "Finite elements in space and time for the analysis of generalised visco-elastic materials. " *International Journal for Numerical Methods and Engineering*, 97.6 (2014):454-472.

Giorla, A. B.r num Nor Numerical Methods in Engineering 97.6 (2014): 454-472.

A Robust Testing Protocol for the Assessment of ASR Reactivity of Concrete

LIONEL SOFIA* — THEODORE CHAPPEX** — CYRIL DUNANT*** — KAREN SCRIVENER*

* *Laboratory of construction materials*
EPFL-STI-IMX, Lausanne 1015, Switzerland
lionel.sofia@epfl.ch, karen.scrivener@epfl.ch

** *TFB SA Technologie et Recherche pour le Béton*
1070 Puidoux, Switzerland
Theodore.chappex@tfb.ch

*** *Engineering Department*
Trumpington Street, Cambridge CB2 1PZ, United Kingdom
cfd30@cam.ac.uk

ABSTRACT: *ASR standard testing methods have been proposed over years. Most of these standards are based on a small group of testing protocol. The conclusions reached from these standards are not always relevant to predict the alkali silica expansion of concretes. In this study, the Swiss Standard SIA 2042 performance test is analysed and some inconsistencies are highlighted. A new protocol is proposed, based upon the acquired experience of the authors' laboratory and the research community. This new procedure aims to avoid the measurement bias introduced by the standard protocol and to be as close as possible to the on-field chemical environmental conditions. Preliminary results are promising, showing a constant expansion rate over more than 400 days. The parameters of the test procedure have to be optimised and results have to be further compared with other test procedures in order to propose a robust testing protocol for ASR performance testing of blended cement concretes.*

KEYWORDS: *ASR assessment expansion, aggregates, pore solution, leaching.*

1. Introduction

Since Stanton's landmark paper in 1940 (Stanton, 1940) for the California road administration, the alkali-silica reaction has been observed worldwide. This reaction is the first non-reinforcement-related cause of concrete degradation in the world and is of particular concern for owners and managers of large mass concrete structures such as dams.

One method to evaluate the potential of concretes to expand deleteriously is to simply measure the expansion evolution of concrete samples (ASTM C1293, ...). As the reaction is extremely slow in the field, these testing methods accelerate the reaction. To achieve this goal, three factors are currently used. The total alkali content of cement is increased. The samples are stored in water vapour conditions. The storage room is maintained at high temperature.

Alkali concentration (Ichikawa, 2009) (Bérubé *et al.*, 2004), temperature (Ulm *et al.*, 2000) (Ben Haha *et al.*, 2007) change the kinetic of the reaction. As it is now known, the wrong use of these parameters can lead to several false positives or false negative in the assessment of ASR reactivity. All these factors must be considered for the design of a test, but the goals of the test will further define how the parameters are chosen. Further, the hydration degree of the sample can be a concern in terms of representativity of field conditions.

2. Materials and Methods

2.1. *Materials and Mix Designs*

2.1.1. *Concrete Sample Preparation*

Following the Swiss standard (SIA M 2042) procedure, concrete prisms (7x7x28 cm) were casted. The aggregates used are reactive aggregates from the Italian Alps (gneiss, micaschist). In accordance with this standard, some sodium hydroxide was added to the batch water until reaching 0.8% Na_2O_{eq} concentration. The cement used was a Portland cement, type CEM I 42 R. The formulation of the concrete is described below:

Cement:	410	[kg/m^3]
Water:	185	[l/m^3]
Aggregates 0/16:	1770	[kg/m^3]
NaOH:	1.552	[kg/m^3]

Additional samples were casted to extract pore solution.

2.2. *Methods for Assessment and Analysis*

2.2.1. *Pore solution Analysis and Simplified Pore Solution*

A batch of concrete was cast in 200 ml sealed containers and stored at 20°C one day. Another batch of concrete was cast in 200 ml sealed containers and stored at 20°C for 28 days. They were pressed in a pore solution extraction device in order to extract the pore solution: the sealed sample is unmoulded and put into the extraction device (5 cm diameter piston in a metallic cell and a container at the bottom to retrieve solution). The sample is loaded with a speed of 100N/min until 1500N. The load is maintained during 10 minutes (Longuet *et al.*, 1973) (Barneyback *et al.*, 1980). This solution is analyzed by ICP (Inductively Coupled Plasma).

2.2.2. *Expansion Tests*

In this study, the influences of three different factors are studied: pre-curing period, storage temperature and storing chemical conditions. 1 day and 28 days pre-curing time are analysed. Most standards prescribe a 1 day pre-curing period. This choice implies that ASR begins while the hydration of cement is far from finished. After one month, cement is almost hydrated, therefore the mechanical properties and the chemical concentration of the pore solution are stabilized. That's why 28 days is also selected as a pre-curing period to neglect the issue of overlapping between hydration and ASR.

Three different storage temperature were used: 38, 50 and 60°C. High temperature ($T \geq 60°C$) is used in standard to accelerate the kinetic of the reaction. However, previous studies have shown that the nature of the gel is dependent on to the temperature. Study of ASR under 38°C enables to get gel characteristics more representative of the on-field ones. Usually, the storage condition prescribed by standard is in water vapour condition. In order to realize this setup, the concrete samples are stored in a sealed container (20°C, HR ≈ 100%). In the bottom of this box few centimetres of water are placed. There is no contact between concrete and water. With this storage condition, alkalis can leach from the concrete to the water in the bottom of the container. In parallel to the standard testing conditions, samples are stored in an alkaline solution to avoid this phenomenon. It is a simplified pore solution made with sodium and potassium alkalis (PS). The Na^+ and K^+ content of the PS testing solution is directly established by the ICP results previously obtained for the pore solution. Therefore, this option enables to have chemical curing conditions close to the on-field ones.

The different testing conditions are summarized in Table 1. For each case, three concrete prisms (7/7/28 cm) are tested to measure the expansion. The case WV-C1-T60 corresponds to the standard SIA M 2042 testing conditions.

	60°C (T60)		50°C (T50)		38°C (T38)	
	1 d. cure (C1)	28 d. cure (C28)	1 d. cure (C1)	28 d. cure (C28)	1 d. cure (C1)	28 d. cure (C28)
Water vapour (WV)	WV-C1-T60	WV-C28-T60	WV-C1-T50	WV-C28-T50	WV-C1-T38	WV-C28-T38
Pore solution (PS)	PS-C1-T60	PS-C28-T60	PS-C1-T50	PS-C28-T50	PS-C1-T38	PS-C28-T38

Table 1. *Testing conditions of the expansion tests*

2.2.3. *Quantification of Alkalis of Storage Solutions*

Samples of storage solutions were collected during the expansion test to measure the alkalis content. These solutions were analyzed by ICP. This analysis enables to quantify the migration of ions. Notably, leaching may be expected in water vapour atmosphere (series WV). Therefore, it is assessed. 3 concrete prisms and 1 litre of distilled water are contained in the storage boxes related to these series. The total molar content of $Na_2O_{eq}^{tot}$ is therefore equal to 224.2 [mmol], it is initially located in concrete. With time, alkalis migrate from concrete to water; this leaching is estimated according to equation [1].

$$leaching = \frac{Na_2O_{eq}^{leach}}{Na_2O_{eq}^{tot}} \qquad [1]$$

where $Na_2O_{eq}^{leach} = \frac{[Na^+]_{ICP} + [K^+]_{ICP}}{2}$ in which $[Na^+]_{ICP}$ and $[K^+]_{ICP}$ are sodium and potassium concentration of water.

3. Results

3.1. *Expansion Tests*

In this paper, the results concerning the impact of the pre-curing period and two different storage conditions (WV and PS) are reported. Expansion curves for the various systems are shown in Figure 1. On these graphs, the horizontal lines are the expansion limit values given by the SIA M 2042 (after 5 months $\varepsilon_{lim, 5\ months}$ = 0.2‰ and $\varepsilon_{lim, 1\ year}$ = 0.3‰).

The main observations are:

– The standard SIA M 2042 testing procedure (series WV-C1-T60) leads to a plateau after about 200 days as often observed.

– The system at 60°C and put in the same conditions as the SIA M 2042 (series WV-C1-T60) shows no significant expansion until 700 days.

– The system cured during 1 day and stored in the simplified pore solution at 60°C (series PS-C1-T60) present an expansion at 400 days 4 times higher than the series WV-C1-T60 at the same time. Moreover, the expansion tends towards a constant rate 150 days after the beginning of the (rate: about 0.01 [m/mm/day]).

– The samples cured during 1 days and immersed in the simplified pore solution at 38°C (series PS-C1-T38) only start to expand 230 days after their casting at a constant rate. This rate is equal to 0.004 [m/mm/day], a value 2.5 time lower than the one of the series cured at 60°C (PS-C1-T60).

– The samples cured during 28 days and immersed in the simplified pore solution at 38°C (series PS-C28-T38) only start to expand 280 days after their casting and have a similar expansion rate than the series PS-C1-T38. It appears a delaying effect between PS-C1-T38 and PS-C28-T38 around 50 days.

– The samples cured during 28 days and immersed in the simplified pore solution at 60°C (series PS-C28-T60) has a similar expansive behaviour than the series PS-C1-T60. The expansion rate is slightly lower, it is equal to 0.007 [m/mm/day].

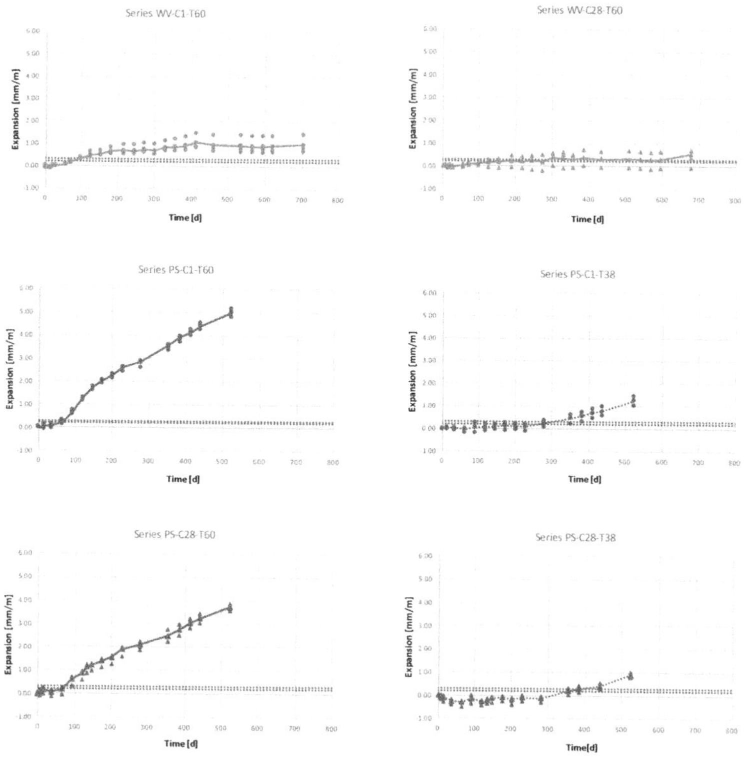

Figure 1. *Expansion over time for some of the studied system (cf. Table 1)*

3.2. *Pore Solution Analysis*

The results of the concrete pore solution after 28 days at 20°C are presented in Table 2. These results were used to create the conditioning solution of the series PS-C28-T38 for sodium and potassium concentration.

3.2.1. *Leaching of Vapour Curing Systems (series WV-C1-T38 and WV-C1-T60)*

The alkali leaching measured for the series WV-C1-T38 and T60 are plotted on Figure 2.

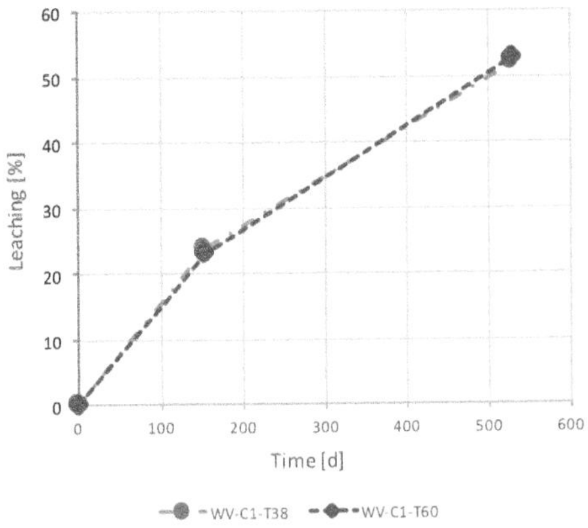

Figure 2. *Na_2O_{eq} leaching over time evolution*

The main observations are:

- There is no impact of the temperature on the results of analysis.
- More than 50% of initial alkaline content is leached in the water.
- It appears no decrease of the leaching rate after 520 days.

The cooling and heating process of measurement protocol may be responsible of this important leaching of alkalis. As the matter of fact, condensation phenomenon may leach importantly during rapid temperature changes.

4. Discussion

As it shown in Figure 2, a least 25% of alkalis are leached in the water, after 150 days, which suggest that almost 75% of free alkalis are available for ASR in the pore solution of concrete. As the mater of a fact, this latter value is overestimated because a certain amount of alkalis is fixed during the hydration of cement. Therefore, the free alkalis content in the pore solution is lower. The observed expansion plateau in series WV-C1-T60 indicates that the reaction is stopped because of a lack of alkalis to sustain this reaction.

When samples are stored in solution no plateau is observed in expansion measurement (series AS and PS). After 50 days, the kinetics of expansion is constant for the series PS.

As already well known, expansion is temperature dependent. It was also observed in this study that systems at 38°C expand less than systems at 60°C.

A delaying effect between PS-C1-T38 and PS-C28-T38 around 50 days is observed. This delay can be explained by a further hydration of the cement for the series PS-C28.

The proposed protocol (series PS) is original notably in the choice of the alkali concentration of the curing solution. This concentration has been chosen to be close to the cement pore solution. It is not possible to have a solution which is exactly the same as it is variable through the hydration process of the cement. To minimise the effect of this variation, we used the results of previous studies in which it is showed that the concentration largely stabilises after one month of cure (Chappex *et al.*, 2012) (Lothenbach *et al.*, 2007). Two types of alkalis are present in the cement pore solution, sodium and potassium. These are known to have different effects in terms of the measured expansion, however, their interactions, if any, are not clear (Leemann *et al.*, 2008). In case of aluminium rich pore solution of certain blended systems, aluminium could also be added to the testing solution. This is being further investigated.

This protocol also has the advantage to respect the mechanical properties of the paste, by curing the samples during 28 days at 20°C before testing. As the expansion of ASR is largely due to the opening of cracks in the microstructure due to the pressure involved by the formation of reaction product, the test protocol should reproduce the relative strengths of the phases, aggregate and paste, as they are in the field. This means in particular that tests where the reaction of the aggregate is very rapid may develop unrealistic damage patterns if the paste has not hardened sufficiently when samples are put into the reactor. Mechanistic simulation of such kinetic effects can be found (Giorla *et al.*, 2013).

5. Conclusion

In this study, some problems with the Swiss standard M 2042 performance testing procedure could be identified. It could be shown that the evolution of the pore solution composition (leaching), is one of the main reason for the flattening of the expansion after only few weeks. A new testing procedure is in course of development and was proposed. The values chosen for this test method are justified using the current knowledge in terms of the chemistry of the reaction as well as the micro-mechanics. It aims to respect the mechanical and chemical properties of pastes. This testing procedure consist of the following protocol:

– Pore solution extraction and analysis of the concrete cured during 28 days at 20°C in sealed containers.

– Samples preparation following the SIA 2042 procedure.

– Casting and curing of concrete samples (7x7x28 mm) during 1 days at 20°C in 95% RH (covered to avoid leaching due to water drops).

– After 1 day, concrete samples are stored in the simplified pore solution at 38° and 60°C (2 different protocols).

– The optimum temperature (between 38°C and 60°C) still has to be defined.

This promising procedure has to be further validated with a number of different aggregates, fast- and slow-reacting alike and several blended cements.

6. References

Barneyback, R.S., Diamond, S., "Expression and analysis of pore fluids from hardened cement pastes and mortars". *Cement and Concrete Research*, vol. 11, no. 2, 279-285, 1980.

Chappex, T., Scrivener, K., "The Influence of Aluminium on the Dissolution of Amorphous Silica and its Relation to Alkali Silica Reaction". *Cement and Concrete Research*, vol. 42, 1513-23, 2012.

Giorla, A.B., "Modelling of alkali-silica reaction under multi-axial load". Doctoral dissertation, EPFL, 2013.

Leemann, A., Lothenbach, B., "The influence of potassium–sodium ratio in cement on concrete expansion due to alkali-aggregate reaction". *Cement and Concrete Research*, vol. 38, no. 10, 1162-1168, 2008.

Longuet, P., Burglen, L., Zelwer, A., "La phase liquide du ciment hydrate". *Revue des matériaux*, 1973.

Lothenbach, B., Winnefeld, F., Alder, C., Wieland, E., Lunk, P. "Effect of temperature on the pore solution, microstructure and hydration products of Portland cement pastes". *Cement and Concrete Research*, vol. 37, no. 4, 483-491, 2007.

Stanton, TE, "Expansion of Concrete through Reaction between Cement and Aggregate". *Transactions of the American Society of Civil Engineers*, vol. 66, 1781-1811, 1940.

CHAPTER 3

Structural Modeling

Modeling of Environmental Conditions and their Impact on the Expansion of Concrete Affected by the Alkali–Silica Reaction

YUICHIRO KAWABATA* — KAZUO YAMADA** — SHOICHI OGAWA***

** Port and Airport Research Institute*
3-1-1, Nagase, Yokosuka, Kanagawa 239-0826, Japan
kawabata-y@pari.go.jp

*** National Institute for Environmental Studies*
10-2, Fukasaku, Miharu, Tamura, Fukushima 963-7700, Japan
Yamada.kazuo@nies.go.jp

**** Taiheiyo Consultant, Co., Ltd.*
2-27-8, Higashi-Nihonbashi, Chuo-ku, Tokyo 103-0004, Japan
Shoichi_Ogawa@taiheiyo-c.co.jp

ABSTRACT. *The trial modeling of environmental conditions such as temperature and water supply and their impact on the simulated expansion of concrete are presented in this paper. Results of the alkali-wrapped concrete prism test (AW-CPT), a new testing method that eliminates alkali leaching and drying, were compared with the results obtained from field exposure tests. The numerical model for prediction using the AW-CPT is then briefly presented, along with the trial modeling of various environmental conditions. The expansion behavior of a concrete block was simulated on the basis of numerical models and the AW-CPT. Results of expansions simulation using AW-CPT were consistent with those of the experimental exposure tests when we provided appropriate environmental inputs. A parametric study revealed that modeling rainfall conditions and setting of an averaging period have critical effects on the simulated expansion.*

KEYWORDS: *alkali–silica reaction (ASR), concrete prism test, field exposure test, modeling, environmental conditions.*

1. Introduction

The ability to predict the expansion of concrete is an important consideration in re-assessment of existing concrete structures that are affected by an alkali–silica reaction (ASR). As such, some advanced models have been developed in order to understand the expansive behavior of concrete structures using numerical tools. These models, however, require calibration of some of their unknown parameters in accordance with the material being tested. The concrete prism test (CPT) is widely used for assessing the potential expansion of concrete and for calibrating the unknown parameters of the numerical model. However, expansion of the concrete under testing conditions is typically smaller than that observed for the field-exposed concrete block (Ideker *et al.*, 2012). Alkali leaching is of concern when performing the CPT (Lindgard *et al.*, 2013); specifically, drying is a serious problem when testing concrete with low alkali content (Yamada *et al.*, 2016). These two issues lead to failure in the evaluation of the potential expansion of the concrete. Avoiding or minimizing these effects is thus essential. To address this, the authors have developed a test method for concrete expansion, the alkali-wrapped concrete prism test (AW-CPT; Yamada *et al.*, 2016). The predicted expansion of the field-exposed concrete block subjected to the AW-CPT has been found to agree well with the experimental results (Kawabata *et al.*, 2016a).

In predicting the expansive behavior of concrete structures affected by ASR, the modeling of environmental conditions must also be taken into consideration. Environmental conditions such as temperature and exposure to water are important factors that affect the expansion of concrete due to ASR. A previous study has revealed that rainfall has an especially significant impact on the expansive behavior of concrete that had been exposed in the field (Kawabata *et al.*, 2016a). Rainwater in contact with the concrete surface is absorbed by the concrete water by capillary suction, which in turn leads to higher water content in the concrete surface. Therefore, modeling rainfall and relative humidity (R.H.), the latter of which is a commonly used as boundary condition, is important. The use of weather statistic data is required in modeling environmental conditions. In order to easily estimate any expansion that may have occurred, environmental conditions may be modeled using data available from weather observation stations (WOSs) near the structure. On the other hand, environmental conditions should be predicted according to the meteorological history of the site when predicting the expansion of the concrete. Multiple methods may be used to average the environmental conditions. Because predicting the daily mean temperature, R.H., and rainfall with high accuracy is not possible, averaging these environmental conditions is necessary for modeling their weekly, monthly or annual trend. The influence of the length of the period over which these conditions are averaged should therefore be examined.

This paper compares the results of the AW-CPT with those of field exposure tests. A numerical model used for predicting the expansion of concrete through the AW-CPT using trial environment modeling conditions is presented. Finally, the impact of environment-modeling conditions on the simulated expansion is presented.

2. AW-CPT and Field Exposure Test

2.1. *AW-CPT*

The AW-CPT was performed on concrete specimens (75 × 75 × 250 mm) possessing a total alkali content of 3.0 kg/m^3 at 20, 40, and 60 °C. The testing protocols of the AW-CPT were described by Yamada *et al.* (2016). The cloth used in the test contained 50 g of 0.73 mol/l NaOH solution. The solution concentration was calculated to simulate the pore solution of concrete having a total alkali content of 3.0 kg/m^3. At each measurement, the weight of the cloth was measured, and pure water (not the alkaline solution) was added to the cloth in order to maintain a constant water content (50 g). The amount of alkali supplied by the wet cloth was estimated to be 0.3–0.4 kg/m^3. Andesite, a highly reactive coarse aggregate with a pessimum proportion of 30 wt%, was mixed with a nonreactive limestone aggregate. Nonreactive limestone sand was used as fine aggregate. The andesite used in this study contained opal and cristobalite. Structures built using this aggregate have been found to undergo severe ASR several years after construction.

Expansion curves obtained from the AW-CPT are shown in Figure 1. All concrete specimens tested showed continuous expansion by more than 0.1% over the test period. Expansion at 40°C was found to be the highest, followed by that at 60 °C, and finally at 20 °C. Even at 20 °C, expansion started after 50 days, eventually matching the expansion at 60 °C. The figure also presents the results of fitting by Equation (2), which is explained in the next chapter 3. Fitting using Equation (1) was difficult because the expansions have not reached a plateau. Actually, we could fit most of the laboratory test results with Equation (1), but the tests were subject to alkali leaching, resulting in an S-shaped expansion curve. Because the concrete structures were massive and alkali leaching was negligible, they showed continuous expansion. A previous work by the authors revealed that concrete subjected to conventional CPTs (namely, AAR-3 & AAR-4; Kawabata *et al.*, 2016a) show little expansion (<0.08%). The expansion of concrete tested at 40°C was found to be especially lower, at <0.03% at half a year, which may be misinterpreted as the concrete being non-expansive. The reason for the lower expansion may be alkali leaching and drying of concrete during the test. This difference between the AW-CPT and conventional CPTs emphasizes the importance of alkali wrapping of concrete prisms (Kawabata *et al.*, 2016a).

Arrhenius plots depicting the temperature dependence of each parameter based on fitting with Equation (2) are shown in Figure 2. Each parameter varies approximately linearly with the inverse of the temperature. The expansion of concrete subjected to various temperatures can thus be simulated by calibrating the parameters using this correlation.

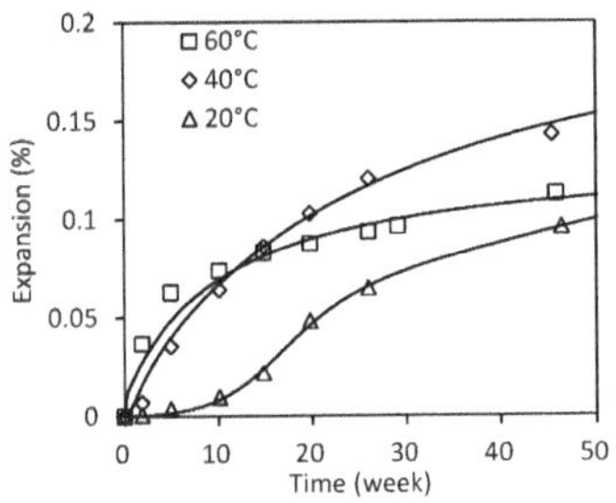

Figure 1. *Expansion of concrete at different temperatures subjected to the AW-CPT*

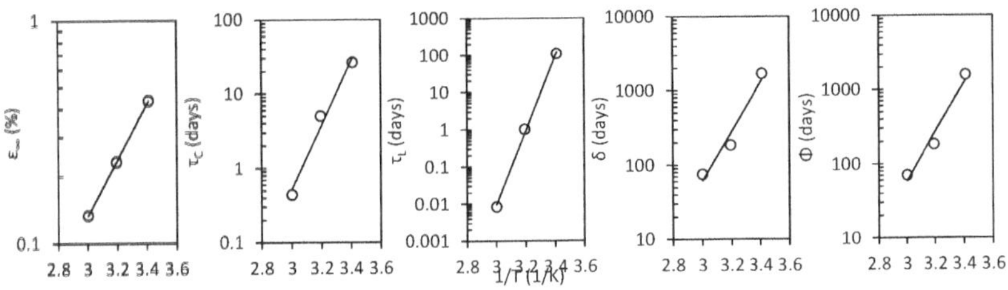

Figure 2. *Arrhenius plot of each parameter of Equation (2)*

2.2. *Field Exposure Test*

Concrete blocks (0.4 × 0.4 × 0.6 m: Fukuoka, 1.0 × 1.0 × 0.5 m: Yokosuka) using aggregates with the same size as those in the aforementioned AW-CPT test were exposed to field conditions in Fukuoka and Yokosuka. The total alkali content in the concrete was determined to be 3.0 kg/m^3. Exposure tests were started on 29th of March, 2011, in Fukuoka and on 19th of February, 2014, in Yokosuka. The meteorological histories of these areas, which were recorded by the Japan Meteorological Agency, are presented in Figure 3. Since R.H. data could not be obtained in the nearest WOS (Miura), data for Yokohama (located 20 km from the exposure site) were used. As shown in the figure, the environmental conditions in Fukuoka and Yokosuka were almost identical.

The expansion behaviors of the concrete blocks are described in Figure 4. In Fukuoka, the expansion of concrete started after 100 days and increased linearly with time over the exposure period. Expansion after 3 years exceeded 0.2%. In Yokosuka, the expansion also started at 100 days, but the expansion rate was higher than that in Fukuoka; expansion exceeded 0.2% after 2 years of exposure. In any exposure site, expansion occurs over a long period, a behavior that cannot be simulated by conventional CPT. Interestingly, the expansion in Yokosuka after 1 year of exposure was almost 0.1% while that in Fukuoka was closer to 0.05%. The degrees of expansion of the field-exposed concrete blocks after 1 year were lower than those predicted by the AW-CPT after 1 year. The field-exposed concrete blocks were subjected to temperatures above 25°C in the summer and below 10°C in the winter. Moreover, the R.H. in the winter was lower than that in summer. The expansion of the field-exposed concrete was therefore significantly influenced by environmental conditions.

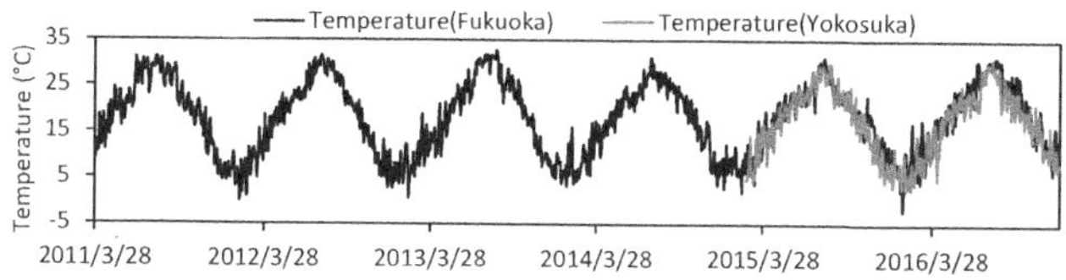

Figure 3. *Environmental conditions of the exposure sites*

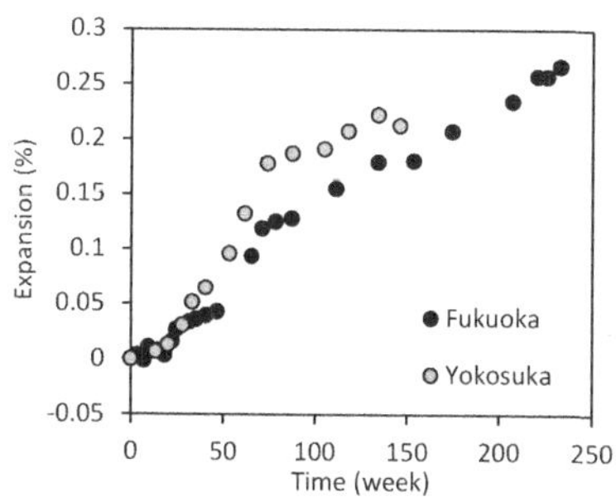

Figure 4. *Expansion of field-exposed blocks*

3. Numerical Model

3.1. *ASR Expansion*

ASR-induced expansion can be formulated as the product of an expansion potential using a kinetic function (Larive, 1998):

$$\varepsilon_t = \varepsilon_\infty \frac{1-\exp\left(\frac{-t}{\tau_C}\right)}{1+\exp\left\{-\frac{(t-\tau_L)}{\tau_C}\right\}}, \tag{1}$$

where t is time (day), ε_t is the expansion at time t (%), ε_∞ is the asymptotic final expansion (%), τ_C is the characteristic time (day), and τ_L is the latency time (day).

Equation (1), which produces an S-shaped curve, was used to simulate ASR expansion. The actual ASR expansion, however, is well known to occur over long periods. Therefore, Equation (2) (Brunetaud *et al.*, 2004) was used to simulate ASR expansion in this study, although this equation was originally devised for expansion due to delayed ettringite formation (DEF):

$$\varepsilon_t = \varepsilon_\infty \frac{1-\exp\left(\frac{-t}{\tau_C}\right)}{1+\exp\left\{-\frac{(t-\tau_L)}{\tau_C}\right\}}\left(1-\frac{\phi}{t+\delta}\right), \tag{2}$$

where ϕ and δ are experimental constants (in days).

ASR expansion is strongly influenced by temperature such that each parameter calibrated by fitting may be calculated as a function of the inverse of the absolute temperature. Each parameter can be expressed according to the following equation derived from the Arrhenius law:

$$\frac{X_{(T1)}}{X_{(T2)}} = \exp\left[\frac{U_X}{R}\left(\frac{1}{T_1}-\frac{1}{T_2}\right)\right], \tag{3}$$

where X and U_X denote each parameter and its activation energy, respectively, R is 8.314 J/(K mol), T_1 is the temperature (K) of the desired environment, and T_2 is the temperature (K) of the laboratory.

Three parameters from Equations (1) and (2), namely ε_∞, τ_c, and τ_L, are moisture-dependent. In further calculations, the saturation degree, s, was used as a relevant descriptor of the moisture content of the material. The moisture dependency can be expressed according to the following equations (Poyet, 2003):

$$\varepsilon_\infty(s)=\varepsilon_\infty(100\%)\left(\frac{\langle s-s_{\infty\_\varepsilon\infty}\rangle^+}{1-s_{\infty\_\varepsilon\infty}}\right)^{m_{\varepsilon\infty}} \tag{4}$$

$$\tau_{c,l}(s)=\tau_{c,l}(100\%)/\left(\frac{\langle s-s_{\infty\_\tau_{c,l}}\rangle^+}{1-s_{\infty\_\tau_{c,l}}}\right)^{m_{\tau_{c,l}}}, \tag{5}$$

where s_{∞_x} is the threshold saturation degree of parameter x below which no expansion occurs, and m_x is the nonlinearity parameter of x ($s_{\infty\_x} = 0.8$ and $m_x = 1.0$ in this study).

3.2. *Shrinkage*

The degree of shrinkage, another important parameter that varies linearly with moisture content, may be written as follows (Torrenti *et al.*, 1999):

$$\varepsilon_{sh(t)}=k_{sh}\times(s - s_{ini}) \tag{6}$$

where k_{sh} is a shrinkage coefficient; s and s_{ini} are saturation degrees at time t and in the initial state, respectively ($k_{sh} = 0.11\%$ and $s_{ini} = 0.97$ in this study).

3.3. *Moisture Transfer*

Moisture transfer in concrete may be calculated by solving the following differential equation:

$$\frac{\delta s}{\delta t} - D_s \frac{\delta s^2}{\delta^2 x} = 0 \tag{7}$$

where D_s is the diffusion coefficient and s is the saturation degree.

The diffusion coefficient is nonlinear and is dependent upon the humidity or saturation degree of the pores. The nonlinear diffusion model that was proposed by Bazant and Najjar (1972) was used in the study:

$$D_s = D_0 \left(\alpha_0 + \frac{1-\alpha_0}{1+\left(\frac{1-s}{1-s_c}\right)^n} \right) \tag{8}$$

where D_0, α_0, n, and s_c are material constants.

On a wet concrete surface, as in the case after rainfall, the concrete skin becomes instantaneously saturated. Capillary suction from the surface then occurs. Therefore, the capillary suction should be simulated in parallel with the diffusion model described above. Since the capillary suction of water favors the filling of larger pores earlier than smaller ones, the pore size distribution significantly affects the capillary suction. Because of the difficulty of modeling the pore size distribution of concrete, the parameters of the diffusion coefficient were varied according to the wetting or drying cycle (D_0 = 100 mm^2/day (wetting) or 25 mm^2/day (drying), α_0 = 0.025 (wetting) or 0.1 (drying), n = 6, and s_c = 0.8 in this study). These values were calibrated to fit the experimental data although the properties of concrete were different (Martin *et al.*, 2013).

3.4. *Environmental Conditions*

Modeling the environmental conditions of the concrete subjected to drying conditions is typically not difficult. Analyzing the properties of shrinkage or creep, for example, simply requires the temperature and R.H. However, liquid water that is supplied by rainfall and reclamation soil, which is difficult to be modeled, must be accounted for when studying structures that are frequently affected by ASR.

Considering the difficulty in modeling the above conditions for this type of system, we adopted a simplified approach in this study. This approach consists of modeling rainfall as a condition occurring when R.H. has a value of 100%; this boundary condition was parametrically varied, as shown in Table 1. Because of the difficulty in modeling capillary action, our study used a simplified approach that can be adapted to the diffusion equation. In order to evaluate the effect of the length of the averaging period on ASR expansion, four different averaging periods were used: 1 day, 1 week, 1 month, and 3 months. Figure 5 shows the modeled temperature variations in Fukuoka for each averaging period of three years.

Case	Boundary conditions
RH	Only R.H. is used
RH+RF1	R.H. is 100% when precipitation is >0 mm
RH+RF3	R.H. is 100% during the day and in the two subsequent days when precipitation is >0 mm (for a total of 3 days)
RH100	R.H. is always 100%

Table 1. *Boundary conditions for the water supply*

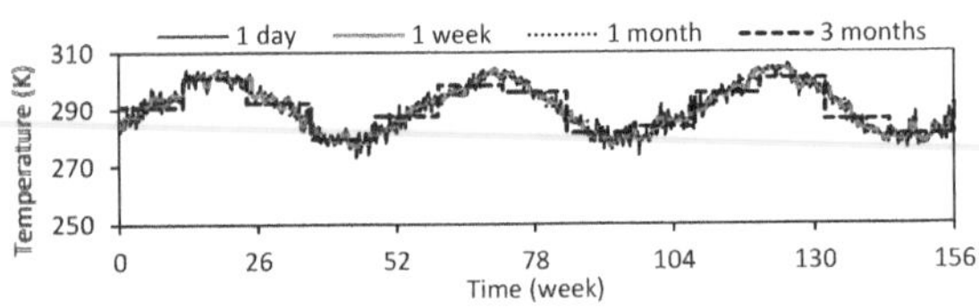

Figure 5. *Modeled temperature variations over different averaging periods (Fukuoka)*

3.5. *Other Assumptions and the Algorithm for the Calculation*

In order to simplify the model, some assumptions were made: (1) The temperature of the concrete instantaneously reaches equilibrium with the measured ambient temperature and solar insolation is negligible. (2) No mechanical calculation is required because the concrete is non-reinforced. (3) Supply or leaching of alkali is absent.

Two-dimensional simulation on the cross-section of concrete block was performed. The incremental expansion of concrete was calculated according to the moisture saturation degree of each element as follows: At time step t, the diffusion coefficient was updated using Equation (7) with respect to the saturation degree at time t-1. The moisture field and resultant incremental expansion was then calculated. The expansion strain was subsequently determined by averaging the strain distribution inside the block, which includes expansion and contraction. The time step was set to 0.2 day, and the calculation was looped over time.

4. Results and Discussion

4.1. *Simulation Using the AW-CPT*

The simulated results using the AW-CPT are presented in Figure 6. Results of simulation with the RH+RF3 conditions in Fukuoka agree well with those of the experiment, whereas simulations with the RH and RH+RF1 conditions produced lower degrees of expansion. This underestimate of expansion may be attributed to insufficient water supply, which results in drying of the concrete block. This in turn prevents expansion when the saturation degree is below the threshold value ($s_\infty = 0.8$). This condition is critically different from the actual conditions (discussed further in section 4.2). Conversely, simulation using the RH100 condition, in which that concrete always expands during exposure, overestimated the experimental results. On the contrary, simulations using the RH100 and RH+RF3 conditions in Yokosuka produced results that are most similar to those of the experiment. The larger size allows the water in the concrete to contribute to expansion, but drying still occurs because of insufficient water supply, thereby preventing continuous expansion at a high rate.

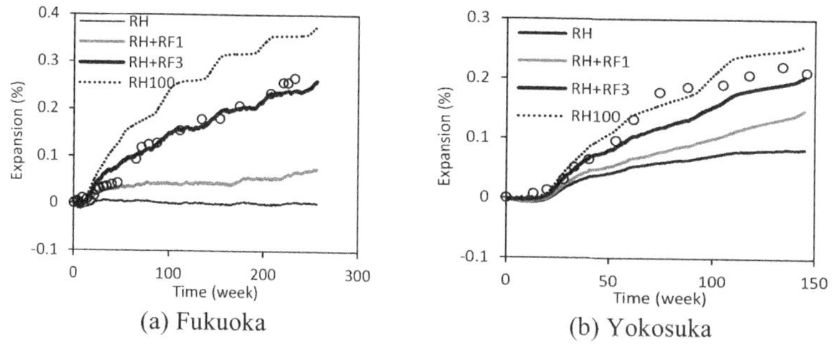

(a) Fukuoka (b) Yokosuka

Figure 6. *Simulated expansions under different water supply conditions (the averaging period is 1 day, and the legend indicates the water supply condition (Table 1))*

Below an expansion of approximately 0.04%, the simulation and experiment results correlated differently than they did elsewhere. The evolution of the transfer properties within concrete due to ASR expansion are critical and therefore must be taken into account, as emphasized in a previous report (Kawabata *et al.*, 2016b). ASR expansion induces microcracks, which allow moisture to move more easily within the concrete.

4.2. *Impact of Modeling Environmental Conditions*

The effect of differing water supply conditions on the simulated expansion of a $0.4 \times 0.4 \times 0.6$ m block is depicted in Figure 7. Legend in the figure shows (averaging period)-(boundary conditions for the water supply). The mean saturation degree, shown in Figure 7 (b), was calculated by averaging the saturation degree of each element within the block. The legend in the figure denotes the various conditions for the averaging period and water supply that were used. Increasing the averaging period was found to reduce the simulated expansion; therefore, the expansion calculated with the 1d-RH+RF3 condition is 1.3 times higher than that calculated with the 3m-RH+RF3 condition. This decrease in expansion may be attributed to a reduction in the mean saturation degree in the concrete. When the condition for the averaging period of the water supply increases, the effect of rainfall decreases, resulting in drying in the winter (the season with lower R.H.). On the contrary, when more precise data were provided for the simulation (as in the case of the 1-day and 1-week averaging periods), the mean saturation degree stays at a higher value.

The effect of the water supply condition on the simulated ASR expansion and mean saturation degree is depicted shown in Figure 8. As explained above, the use of the RH or RH+RF1 condition leads to a reduction in the mean saturation degree of the concrete. On the contrary, the mean saturation degree remained the same in the case of the RH+RF3 condition, resulting in a higher expansion rate.

The obtained results emphasize that precise environmental conditions are required as inputs in order to obtain better results. In this calculation, temperature dependency of ASR expansion was modeled by Arrhenius law which is likely to be sufficient to simulate the actual expansion. It should be noted that rainfall, in particular, has a critical impact on the mean saturation degree of the concrete and thus the ASR expansion. When using a longer averaging period to simulate concrete expansion, it is important to consider that it may likely result in an underestimate of the actual expansion. Therefore, an accurate model of moisture transfer during wetting and drying must be developed.

Various exposure tests have been performed in various parts of the world. In the PARTNER project, for example, 0.3 × 0.3 × 0.3 m blocks have been used (Lindgard *et al.*, 2010), while blocks with dimensions of 0.4 × 0.4 × 0.7 m have been used in experiments conducted in the United States (e.g., Moffatt *et al.*, 2016). A 3.0 × 3.0 × 3.0 m block was also fabricated for the Mactaquac Dam (Moffatt *et al.*, 2016).

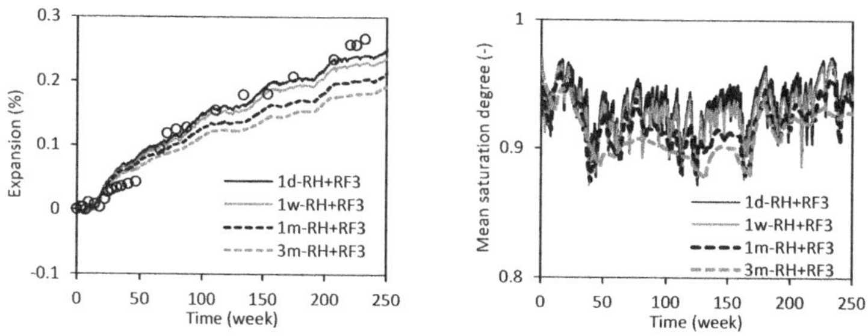

(a) Expansion (b) Mean saturation degree

Figure 7. *Effect of the averaging period on the simulated expansion under different water supply conditions*

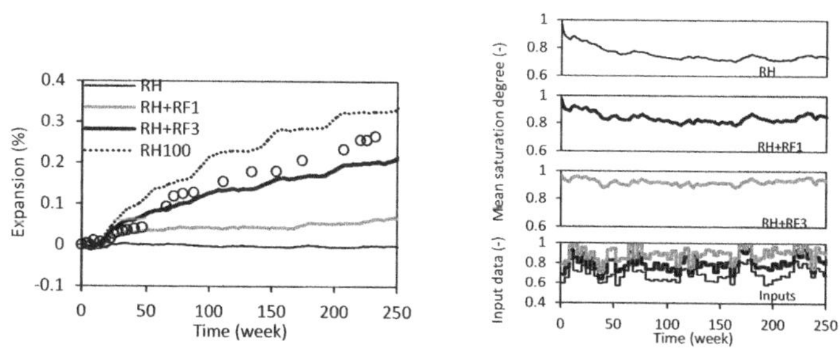

(a) Expansion (b) Mean saturation degree

Figure 8. *Effect of water supply on the simulated expansion (1 month-averaging)*

From the simulated results as shown in this paper, it can be pointed out that environmental conditions influence smaller concrete blocks more than they do larger blocks. The smaller blocks had a more sensitive response to the changing environment, while the largest block maintained almost the same value of mean saturation degree. Therefore, additional caution should be taken when interpreting the results of smaller concrete blocks that have been exposed in the field.

5. Conclusions

This paper presents the results of trial modeling of environmental conditions such as temperature and water supply, as well as their impact on the simulated expansion of concrete. The following conclusions may be drawn from this study:

– Results of the prediction of expansion using the AW-CPT, a new testing method, are consistent with the experimental exposure test results when precise environmental inputs are provided.

– Rainfall as well as temperature has a significant impact on the expansion behavior due to ASR. Therefore, a model accounting for rainfall and moisture transfer should be developed in order to produce simulation results that are more consistent with those obtained from the field.

– A longer averaging period reduces the effect of the R.H. boundary condition used in the simulation, resulting in less expansion. Therefore, an appropriate averaging period should be used in simulation using the AW-CPT in order to accurately describe the actual ASR expansion.

6. References

Bazant, Z. P. and Najjar, L. J., "Nonlinear water diffusion in nonsaturated concrete", *Mat. Struc.* Vol. 5, pp. 3-20, 1972.

Brunetaud, X., et al., "Delayed ettringite formation: Suggestion of a global mechanism in order to link previous hypotheses", *7th CANMET/ACI Int. Conf. on Recent Advances in Concrete Technology*, pp. 63-76, 2004.

Japan Meteorological Agency: http://www.jma.go.jp

Ideker, J. H., *et al.*, "The Importance of Outdoor Exposure Site Testing", *Proc. of 14th Int. Conf. on Alkali-Aggregate Reaction*, 051412-IDEK, 2012.

Kawabata, Y., et al., "Correlation between laboratory expansion and field expansion of concrete: Prediction based on modified concrete expansion test", *Proc. of 15th Int. Conf. on Alkali-Aggregate Reaction*, 15ICAAR2016_034, 2016a.

Kawabata, Y., et al., "Modelling of evolution of transfer properties due to expansion of concrete induced by internal swelling reaction", *Proc. of 15th Int. Conf. on Alkali-Aggregate Reaction*, 15ICAAR2016_033, 2016b.

Larive, C., "Apports Combinés de l'Expérimentation et de la Modélisation à la Compréhension de l'Alcali Reaction et de ses Effets Mécaniques", Laboratoire Central des Ponts et Chaussées, OA28, 1998. (in French)

Lindgard, J., et al., "The EU "PARTNER" Project – European standard tests to prevent alkali reactions of aggregates: Final results and recommendations", *Cement and Concrete Research*, Vol. 40, pp. 611-635, 2010.

Lindgard, J., et al., "Alkali-silica reaction (ASR) – performance testing: Influence of specimen pre-treatment, exposure conditions and prism size on alkali leaching and prism expansion", *Cement and Concrete Research*, Vol. 42, pp. 223-243, 2013.

Martin, R.-P., et al., "Importance of considering the coupling between transfer properties, alkali leaching and expansion in the modelling of concrete beam affected by internal swelling reactions", *Construction and Building Materials*, Vol. 49, pp.23-30, 2013.

Moffatt, E.G., et al., "Remediation strategies intended for the reconstruction of the ASR-induced Mactaquac dam", *Proc. of 15th Int. Conf. on Alkali-Aggregate Reaction*, 15ICAAR2016_069, 2016.

Poyet, S., "Etude de la dégradation des ouvrages en béton atteints par la réaction alcali-silice: approche expérimentale et modélisation numérique multi-échelles des dégradations dans un environnement hydro-chemo-mécanique variable", Université de Marne la Vallée, 2003. (in French).

Torrenti, J.-M., et al., "Modeling concrete shrinkage under various ambient conditions", *ACI Mat. J.*, Vol. 96, pp. 35–39, 1999.

Yamada, K., et al., "Importance of alkali-wrapping in Concrete Prism Tests", *Proc. of 15th Int. Conf. on Alkali-Aggregate Reaction*, 15ICAAR2016_084, 2016.

Nonlinear Finite Elements for the Assessment of Hydraulic Concrete Structures Affected by Alkali-Aggregate Reaction: A Study Case

MAHDI BEN FTIMA* — PIERRE LÉGER* — FATEH BOUSSAHA**

* *École Polytechnique de Montréal*
P.O. Box 6079, Station Centre-ville, Canada
mahdi.ben-ftima@polymtl.ca, pierre.leger@polymtl.ca

** *Hydro-Québec – Expertise en barrages*
75, boul. René-Lévesque ouest, Montréal, Canada
boussaha.fateh@hydro.qc.ca

ABSTRACT. *The structural stability of several AAR affected concrete hydraulic structures is very often impaired by the presence of severe tridimensional discrete cracks induced by AAR. A methodology to assess a sliding safety factor (SSF) and structural safety margin for this type of structure is presented herein using two complementary numerical models: (i) a quasi-static explicit continuum nonlinear finite elements smeared crack model (QSE-FEM-S) and (ii) a quasi-static explicit nonlinear finite elements discrete crack model (QSE-FEM-D) using the strength reduction approach. The QSE-FEM-S is using multi-physics thermal, hygral, mechanical formulations including an AAR kinetic model calibrated from field and material data. The QSE-FEM-S is essentially a "damage analysis" (i) to identify dominant crack patterns forming potential failure planes, and (ii) to compute the initial condition to perform a "stability analysis" using the discrete crack formulation (QSE-FEM-D). The proposed methodology is robust and leads classical engineering scalar stability indicators (i.e. the SSF) that are easier to interpret than smeared "damage" field. The application of the methodology is presented through the sliding stability assessment of an existing hydraulic structure.*

KEYWORDS: *concrete hydraulic structure, finite elements; quasi-static explicit analyses; alkali-aggregate reaction, smeared crack, discrete crack, sliding safety assessment.*

1. Introduction

Several concrete dams and spillways have been subjected to severe tri-dimensional discrete cracks induced by concrete expansion due to alkali aggregate reaction (AAR) producing an assembly of unreinforced cracked mass concrete blocks resting one on top of the others (Fig.1c). Typical examples have been reported in the literature such as Beauharnois dam, Canada, Fontana dam, USA, La Tuque dam, Canada, Chambon dam, France and Temple-sur-Lot spillway piers. (Vuillet et al. 2016). This paper presents a global methodology to assess the stability of cracked hydraulic structures affected by AAR. Two complementary finite element models (FEM) have to be developed within the framework of the quasi-static Abaqus explicit nonlinear equation solver. In a first stage, the FEM is using multi-physics continuum mechanics, smeared cracks and a user defined AAR constitutive model based on the work of Saouma and Perotti 2006. It is used to predict the evolution of displacements, crack patterns and emerging failure mechanism in conjunction with field monitoring and field investigations. In the second stage the FEM model is using a discrete crack model to assess sliding stability along a priori postulated 3D failure planes defined from the first stage. The second stage is to compute classical scalar sliding safety factor (SSF) using the material shear strength reduction method to detect the critical friction coefficient that will induce sliding instability. The illustration of the proposed two stages methodology is done through a case study adapted from an existing AAR affected cracked hydraulic structure to evaluate its sliding stability.

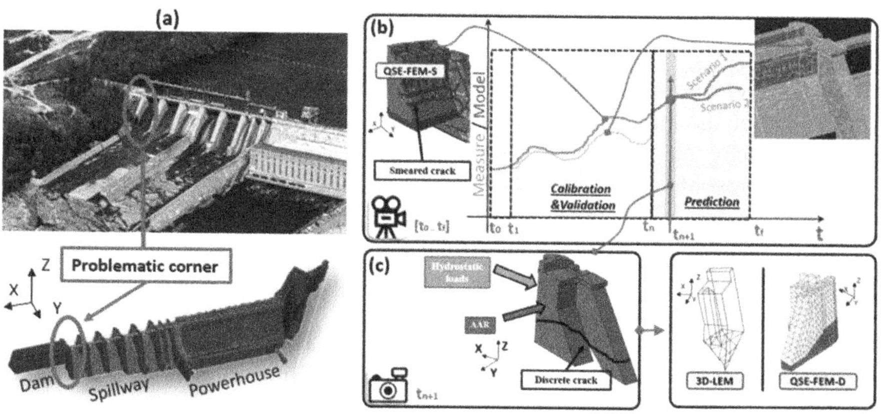

Figure 1. *(a) Description of the structural problem; Assessment methodology: (b) continuous and (c) instantaneous*

2. Methodology to Assess Hydraulic Concrete Structures Affected by AAR

The assessment of hydraulic structures affected by AAR is a challenging task that presents numerous aspects. From the structural point of view, the AAR is a continuous and slow process, driven by "imposed" internal deformations of the overall hydraulic facility leading to significant displacements, distortions and concrete cracking. This behaviour of AAR affected hydraulic structures is highly nonlinear and encompass multi-scale, multi-physics coupled processes including (i) mechanical (gravity G and water pressure P), (ii) chemical (AAR), (iii) thermal (T) and (iv) hygral (H). While withstanding environmental variations (H, T), and accelerated degradation (induced by AAR), the main structural function of the facility is to retain the upstream water (P, including hydrostatic P_H and uplift pressures P_U). Two different families of actions interact on the structural system: (i) actions driven by *loads* (G and P) and (ii) actions driven by *deformations* (T, H and AAR). The deformations induced by AAR are at least of an order of magnitude larger than those induced by temperature or shrinkage effects. The AAR process results over time in a network of localized cracks due to the low steel reinforcement ratio, typical for massive hydraulic structures. Once these new cracks form, water pressure can build up over time and induce stability problems of isolated concrete blocks (Figure 1). From a numerical point of view, no consensus has emerged from the literature and from practicing engineers to address this problem even when considering solely the concrete constitutive model mechanical aspects. Uncertainties exist for the estimation of nonlinear input parameters (e.g. tensile strength, fracture energy …) mainly due to size effects, transient evolution of microcracks and material damage forming discrete cracks of structural significance and to the difficulty to assess these parameters for an existing concrete structure. Furthermore, difficulties are known to arise when estimating kinetic parameters of the AAR. Calibrations at the structural level using available monitoring displacement data from pendulum measurements are generally used to fine-tune these input parameters. Most of the available chemo-mechanical constitutive models are known to provide satisfactory predictions for the displacement field of hydraulic facilities, but none of them is known to provide robust predictions of the residual strength and the related structural safety margin. The lack of experimental work at the element/structural levels to validate these chemo-mechanical numerical models is an obvious argument that supports this fact. On the other hand, more consensual views seems to exist on the constitutive model and mechanical properties of the concrete to concrete interfaces (e.g. Mohr-Coulomb), and on the analytical/numerical methods to assess the stability of rigid or deformable bulk material components (e.g. stability of dams at concrete lift joints). Computation of the sliding stability safety factor of hydraulic structures is a common task in engineering practice.

Taking into account these considerations, a methodology based on the authors past experience is suggested in this work. The methodology, schematically depicted in Figures 1.bc, is based on two parallel processes: (i) a *continuous damage* assessment process and (ii) an *instantaneous or 'snapshot' residual strength and stability* assessment process.

2.1. *Continuous Displacement and Damage Assessment Process (Figure 1.b)*

The continuous damage assessment process is based on a nonlinear numerical *companion model* of the existing structure or substructure. This model shall be ideally able to represent the major past historical events of the facility from the first day of concrete pouring t_0, to the reservoir impounding of the facility, up to the present day t_n. The first weeks, or months are very important in the life of hydraulic structures because they are responsible of more than 80% of cracks in a normal structure due to hydration process (i.e. without AAR pathology). Fine-tuning of this companion model is a *continuous learning process* for the model, that can be based on the monitoring data available from the time t_1 up to present time t_n. This *calibration & validation* process allow: (i) a good interpretation of past monitoring data and a better comprehension of the behaviour of the facility, (ii) optimization and rationalization of the location of future sensors. Fine-tuning essentially concern: (i) the kinetic parameters for creep and AAR processes, (ii) the nonlinear parameters of the bulk mass concrete material, (iii) the nonlinear parameters of the concrete-concrete and concrete-rock interfaces, and (iv) more generally the boundary conditions of the numerical model. Laboratory testing of sample materials or field-testing, such as localised stress release by cutting steel bars or performing a cut in concrete, could be used for this purpose and would be ideally guided by the numerical model. Contrarily to several past studies, quantitative data complementary to displacements, could be used to evaluate the performance of the model with respect to the existing facility (Figure 1.b). These data would better represent the context of nonlinear multi-physical analyses: crack length, crack width, piezometric pressure in evolving cracks and measurements of downstream seeping water flows.

The same companion model can be used for *prediction* purposes from the present time t_n to the anticipated end-of-life time of the facility t_f (whether functional or physical). According to *V&V* principles (Thacker et al. 2004), the evolving calibration and validation process (from t_1 to t_n) will help to assess the *predictive capability* of the numerical model. The prediction process allows: (i) detection of potential future problems (e.g. at t_{n+1}), (ii) a better orientation of maintenance operations, and (iii) comparison of alternative major repair scenarios and long term strategic planning (e.g. estimation of the functional end-of-life time).

Nonlinear analyses for this continuous process shall be based on 3D finite element models using appropriate multi-physic concrete representation. The related constitutive models should at least include the chemo-mechanical coupling and consider the influence of environmental conditions (T and H). According to the authors past experience, the feasibility of these complex analyses is a real industrial issue because large models have to be developed for the entire facility (or strategic substructures of the facility), with an adequate mesh refinement to be able to visualize in a comprehensive manner crack patterns. The specific case of hydraulic structures with mass concrete (i.e. with little or no steel reinforcement) is particularly challenging from the numerical point of view due to concrete softening (Ben Ftima 2014). The Quasi-Static Explicit FE method using smeared cracking QSE-FEM-S has been found to be an appropriate solution to this feasibility problem. It has been successfully applied in the industry of hydraulic structures in Quebec during the past ten years (e.g. Ben Ftima and Massicotte 2015). The use of QSE-FEM-S for this continuous process is adequate to identify the potential failure planes and related dominant cracks within the concrete body of the facility. Experience has shown that these cracks are not always along the concrete lift joints. QSE-FEM-S is a powerful tool that has, however, some limitations such as the aleatory and knowledge uncertainties in estimating the fracture parameters of the AAR affected bulk concrete material and epistemic uncertainties for the predictive capability of the constitutive laws in estimating the residual safety margin at time t_n. To overcome these weaknesses, the suggested methodology is to combine the results of QSE-FEM-S (smeared crack model) with in-situ investigations (visual/submarine camera inspections and concrete drilling) to identify potential failure 3D crack surfaces or more generally potential failure mechanisms. Extrapolation of specific data measurements at particular location to a complete and critical 3D crack interface pattern is certainly required in this phase through expert engineering judgement. Instantaneous assessment processes can then take place to evaluate the safety factors for the identified potential failure mechanisms. Details of the QSE-FEM-S tool are presented in section 3.

2.2. *Instantaneous Residual Strength and Stability Assessment Process (Figure 1c)*

Contrarily to the continuous structural health monitoring (SHM) and related continuum mechanic models, the "instantaneous" residual strength and stability assessment is a '*snapshot*' estimation of the stability of the structure or a substructure for an identified critical '*timeframe*' t_{n+1} and an identified potential failure mechanism. For the example considered in this study (Figure 1), a 3D potential failure surface was identified by site investigations, and raised the stability problem of an upper wedge concrete block. The assessment of safety factors and evaluation of potential major repair scenarios is then required in this phase and can be viewed as the "*true*" engineering phase. The 3D geometry of the discrete crack and particularities of the loading combinations (e.g. 3D loading due to AAR thrust from the right side dam, presence of uplift pressure in cracks …) makes the problem particularly challenging. Vuillet et al. 2016 have recently developed specific numerical tools and strategies to assess the sliding safety factor for these cases. The Quasi-Static Explicit FE analyses with Discrete crack (QSE-FEM-D) and an extension to the conventional limit equilibrium method (3D-LEM) are two tools that have been developed within the overall proposed assessment methodology. Details of these tools are presented in section 4.

3. Quasi-Static Explicit FE Using the Smeared Crack Approach: QSE-FE-S

3.1. *The Quasi-Static Explicit Approach*

In the quasi-static explicit approach, the nonlinear problem is solved dynamically considering Newton's second law of motion. Conventional nodal forces are converted into inertia forces by assigning lumped masses into the nodes. The dynamic equilibrium equations are written in terms of inertia forces, where $\boldsymbol{M}$ is the lumped mass matrix of the model, $\boldsymbol{P}$ is the external load vector, and $\boldsymbol{I}$ is the internal load vector:

$$\boldsymbol{M}\ddot{u} = \boldsymbol{P} - \boldsymbol{I} \tag{1}$$

With comparison to the conventional implicit approach, no iterations are performed. The algorithm advances explicitly in time using a very small time increment to ensure stability. The loads are also applied slowly enough to minimize the kinetic energy with respect to the internal strain energy of the system. This approach has been initially developed for the industry of metal forming at the beginning of the nineties (Prior 1994), and recently applied to the case of civil engineering concrete hydraulic structures (Ben Ftima 2014). Advantages of this approach were shown with respect to the conventional implicit approach using Newton-Raphson iterations, for the specific case of highly nonlinear problems involving large concrete structural models with concrete softening and complex contact interactions.

3.2. *User Defined Material AAR Model for QSE-FE-S and Computational Multi-Physical Framework*

A user-defined subroutine was developed under Abaqus-Explicit using the VUMAT interface (Hibbitt et al. 2016). As shown in Figure 2, the model was originally implemented as a purely mechanical constitutive law EPM3D (Massicotte et al. 2012) that has undergone a rigorous V&V process over the last ten years and was used in the industry field of hydraulic structures and bridge engineering (Ben Ftima and Massicotte 2015, Massicotte et al. 2015).

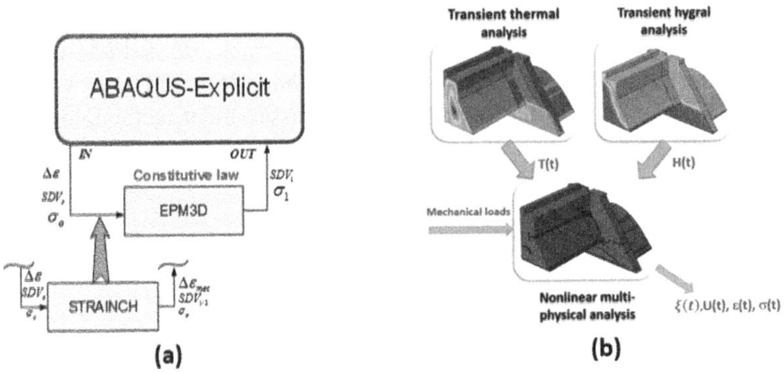

Figure 2. *Numerical framework for the QSE-FE-S: (a) user-subroutine implementation; (b) multi-physical approach (Ben Ftima et al. 2016)*

More recently (Ben Ftima et al. 2016), a multi-physical extension of this law has been performed using the decomposition of the total strain into mechanical, thermal, creep, shrinkage and AAR components. Similarly to the work of Pan et al. 2013, the total strain increment $\Delta\epsilon$ is decomposed into mechanical $\Delta\epsilon_{mec}$, thermal$\Delta\epsilon_{th}$, creep $\Delta\epsilon_{cr}$, shrinkage $\Delta\epsilon_{sh}$and AAR strains $\Delta\epsilon_{aar}$ according to the following equation in the incremental form:

$$\Delta\epsilon(t,T,H) = \Delta\epsilon_{mec}(t,T,H) + \Delta\epsilon_{th}(t,T) + \Delta\epsilon_{cr}(t,\sigma_0) + \Delta\epsilon_{sh}(t,H) + \Delta\epsilon_{aar}(t,\xi,T,H,\sigma_0) \tag{2}$$

The AAR strain is computed according to Saouma and Perotti 2006 anisotropic model and to Larive 1998 model for the AAR kinetics. In parallel to this extension, a multi-physic framework was developed in Abaqus based on three different successive finite element analyses: (i) thermal implicit, (ii) hygral implicit and finally (iii) nonlinear multi-physic explicit analyses (Figure 2b). An innovative formulation to address the problem of time scale difference between the implicit and explicit approaches was developed for this purpose.

4. Stability Assessment Using Discrete Quasi-Static Explicit FE and Extended Limit Equilibrium Methods (QSE-FE-D and 3D-LEM)

Once a kinematically admissible failure mechanism, and related failure surfaces, have been postulated from (i) field monitoring and (ii) field investigations in complementarity with (iii) the QSE-FE-S rheological model, a limit equilibrium approach is used to compute the sliding safety factor (or other indicators). In the proposed methodology, Vuillet et al. 2016 are using first a 3D extension (labelled 3D-LEM) of the well-known 2D Limit Equilibrium Method (2D-LEM) using the horizontal equilibrium of series of rigid body wedges in contact with each other subjected to applied external forces. The 2D-LEM method was used in 2009 for the stability assessment of the application example considered in section 5. In Vuillet et al. 2016 work, a MATLAB code was developed by implementing a progressive reduction of the concrete-concrete sliding interface friction coefficient and incrementing the potential angular sliding direction. The 3D-LEM allows a first approximation to establish a range of plausible values for the critical friction coefficient inducing sliding and related sliding direction. The 3D-LEM is used as a preliminary analysis tool to define the initial friction coefficient susceptible to induce sliding in the detailed QSE-FE-D model and to locate a control point to monitor the potential sliding displacements at the end of the sliding wedge. In the QSE-FED-D model, a bracketing procedure is used by performing series of FE simulations to refine the identification the critical friction coefficient using the displacement of the control point and the ratio of the kinetic energy to the strain energy as failure criteria.

5. Application

5.1. *Geometry and Material Properties*

Figure 3a shows the geometry and the FE model of a substructure extracted from the facility in Figure 1a.

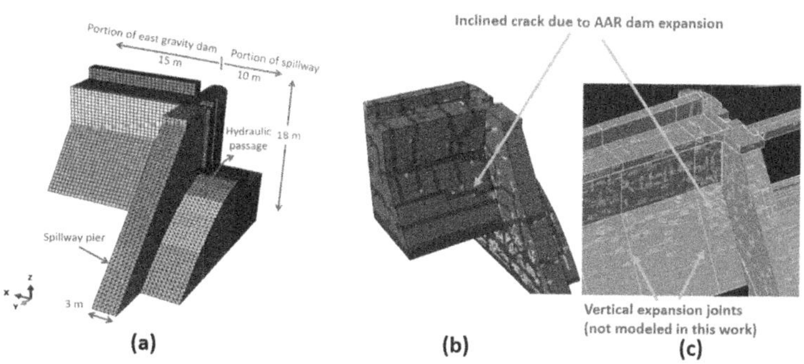

Figure 3. *FE model of the hydraulic structure and results of QSE-FE-S (Ben Ftima et al. 2016): (a) Geometry and mesh; (b) Predicted damage pattern (damaged elements in blue) and (c) real cracking pattern record*

Only portions of the east gravity dam and the spillway are considered. The pier of the spillway is subjected to the thrust action of the dam due to swelling AAR and behaves essentially as a cantilever beam. Fixity condition is assumed at the bottom faces of the model and plain strain conditions are imposed at the lateral vertical faces of the dam and the spillway chute. Hydraulic loads are applied at the upstream faces with the normal closed condition of the spillway gates. The upstream water level is located 1.2 meters below the crest of the dam. Table 1 gives some input data used for multi-physic analyses. For more details, the reader is referred to Ben Ftima et al. 2016. Preliminary input data were used for the AAR kinetic parameters, and the use of available monitoring data to calibrate these parameters is planned in future work.

Property	Symbol	Value	Unit
Mass density	ρ	2350	kg/m^3
Young modulus	E	25000	MPa
Compressive strength	f_c'	25.0	MPa
Tensile strength	f_t'	2.3	MPa
Poisson's ratio	ν	0.18	-
Mode I fracture energy	G_F	0.2	kN/m
Long term volumetric AAR strain	$\varepsilon_{v\infty}$	0.002	-

Table 1. *Some input data used for the multi-physical analysis in QSE-FE-S*

5.2. *Results of QSE-FE-S*

The computational time for the multi-physic model (with more than 120 000 nodes) was around eight hours on a standard quad-core laptop. Figure 3.b shows the final damage (crack) pattern evolution after a total analysis period of 100 years. The cracked elements are blue-colored whereas non-cracked elements are red-colored. The actual cracking pattern of the existing structure is shown in Figure 3. The similarity between computed and actual patterns is interesting though very preliminary data was used. This generalised crack pattern caused by AAR is closely monitored and does not compromise stability under the hypotheses formulated in the preliminary analyses reported in section 5.3. The tendency related to the apparition of an inclined crack near the dam- pier intersection is noticeable in both the real and predicted (computed) patterns. Parametric studies showed that the inclination of this crack in the numerical model (and more generally the overall cracking pattern) matched better the real records when larger portion of the east gravity dam and the vertical expansion joint between east gravity dam and spillway were considered in the model. Site investigations based on visual inspection of crack contours for the downstream face, or camera inspection with a submarine for the upstream face and concrete vertical drilling from the dam crest allowed to identify a potential 3D cracking surface in the spillway pier and portion of the dam as shown in Figure 1.b. This topology was then used as input to the stability assessment phase using 3D-LEM and QSE-FE-D tools.

5.3. *QSE-FE-D and 3D-LEM Analyses*

The stability analysis considered herein is performed for the timeframe just before the repair of the structure, which consisted of anchoring the upper block with vertical passive steel anchors across the crack surface. In this work, stability of the upper block is considered using the loading combination schematically shown in Figure 4. Only actions induced by loads were considered in the stability problems: G, P_H and P_U. It is assumed in this study that actions induced by deformations, and more particularly the AAR thrust from the east gravity dam, do not contribute to the potential instability of the upper block. This hypothesis, though intuitive, will be validated in future studies. In lieu of this AAR thrust, a conservative hydrostatic pressure block is assumed at the vertical east side of the model, coincident with the spillway to dam expansion joint (Figure 4.b). The mesh of the model used in both 3D-LEM (for the top and bottom faces of the upper block) and QSE-FE-D tools is shown in Figure 4.c. Zero cohesion is assumed for the crack interface with a friction coefficient of 1.13 (corresponding to a 48.5 deg friction angle).

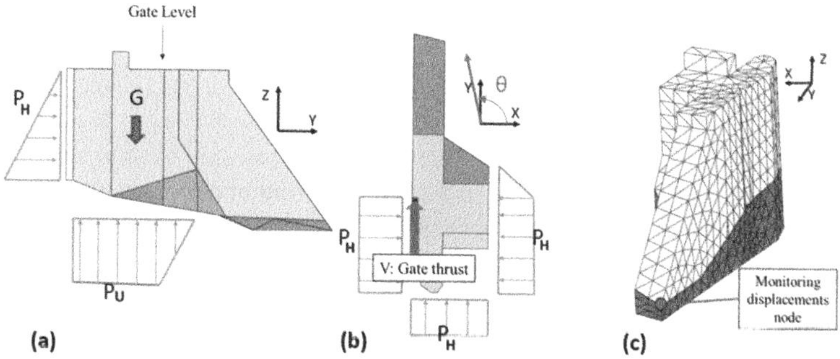

Figure 4. *(a) Model, (b) Load conditions of the identified potentially instable concrete wedge and (c) FE mesh for the QSE-FE-D analysis*

Using these hypotheses, the computed safety factors and potential sliding directions are given in Table 2. Very close results were obtained with both methods for the sliding direction. This direction was close to the upstream/downstream direction as shown in Figure 4.b. Sliding safety factors (SSF) relatively close to each other were obtained with both tools, with a lower bound tendency of the 3D-LEM tool as discussed in Vuillet et al. 2016.

3D-LEM		QSE-FEM-D		
$SSF_{LEM\text{-}D}$	$\boldsymbol{D}_{LEM\text{-}D}$ θ (deg)	$SSF_{FEM\text{-}D}$ Criterion C_{Eng}	$SSF_{FEM\text{-}D}$ Criterion C_{Disp}	$\boldsymbol{D}_{FEM\text{-}D}$ θ (deg)
1.66	106	1.94	1.78	105

Table 2. *Results of the stability analysis*

6. Conclusions and Perspectives

The structural stability of several AAR affected concrete hydraulic structures is very often impaired by the presence of severe tridimensional discrete cracks induced by AAR. A global methodology has been developed to assess a sliding safety factor (SSF) of this type of structure using two complementary numerical models: (i) a quasi-static explicit continuum nonlinear finite elements smeared crack model (QSE-FEM-S) and (ii) a quasi-static explicit nonlinear finite elements discrete crack model (QSE-FEM-D) using the strength reduction approach. The proposed global methodology is robust because the use of explicit nonlinear equation solvers alleviate several convergence difficulties often experienced with implicit solvers applied to highly nonlinear bulk material and opening-closing, stick-slip interface contact problems encountered in AAR affected hydraulic structures. Most importantly, we believe that the proposed methodology is a significant step forward to compute residual strength and scalar stability indicator (i.e. SSF) that are meaningful to the engineering profession and could facilitate the decision-making process and justification of structural strengthening actions or mitigations, if necessary.

7. References

Ben Ftima, M., Sadouki, H. and Brühwiler, E., "Development of a computational framework for the use of nonlinear explicit approach in the assessment of concrete structures affected by alkali-aggregate reaction", *Proceedings, 9th International Conference on Fracture Mechanics of Concrete and Concrete Structures, FRAMCOS-9*, V. Saouma, J. Bolander, and E. Landis (Eds), May 22-25, Berkeley, California, USA, 11 pp., 2016

Ben Ftima, M. and Massicotte, B., "Utilization of nonlinear finite elements for the design and assessment of large concrete structures, part II: Applications", *ASCE Journal of Structural Engineering*, Vol. 141, No. 9, 2015.

Ben Ftima, M., "Utilisation de la méthode des éléments finis non-linéaires pour la conception des structures en béton armé: application aux structures massives", PhD Thesis, École Polytechnique de Montréal, Canada, 2014.

Hibbitt, H.D., Karlson, B.I. and Sorensen, E.P., ABAQUS version 2016, finite element program. Hibbitt, Karlson and Sorensen, Providence, R.I., USA, 2016.

Larive, C., "Apports combinés de l'expérimentation et de la modélisation à la compréhension de l'Alcali-Réaction et de ses effets mécaniques", PhD Thesis, LCPC, Paris, France, 1998.

Massicotte, B., Ben Ftima, M. and Conciatori, D., "Établissement d'une méthode d'évaluation des ponts utilisant les analyses non-linéaires: application à des ouvrages existants", Research report, École Polytechnique Montréal, 2015.

Massicotte, B., Nour, A., Ben Ftima, M., Yildiz, E., and Conciatori, D., "EPM3D v 2.0, A user-supplied constitutive model for the nonlinear finite element analysis of concrete structures". Research report, École Polytechnique de Montréal, 2012.

Pan, J., Feng, Y., Jin, F. and Zhang, C., "Numerical prediction of swelling in concrete arch dams affected by alkali aggregate reaction." *European Journal of environmental and Civil Engineering*, 17(4): 231-247, 2013.

Prior, A., "Applications of implicit and explicit finite element techniques to metal forming." *Journal of Materials Processing Technology*, 45(4): 649-656, 1994.

Saouma, V., and Perotti, L., "Constitutive model for alkali aggregate reactions." *ACI Materials Journal*, 103(3):194–202, 2006.

Thacker, B.H, Doebling, S.W., Hemez, F.M., Anderson, M.C., Pepin, J.E. and Rodriguez, E.A., (2004). "Concepts of model verification and validation." Los Alamos National Laboratory report LA-14167-MS, 2004.

Vuillet, F., Ben Ftima, M. and Léger, P., "Stability of cracked concrete hydraulic structures by nonlinear quasi-static explicit finite element and 3D limit equilibrium method". Submitted to *Computers & Structures*, 23/09/2016.

Macro-Modelling of AAR-Affected Hydraulic Structures

Assessment of Parameters Influencing the In-Situ Concrete Growth

VLADIMIR GOCEVSKI* — EMRE YILDIZ**

** Hydro-Quebec Equipment*
855 St-Catherine Est, 10th floor, Montreal, PQ, H2L 4P5, Canada
gocevski.vladimir@hydro.qc.ca

*** IDAE s.e.n.c.*
204 rue Saint-Sacrement, Suite #300, Montreal, PQ, H2Y 1W8, Canada
emre@idae.ca

ABSTRACT. *The evaluation of three-dimensional, anisotropic rate of swelling is essential for obtaining accurate analytical results, of the behavior of the AAR-affected structures, corresponding to the in-situ observations. In this paper, a practical engineering approach is presented addressing the influence of all contributing parameters, in the numerical model based on the plasticity of brittle materials. In particular, the following will be discussed in more details: (a) the effect of non-uniform moisture distribution on the rate of swelling throughout the massive concrete elements; and (b) the methodology of evaluation of internal and external confinement in massive concrete structures, the effect of confinement on rate of concrete swelling and its distribution throughout the entire structure. The practical application of the methodology in the analysis of hydroelectric power plant structures will be presented.*

KEYWORDS: *alkali-aggregate reaction, concrete, moisture, internal and external confinement.*

1. Introduction

During the last 40 years, extensive research has been conducted in studying the behavior; the characterization and the modelling of material of ASR-affected concrete structures. Based on the initial research, to detect the reaction in the concrete and to judge its intensity, testing procedures have been formalized in many national standards. In addition, the attempts have been made to relate the results obtained from standardized accelerated concrete tests on small samples to the in-situ behavior of AAR-affected concrete. The difficulty to obtain realistic in-situ concrete properties, directly from AAR accelerated small sample tests and consequently to determine the behavior of the concrete in real structures, lead to the conclusion that obtaining the required properties directly from the affected structure is a more feasible approach. The understanding of the in-situ concrete behavior is essential for an adequate analytical approach in the evaluation of an existing structure and to provide answers to the following questions: (a) Has the AAR deteriorated the safety of the structure? (b) Are the structural elements resistant enough to sustain the applied loads? (c) Can the displacements caused by the concrete growth be accommodated to ensure uninterrupted service of the plant?

In this paper, an elaborate but practical engineering approach for the analysis of an AAR-affected hydroelectric and nuclear power plant concrete structures is briefly presented. Using this approach has led to appropriate evaluations and cost effective interventions for the last 30 years. The experiences gained in actual projects improved the methodology of defining and evaluating the important parameters of AAR modeling. Some of those parameters will be discussed here in this paper.

2. Numerical Model Developed and Used by Hydro-Quebec

Hydro-Quebec's first AAR model was developed in 1992 in an implicit scheme, implemented as a user material within commercially available COSMOS/M finite element software. The earlier versions of the code were able to adequately simulate the nonlinear behavior of the concrete as well as the AAR strains. Since then, the model is under continuous improvement. The latest versions of the code has the most important AAR parameters implemented, such as the anisotropic swelling, anisotropic confinement, relative humidity dependence, time dependence of both ASR and concrete materials etc. In 2012, the code has been moved to Abaqus Explicit [Hibbitt 2016], commercially available finite elements software. The use of explicit analysis provides a significant advantage in dealing with large models with strongly nonlinear behavior, such as AAR-affected hydraulic structures.

3. Discussion on Primary Parameters Affecting the Results Obtained by Hydro-Quebec Equipment (HQE) Numerical Model

3.1. *Maximum AAR Expansion Rate of the In-Situ Concrete*

The maximum free AAR expansion rate may be defined as the largest rate possible when the relative humidity and temperature conditions are "optimal" for the expansion, and there are no stresses applied to the material.

The estimation of the maximum free expansion rate should be based on the interpretation of the in-situ measurements. It should also be based on the inspection and survey reports, to predict the initiation time of the reaction. The evaluation of the in-situ rate is hence a trial and error process, comparing structural analysis results to measurement data. This procedure can be accomplished in a few steps:

The first step involves an initial estimation of the maximum free expansion rate of the concrete at 100% relative humidity and with no confinement effects. The second step is the calculation of the steady-state relative humidity distribution within the structure. This distribution represents the average humidity distribution during the service life of the structure. In the third step, the effect of the reinforcement on the AAR expansion rate is evaluated, if the reinforcement is not explicitly modelled. The confinement due to boundary conditions and its effects on volumetric expansion and on the anisotropic distribution of the expansion is then evaluated by the software during the analysis. The analysis also includes the effects of micro and macro-cracking on the rate of expansion.

At the end of this process, the calculated displacements should match closely to the in-situ measurements in three directions for a representative number of instrumented points on the structure.

3.2. *Effect of Reaction Duration on Free Expansion Rate of the Concrete*

Based on experimental evidence, the alkali-aggregate reaction sets on when the concentration of the built-up gel exceeds a particular threshold value. In the formulation, this is accounted for by introducing the initiation time. Figure 1a presents the incorporated exponential law for the variation of the AAR rate with time.

As discussed earlier, the expansion of concrete does not start until the concentration of reactive material reaches a certain level. This is introduced by the 'Reaction Start Time' parameter in Hydro-Quebec's model.

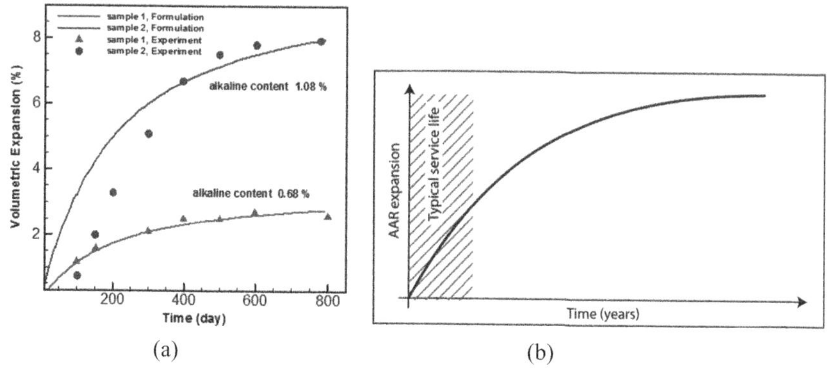

Figure 1. *(a) Incorporation of the exponential law, (b) AAR expansion over time vs. life span of a structure*

As the constituents of the chemical reaction of ASR are consumed, the rate of reaction slows down. The expansion amount over time can be considered as an exponential curve to simulate the slowing of the reaction. However, for the periods of time corresponding to a life span of a structure, the change in the rate is negligible, as shown in Figure 1b. For that reason, the free AAR rate can be considered constant for structural evaluation purposes.

3.2.1. *The Distribution of RH Throughout the Structure and its Effect on the Rate of Free Expansion*

The relative humidity is a major factor defining the rate of chemical reaction causing the expansion of concrete in AAR-affected structures. Many tests in the literature have shown that the amount and the rate of the alkali-aggregate reaction are dependent upon the water content in the concrete and below a certain value of relative humidity, the reaction would not occur.

While there are some empirical relationships provided in the literature for the AAR expansion rate in function of the relative humidity within the concrete, they are all based on small sized tests in accelerated conditions. Hydro-Quebec's experience shows that these tests tend to underestimate the expansion rate.

Three curves, presented in Figure 2, are being used to introduce the RH effect in HQE model. The choice of the curve to use for a given type of concrete is based on the observations on the studied structure.

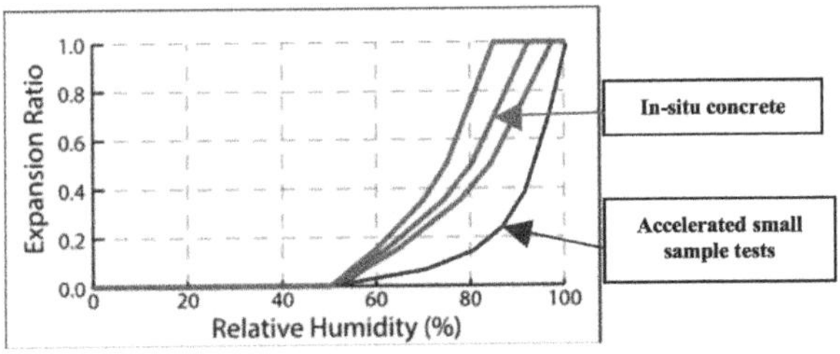

Figure 2. *Variation of the ASR expansion in function of the relative humidity*

The effect of the relative humidity is introduced explicitly into Hydro Quebec's methodology where its values are obtained from a steady-state humidity transfer analysis done prior to AAR modeling and its effect is calculated at each integration point, according to the curve presented above. An example of this evaluation is provided in Figure 3.

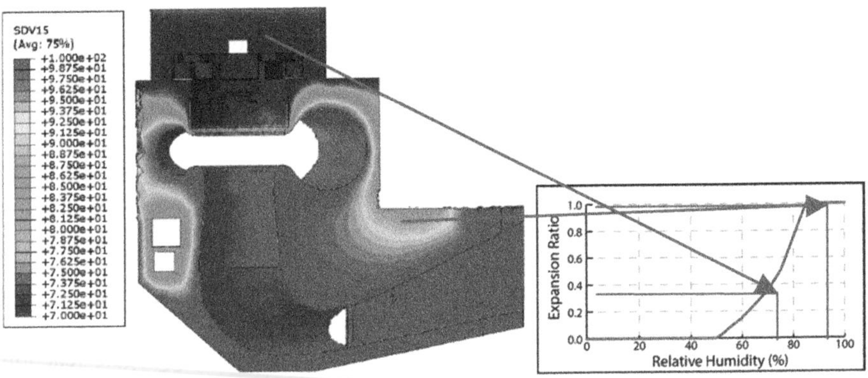

Figure 3. *Effect of RH on the AAR expansion*

3.2.2. *Effect of the Reinforcement (Internal Confinement)*

The internal confinement results from the internal forces created by the reinforcement steel subjected to the concrete AAR expansion strains. While the expansion develops, the reinforcement steel is subjected to tensile deformations imposed by the concrete. As a reaction, the reinforcement steel resists to these tensile deformations by applying a compressive force on the concrete.

When the AAR expansion becomes significantly large, the reinforcement steel starts to yield and the compressive force applied to the concrete remains constant, in a yielding plateau, or changes slightly with deformation, in a strain hardening stage.

Therefore, the confinement forces increase linearly as the AAR expansion starts but then they are stabilized once the deformation reaches the yielding strains.

The reinforcement steel effect may be introduced explicitly by modelling the reinforcement steel or implicitly by introducing its effect as a volumetric expansion reduction that is in function of the volumetric steel ratio. If introduced implicitly, the volumetric reinforcement ratio should be calculated for a given concrete volume.

The effect of internal confinement has been demonstrated by many tests in the literature. Jurcut [Jurcut 2015] presented a summary of tests showing the variation of AAR expansion in function of the reinforcement ratio, as shown in Figure 4a. A simple model has been analyzed using the Hydro-Quebec's model to study the effect of the reinforcement ratio on AAR expansion. The model used is a simple rectangular prism with a reinforcement placed in the center of it, as shown in Figure 4b. The area of the reinforcement is varied to obtain different reinforcement ratios.

The concrete is then subjected to AAR expansion. At the end of the analysis, the displacements of the concrete block are used to calculate the average yearly AAR expansion rate. The simulation is repeated for three different free expansion rates: 50, 100 and 150 µε/ year. As can be seen in Figure 4c, for all three cases, the results are within the experimental results range.

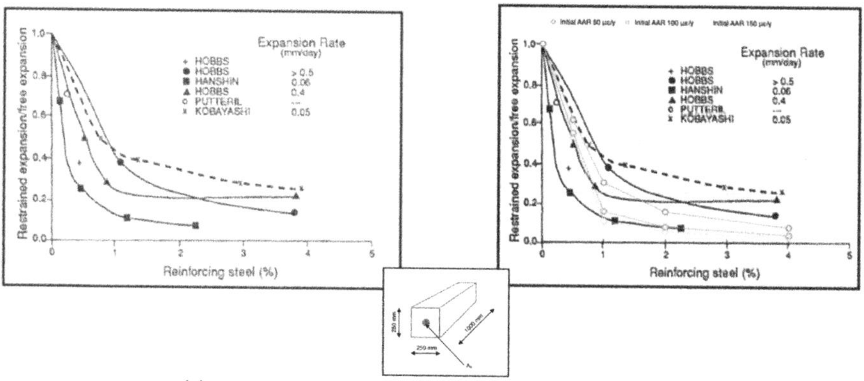

(a) (b) (c)

Figure 4. *(a) Test results on reduction of AAR expansion with the steel ratio (b) Geometry of the model used for validation (c) Comparing analysis results to experimental values*

The "simplified curve", obtained as the average of the curves acquired from the tests and those found numerically from the model (Fig. 4b and 4c), defines the ratio of the restrained over free expansion as a function of the reinforcement steel percentage in a concrete section is presented in Figure 5.

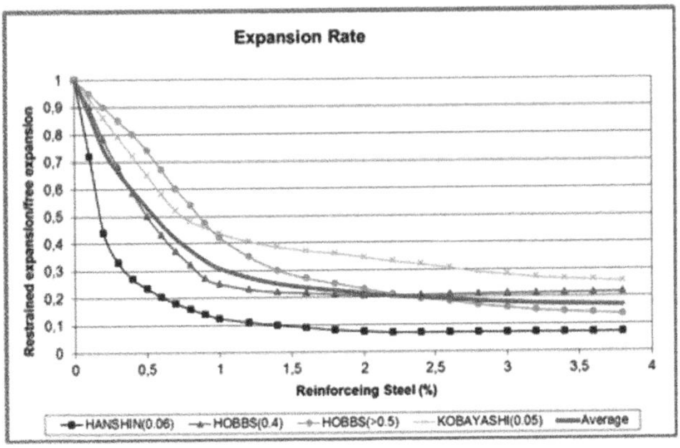

Figure 5. *Simplified Curve on reduction of AAR expansion with the steel ratio obtained from available test results and numerically calculated*

3.2.3. *Effect of External Confinement*

External forces applied to a given volume of concrete cause external confinement. These forces may be caused by the permanent and live loads applied to the structure, such as the dead load or by the forces created by prestressing or by post-tensioning of concrete. They may also be created by the AAR expansion itself, as the expansion of the concrete faces a reaction from the boundary conditions. The confinement forces that are created by the expansion will continue to increase as the AAR progresses. The stress levels where the confinement effect begins and where the confinement reaches a full stop may vary with the concrete properties and AAR aggressiveness. As a simplified approach, it can be assumed that the confinement effect varies linearly between these two thresholds, as presented in Figure 6a with a continuous line. Many experimental results show a parabolic tendency presented as a dashed line. The assumption of a linear effect is slightly conservative than the actual behavior. The effect of confinement on the rate of concrete expansion of a powerhouse is shown in Figure 6b.

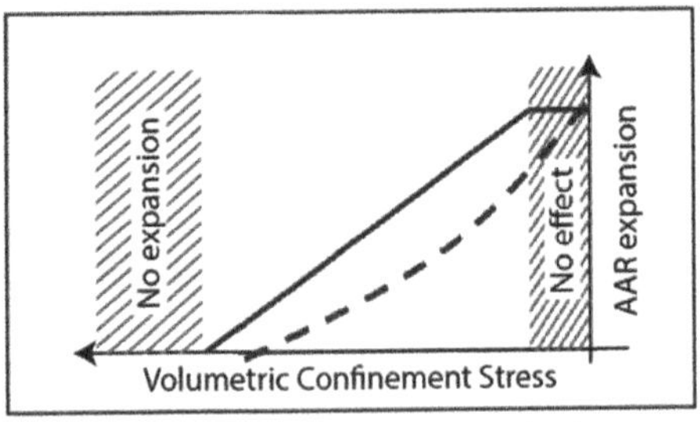

Figure 6a. *Confinement effect in function of volumetric stress*

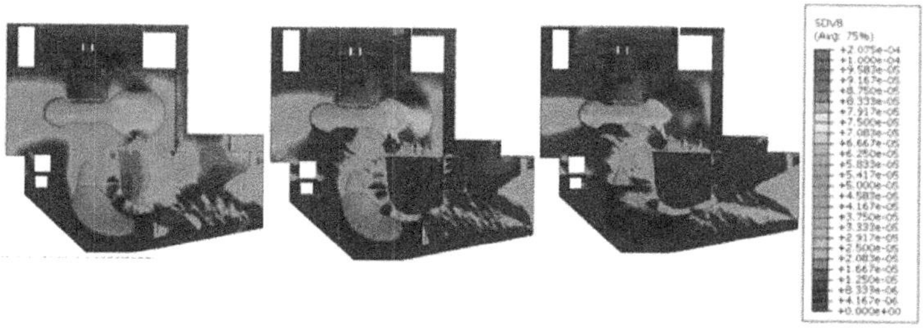

(a) (b) (c)

Figure 6b. *Variation of in-situ rate of expansion in right-left bank direction;(a) 5, (b) 18, (c) 50 years after construction respectively*

3.2.4. *Confinement Effect on the Volumetric Expansion and the Anisotropy of the Swelling*

The external confinement reduces the volumetric expansion of the concrete and creates anisotropy of the swelling. As demonstrated by the tests performed by Sherbrooke University under the supervision of Hydro-Quebec, a sufficiently high confinement stress would stop the AAR expansion completely. Similar conclusions have been drawn from multiple on-site observations realized by Hydro-Quebec's experts.

The effect of external confinement is included in Hydro Quebec's model through a volumetric expansion reduction and an anisotropic distribution of the expansion. The upper and lower stress thresholds are parameters that need to be defined by the user for a given concrete material.

The expansion of AAR is not of equal amplitude in all directions. Its rate varies with a number of confinement stresses in each direction. As demonstrated in the tests performed by University of Sherbrooke, while the expansion in one direction is reduced with the confinement stress in a given direction, it would not slow down equally in the other directions where no such stress exists. It is, therefore, necessary to apply an anisotropic distribution of the volumetric expansion, in function of the stress state in each direction.

As an example, a thin-walled circular member of a nuclear power plant containment building is studied. The section is pre-stressed in vertical (z) and horizontal (x) in plane axis. The stress state is presented in Figure 7.

STRESS STATE DATA

PARAMETER	VALUE
Stress XX	-5.4 MPa
Stress YY	-2.0 MPa
Stress ZZ	-4.6 MPa

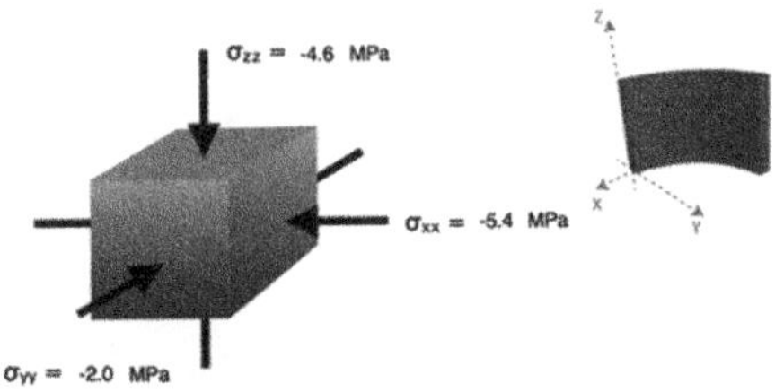

Figure 7. *Stress state of the thin-walled circular member*

In this example, the free expansion rate at 100% relative humidity was estimated to be 140 με/year. With the effect of the average relative humidity obtained from 15 in-situ locations of the reactor building, the AAR expansion without any confinement effect is evaluated to be 50 με/year.

The calculations of confinement effects start by defining the volumetric confinement parameter. This parameter determines at what stress level the AAR expansion would start lessening and at what level it would stop completely. These stress levels depend on the concrete mix and the reinforcement ratio. They can be estimated based on the stress/expansion relations obtained from laboratory testing of small samples and in-situ stress measurements using overcoring and instrumented concrete cylinders inserted in the structural concrete. In this case, the variation of volumetric confinement is as shown in Figure 8. The stress level where the expansion would come to a stop is estimated to be larger than most cases of massive concrete of hydraulic structures. At this moment, based on our knowledge, there are not enough tests to confirm the real reason of this difference between massive and structural concrete.

MATERIAL RELATED DATA

PARAMETER	VALUE
Confinement Starts	-3.0 MPa
Expansion Stops at	-18.0 MPa

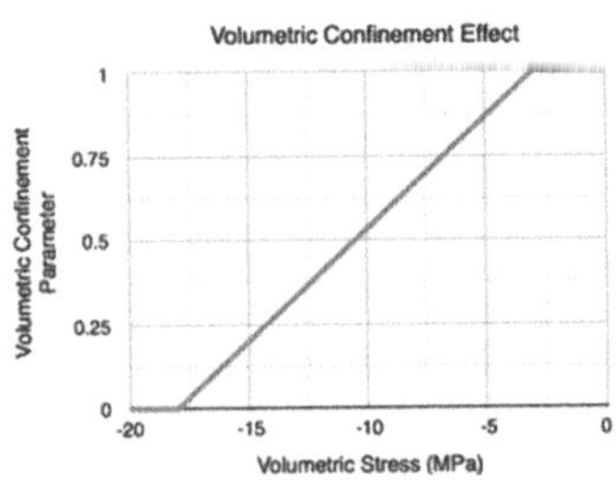

Figure 8. *Variation of volumetric confinement parameter in function of volumetric stress*

The stress levels are defined with negative values for compressive stresses, as it would in a finite element analysis. As shown in Figure 9a, this volumetric stress of this example would reduce the expansion to 40% of the expansion rate with no confinement. The in-situ volumetric expansion, with the effect of volumetric confinement and the humidity, is hence estimated to be 20 με/year.

In the next step, the anisotropic distribution of this volumetric expansion should be calculated. For this, the uniaxial confinement parameter will be defined. Based on Hydro-Quebec's experience, the uniaxial confinement stress levels can be estimated in most cases to be the third of volumetric confinement levels. The curve used for this application and the uniaxial confinement parameters for each direction, according to the magnitude of stresses, is presented in Figure 9b.

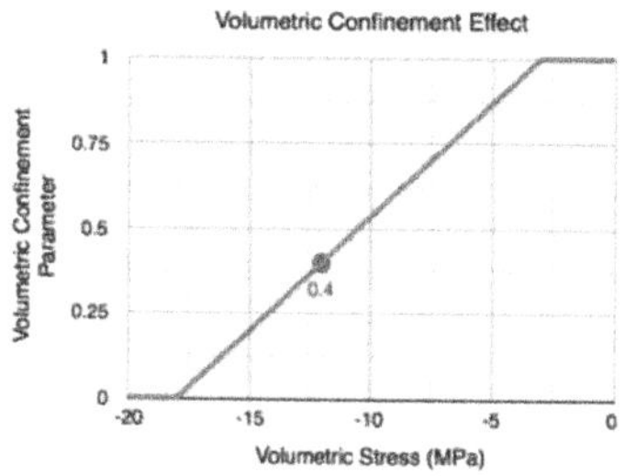

(a)

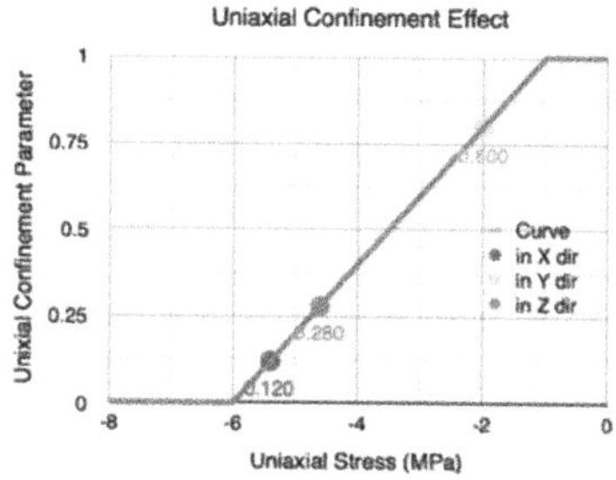

(b)

Figure 9. *(a) Reduction of volumetric expansion of 40%; (b) evaluation of the uniaxial confinement parameters for each direction*

The values obtained for uniaxial confinement parameters are then used to calculate the percentage of expansion in each direction. The reduced volumetric expansion is then distributed in each direction, according to the calculated percentages in Figure 10a. The final results of the uniaxial expansion rates are presented in Figure 10b. The obtained expansion rates match closely to the observed ones.

Uniaxial expansion distribution ratios

PARAMETER	Uniaxial Confinement Parameters	% of AAR expansion (UCP/Sum)
Direction XX	0.120	10%
Direction YY	0.800	67%
Direction ZZ	0.280	23%
Sum of Parameters	1.200	

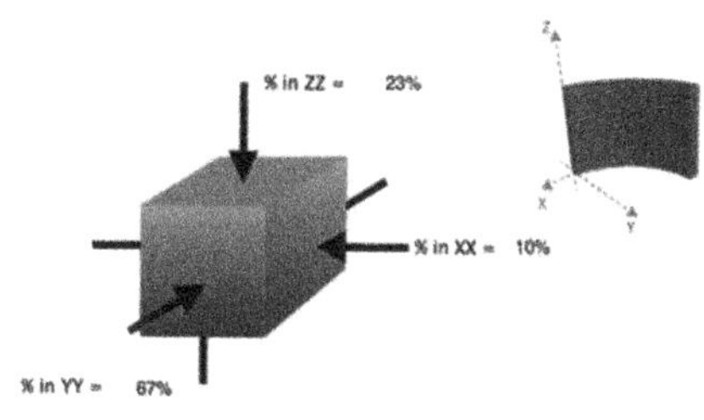

(a)

Uniaxial expansion rates

PARAMETER	% of AAR expansion (UCP/Sum)	AAR expansion rate
Direction XX	10%	2.0 με/year
Direction YY	67%	13.3 με/year
Direction ZZ	23%	4.7 με/year

(b)

Figure 10. *(a) – Calculation of percentages of expansion. (b) Distributed expansion rates in each direction.*

While there are other methods in the literature to calculate the anisotropy of the swelling, in Hydro-Quebec's experience, this method is more efficient, simpler than others, and provides accurate results.

3.2.5. *The Effect of Development of Micro- and Macro-Cracking on the Rate of Displacements*

To assure the safety of structures from the initial period of the operation to the end of the service life, the movements of the structure is surveyed regularly. It is, therefore, a standard procedure to install surveying targets at the crest of dams, spillways, intake structures and generating station floors during the construction of hydroelectric power plants. The movements are surveyed twice a year. These survey results are very useful for evaluation of in-situ vertical displacement rates in case of AAR-affected hydro structures. Some structures are equipped with instruments such as extensometers, pendulums, crack meters and so on. The interpretation of the measurement data should be realized with care. Most of the instruments function by measuring displacements between two points and measuring the average displacement rate between these points.

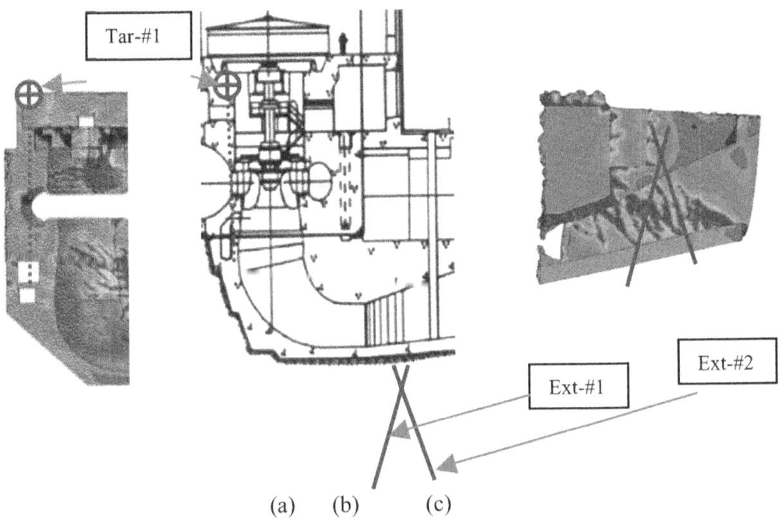

Figure 11. *(a) moderate macro-cracks through the entire height of the unit; (b) The position of instruments in the unit of the powerhouse; (c) extensive cracking and deformation in the pier of the draft tube, over a region of 2.0 m*

For power plants where the surveys are realized at the crest of the intake structure, at the turbine floor or the generating units' floor of the powerhouse, the obtained measurements represent the vertical displacement between the target point and the vertical projection of this point on the contact surface of the structure with the rock foundation. Therefore, we are measuring average displacements over a certain length and the derive rates are primarily average displacement rates. These should not be confused with the AAR rate as the displacement rates are indeed influenced by concrete expansion but also by many other factors.

As an example, the instruments' readings of a powerhouse will be studied. The location of the instruments is presented in Figure 11b.

The crack openings are approximately estimated using the plastic strains over the length of cracked area obtained from the finite element model. The crack opening values are shown in the direction of the instruments. Based on these values, the average concrete expansion rates can be approximately separated from the average displacement rates. It can be seen that the concrete expansion rates in extensively damaged (cracked) areas are 40 to 60% lower than average displacement rates.

The effect of macro-cracking on the measurements for 25 years is summarized in the table below.

Instrument	Length (m)	Measured displacement (mm)	Estimated crack movement (mm)	Corrected displacement (mm)	Rate without correction (μm/m/y)	Rate with correction (μm/m/y)
Ext – 1	16.0	41.73	26.0	15.73	104.32	39.92
Ext – 2	14.0	21.77	10.0	11.77	62.20	33.63
Tar – 1	17.5	17.75	5.5	12.25	40.57	28.00

Table 1. *Example of correction of instrument readings for cracking*

4. Examples on Modelling Large Structures with Complex Geometry

The numerical model and the methodology developed by Hydro-Quebec were used for evaluation and refurbishment of more than 25 AAR-affected structures. Few examples of the analysis are results of a powerhouse shown in the figures below.

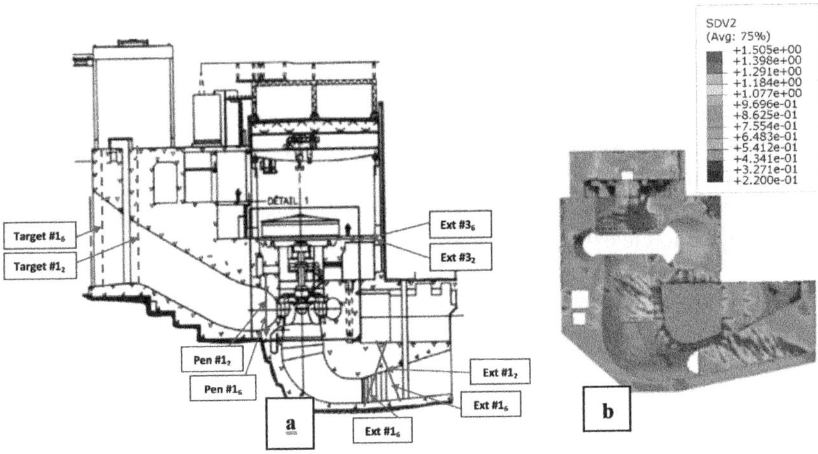

Figure 12. *(a) Location of the instruments; (b) Damage distribution*

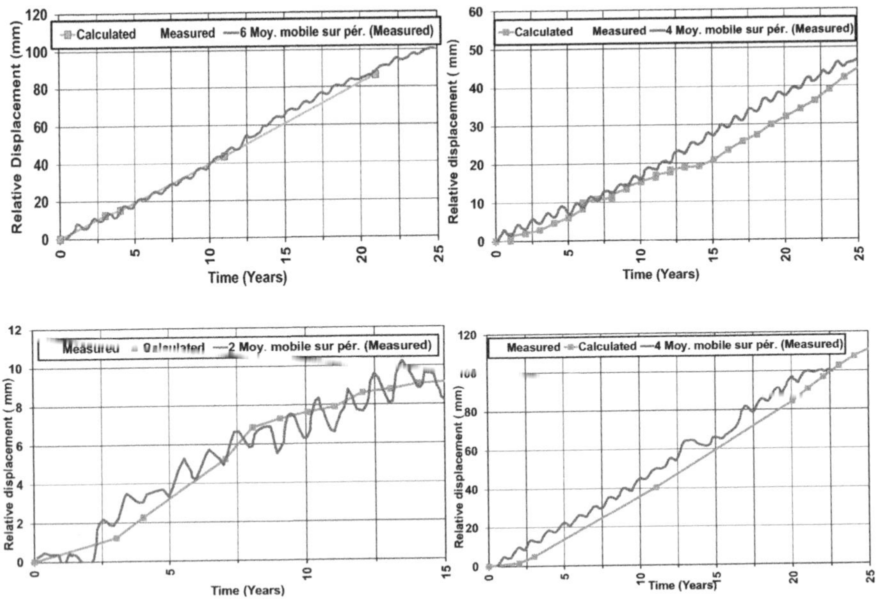

Figure 13. *Two examples only: comparison of measured and calculated displacements*

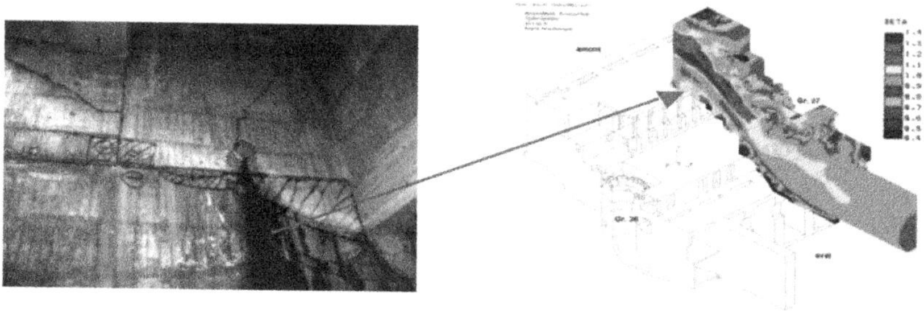

Figure 14. *The damage is expressed with the damage parameter β for the year 2010*

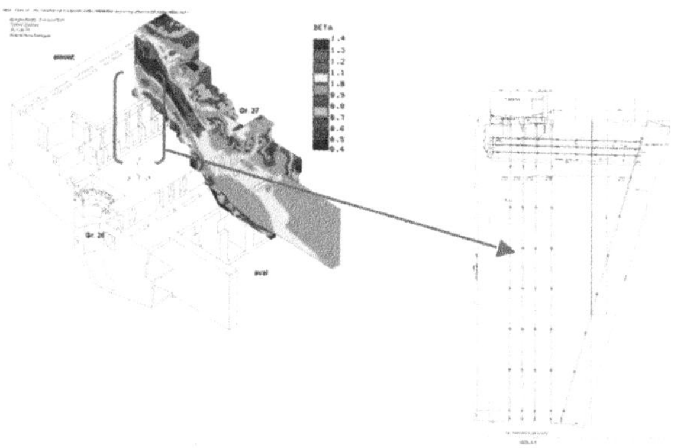

Figure 15. *Damage predicted for the year 2033; Pier strengthened in 2010 by 35M rebars*

The results presented here are a few examples taken from the results of a life extension study of a hydroelectric power plant severely affected by AAR. The study was able to provide answers to the most important, from the structural point of view, questions: (a) is the structure safe? (b) what is the life span for an uninterrupted production?, and (c) when is a suggested refurbishment economically justifiable?

5. Conclusions

In this paper, an elaborate but practical engineering approach, used by HQ for analysis of AAR-affected hydroelectric and nuclear power plant structures is briefly presented.

The modelling of AAR expansion defers from those frequently used by different models today. It is based on the in-situ obtained concrete expansion parameters, and uses a numerical formulation of nonlinear plasticity of brittle materials.

The major contribution presented in this paper is the introduction of a successful application of a simple engineering approach in evaluating the effect of primary parameters on the in-situ variability of the rate of expansion and their influence on the obtained results in the analysis of ASR-affected structures.

6. References

Gocevski, V., "Presentations with Bibliography of more than 60 papers and reports prepared by HQ personnel and their collaborators in the field of ASR of concrete - research and application", In the *proceedings of the:(a) ICOLD workshop, Fontana NC, USA, 2013, (b) NEA/CSNI/ IAGE CAPS Workshop, Gaithersburg, MD, USA June 29- July 1, 2015, (c) Meeting RILEM TC 259-ISR Boulder, CO, 6-7 June 2016, (d) Meeting EPRI- Charlotte NC, 7 December 2015.*

Hibbitt, H.D., Karlson, B.I. and Sorensen, E.P., *ABAQUS version 2016, finite element program*, Providence, R.I., USA, 2016.

Jurcut, A.C., *Modelling of Alkali-Aggregate Reaction Effects in Reinforced Concrete Structures*, University of Toronto, Canada, 2015.

AAR and DEF Structural Effects Modelling

ÉTIENNE GRIMAL* — PIERRE MORENON*,** — ALAIN SELLIER** — STÉPHANE MULTON** — ÉRIC BOURDAROT*

** EDF - CIH*
Savoie Technolac, 73373 Le Bourget du Lac Cedex, France
etienne.grimal@edf.fr, eric.bourdarot@edf.fr

*** LMDC Toulouse*
155 avenue de Rangueil, 31400 Toulouse, France
morenon@insa-toulouse.fr, alain.sellier@insa-toulouse.fr, stephane.multon@insa-toulouse.fr

ABSTRACT. *The internal swelling reaction (ISR) affects a number of Electricité de France's structures. In order to manage operating dams, a methodology based on monitoring, laboratory testing, modeling and remedial works is provided. This paper presents the model use to compute the behavior of dams affected. This model is able to consider different pathologies (AAR, DEF...). Modeling the behavior of ISR damaged concrete is made complex by the large number of elementary physical phenomena to be taken into account (concrete reactivity, thermal activation, moisture dependence, concrete creep and shrinkage, and damage interaction). The model is based on elementary physical principles that lead to the formulation of a visco-elasto-plastic orthotropic damage model including chemical pressure induced by ISR in a poro-mechanics framework. The anisotropic plasticity and damage alow to realistically model the strong cracking and swelling anisotropy observed on affected structures. It considers also the anisotropic stiffness recovery due to oriented crack reclosing during cyclic loading.*

KEYWORDS: *AAR, DEF, numerical model, dams behaviour.*

1. Introduction

The internal swelling reactions (alkali aggregate reaction or delayed ettringite formation) impact the long term behaviour of massive concrete structure such as dams. In order to manage the operation of its dams, Electricité de France (EDF) needs to reach a better understanding and a better prediction of swelling phenomena.

Indeed, EDF developed a methodology based on the following various phases:

– Swelling Detection: In this step, the swelling is detected by visual inspection and the monitoring. The origin of swelling (AAR or DEF) need to be confirmed by laboratory tests.

– Supervision and monitoring: After the detection, the visual inspections are increased and specific auscultation apparatuses are generally installed.

– Impact on operations: The swelling will cause structural dysfunction such as the blockage of the spillways, deformation of the production groups, a cracking which may lead to an increase in interstitial pressures.

– Consequences on the safety level: In this step, it is difficult to know the impact of swelling on structure affected. Numerical modeling is needed to analyze and to understand the behavior of the structure. To calibrate the model, monitoring and laboratory tests are used.

– Possible remedies: Once the behavior of the structure is understood, remedies may be proposed. The numerical model allows to design and to analyze the effect of the work.

– Long-term evolution: The last step to keep in service a structure affected by swelling is to evaluate the long-term behavior of the structure. For this the numerical approach is a great help.

As is shown in this methodology, laboratory testing, investigation and modeling are central to the management of structures affected by swelling.

In this context, EDF and LMDC Toulouse have been developing a numerical model integrated in a finite element computer code in order to simulate AAR (Grimal et al., 2010) and DEF (Salgues et al., 2014 ; Morenon et al., 2017) affected structures.

The constitutive model is based on a global rheological scheme which considers the porous pressures, by drying or other intra-porous phenomena (such as AAR), the external loading, and the concrete matrix as a representative elementary volume where the coupling between the swelling, creep and anisotropic damage takes place (Grimal et al., 2008).

The model was compared to experiments, to check its robustness (Grimal and al. 2007, 2008). It is now used to analyze the behavior of long-term structures.

2. Model Description

The following models are developed in Code_Aster. These models are also explained in (Morenon, 2017)

2.1. *Rheological Model*

The concrete behaviour is modelled in the poro-mechanical framework. The model contains a damage model (inspired from Grimal et al., 2008), and a rheological model regularly improved (Sellier et al., 2016), in order to consider realistic interactions between ASR, creep, shrinkage and damage. In figure 1 Pg is the pressure induced by AAR and Pw the capillary pressure causing shrinkage in non-saturated conditions.

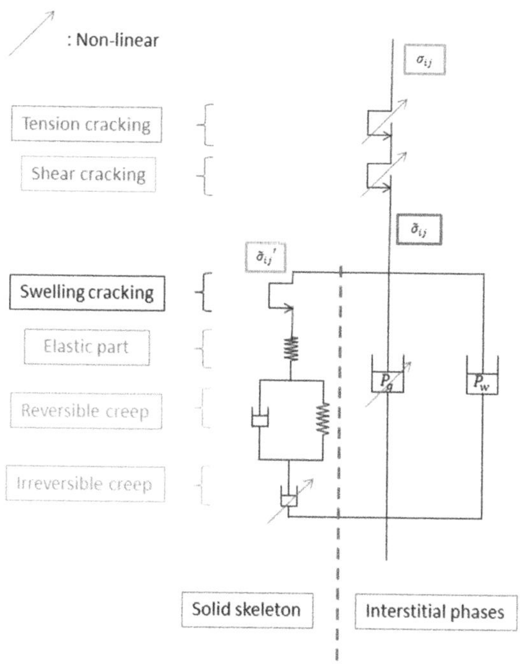

Figure 1. *Rheological scheme of the model*

2.2. *ASR Modelling*

The gel pressure Pg depends on the ASR advancement A_{asr} which varies according to equation (1) from 0 (before the start of the reaction) to 1 (when the reaction ends). Its evolution principally depends on the temperature and the humidity of the material (equation 1). $\tau ref\ asr$ is a characteristic time which gives a mean representation of the different mechanisms of ASR-gels production (diffusive phenomenon and silica reactivity). It can be obtained by calibration on measured expansion. $Aasr\ \infty$ is the maximal advancement value reachable according to environmental conditions. It cannot exceed the saturation degree because only wet reactive sites can interact in ASR-gels development (Kim et al., 2014).

$$\frac{\delta A^{asr}}{\delta t} = \frac{1}{\tau_{ref}^{asr}} C^{T,asr} C^{W,asr} (A^{asr\,\infty} - A^{asr}) \tag{1}$$

Temperature effect ($C^{T,asr}$) is managed by an Arrhenius' law (equation 2), where E^{asr} is the chemical activation energy (≈ 40,000 J/Mol) (Perruchot et al., 2003) and T_{ref} the reference temperature for which τ_{ref}^{asr} has been set.

$$C^{T,asr} = \exp\left(-\frac{E^{asr}}{R}\left(\frac{1}{T} - \frac{1}{T_{ref}}\right)\right) \tag{2}$$

When the material is not saturated, reaction kinetics is slowed down. The parameter $C^{w,asr}$ manages this phenomenon (equation 3). S_r represents the saturation degree and $S_r^{th,asr}$ the saturation degree threshold below which the reaction stops

$$C^{W,asr} = \begin{cases} \dfrac{S_r - S_r^{th,asr}}{1 - S_r^{th,asr}} & if \quad S_r > S_r^{th,asr} \\ 0 & if \quad S_r \leq S_r^{th,asr} \end{cases} \tag{3}$$

The fraction of gel produced by the reaction is ϕ_g. It is the product between the advancement and the maximum gel potential ϕ_g^{∞} (equation 4).

$$\phi_g = \phi_g^{\infty} . A^{asr} \tag{4}$$

2.3. *ASR Mechanical Effects*

The gel produced applies a pressure P_g (equation 5) on aggregate and concrete (figure 3). It mainly depends on the accessible porosity (ϕ_g^v) (Sellier et al., 1997), the initial one or the one created by strains (elastic or plastic).

$$P_g = M_g\left(\phi_g - \left(\phi_g^v\left(\frac{P_g}{P_g^{lim}}\right) + b_g tr(\varepsilon - \varepsilon^{p,g}) + tr(\varepsilon^{p,g})\right)\right) \quad (5)$$

$b_g tr(e)$ is the variation of the porosity due to concrete strain. $tr(e^{p,g})$ is the variation of the porosity due to plastic ASR-strains. M_g is the gel-matrix modulus. ϕ_g^v is a fraction of gel non-efficient to create expansion under the characteristic pressure P_g^{lim}. This last one is the pressure to reach to micro-crack the solid skeleton (equation 6) in a specimen free of stresses. It usually depends on the traction strength $\tilde{R}_I^t$ of concrete and on k_g which represents the stress concentration factor (which depends on aggregate shape). This parameter fits the local strength to intra-porous gel pressure.

$$P_g^{lim} = \frac{\tilde{R}_I^t}{k_g} \quad (6)$$

P_g expression (equation 5) is managed in two main parts. When $P_g > P_g^{lim}$, the gel spreads on the connected porosity under pressure. The second part of the expression is the volume which represents the porosity created by strains which absorb also a part of the gel. The entire volume created by strains (except the ASR plastic volume) is affected by the Biot coefficient bg which comes from poromechanics considerations (Biot, 1972) (Poyet, 2003). Grimal's model gives calibration of b_g lying between 0.1 and 0.4. For plastic strains induced by the ASR ($\varepsilon_{p,}$), b_g=1 because cracks created by the gel are supposed to be always accessible for the gel and then totally filled by it (Figure 2).

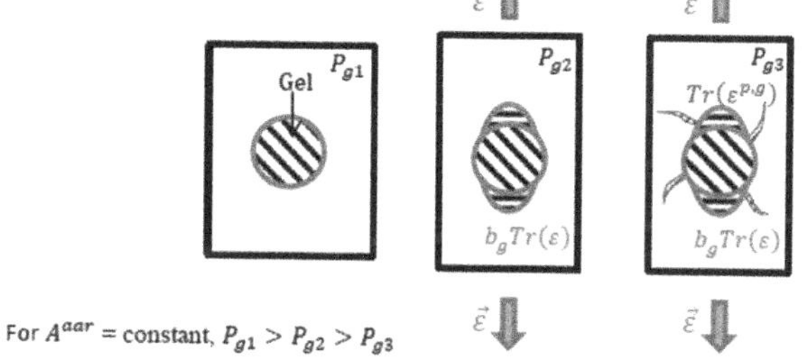

Figure 2. *Strains effect on gel pressure*

The concrete behaviour is modelled in the poro-mechanical framework. The model contains a damage model (Cousy, 1995), and a creep model which considers a realistic interaction between ASR and structural behaviour (Grimal, 2007). Creep model Burger chain is explained in (Grimal et al., 2008) and recents improvements are given in (Sellier et al., 2016). Concrete shrinkage during the same period can also be taken into account as illustrated in figure 3.

The non-linear mechanical behaviour is described by using an anisotropic plastic criteria.Damages depend on plastic strains (Sellier, 2016). In order to calculate plastic strains, there are two criteria groups: one to manage shear cracking (Drucker Prager criteria) which is isotropic, and one to operate traction behaviour (Rankine criteria) which is orthotropic. This last one separates structure macro-cracks, reclosing macro-cracks and intra-porous pressure micro-cracks (ASR).

P_g (equation 7) is used to determinate micro-cracking due to ASR: criterion f_{Ig} combines the stress state $\tilde{\sigma}_I$ and the gel pressure in each principal tensile stress direction (I).

$$\begin{cases} f_I^g = k_g P_g & if \quad \tilde{\sigma}_I \geq \tilde{R}_I^t \\ f_I^g = k_g P_g + \tilde{\sigma}_I - \tilde{R}_I^t & if \quad \tilde{\sigma}_I < \tilde{R}_I^t \end{cases} \quad \text{with } I \in [I,II,III] \tag{7}$$

3. Model Application on the Chambon's Dam

In order to study the swelling phenomena, the version of the model integrated in Code_Aster (BETON_RAG) was applied on Chambon's dam.

The Chambon is a concrete gravity dam located in the French Alps on the Romanche river and operated by Electricité de France since 1945. It was build between 1929 and 1935. It was constructed with a cyclopean concrete (150 to 250 kg/m^3 cement ratio). Its height is 137 m above the subglacial deep narrow channel (88 m above river bed). Its crest is 294 m long and 5 m wide.

First disorders were observed as soon as 1958 with cracks mainly located on lift joints on upstream and downstream faces, as in galleries. In parallel, first monitoring results indicated an atypical mechanical behavior with an upstream tilting of the crest on the left curved part. Several laboratory tests performed in the 1967-1996 period indicated AAR signs mainly due to aggregates. They pointed also out the heterogeneity of the location, intensity and potential expansion of reactive sites inside the dam.

Several remedial works campaigns were performed during the 1990-1997 period (crack grouting, new spillway, PVC membrane on the upstream face, slot cuttings). All this work made it possible to ensure the safety of the structure. After 10 years, in 2007 decision was made to analyze the dam behavior with the new model. The aim of the new approach is to describe the cracking, the displacements and the efficiency of remedies works made on 1990th and to estimate the long-term evolution of the dams.

3.1. *Hypothesis and Data*

First of all, dam's structure is composed by different type of concrete, which are taken into account on the mesh.

The dam's temperature is assumed to remain uniform and constant during all the computation. The picked temperature is the average dam temperature since its building that is 8°C.

Chemical reaction is assumed always possible due to high alkali and silica contain in the aggregate.

3.2. *Fitting Procedure*

To perform relevant calculations, all phenomena are characterized in laboratory (mechanical properties of the concrete and AAR). Experimental tests are not presented in this paper but a description is done in (Grimal, 2008).

Measurements should be made of strength, modulus and creep characteristic but also AAR kinetic. The method has been successfully tested for several dam as Temple-sur-Lot dam and Chambon dam. The results show that the fitting of the amplitude, using one significant displacement combined with the laboratory determination of the chemical kinetic parameters, allows to be assessed and other displacements, damage and stress fields of the dam are used for the fitting validation. Once the parameters validated, a prediction of the dam's displacements, stresses evolution and damage fields is made for the coming decades. It shows that, although most of the displacement has already occurred, displacements and damage will continue to occur for several decades. This non-linear model allows also testing the influence of different repairs on the behaviour of the structure.

The global methodology to fit the chemical model is summarized in figure 3. Two complementary stages are recommended: first laboratory tests on drilled specimens allows to obtain kinetic parameter (τ^{AAR} in eq 3), next a finite element inverse analysis of displacement and crack pattern allows to fit maximal swelling amplitude (Vg in eq 2).

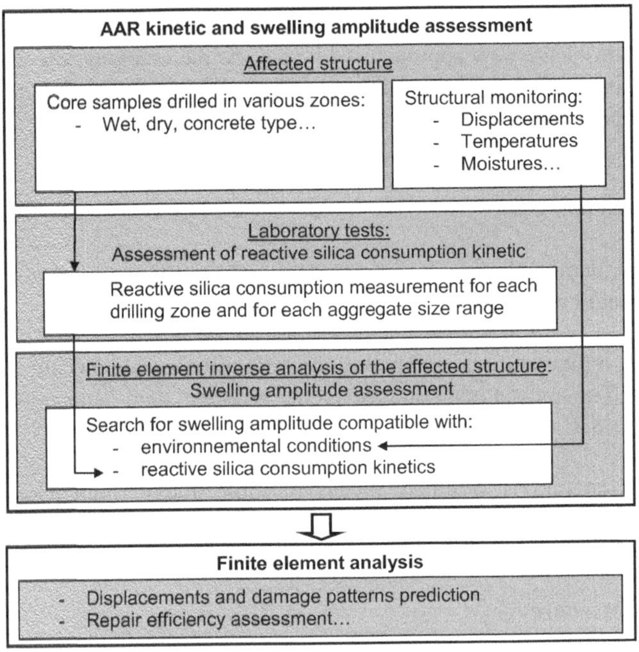

Figure 3. *Global methodology summary*

3.2.1. *Kinetic of AAR*

As mentioned, the objective of the first fitting stage is to obtain parameter τ^{AAR} of eq 3. Based on previous works on the effect of the size of reactive aggregates on the swelling capability of mortars (Poyet et al., 2006), an original procedure is used for assessing chemical advancement for each aggregate size of the concrete. This process splits the problem into two sub-phases:

Aggregates of the affected concrete are first extracted by chemical attack and sifted (Gao et al., 2013). Then the residual reactive silica content is assessed for each reactive aggregate size using a specific procedure based on a comparative study of swelling tests of mortars (Gao et al., 2013). Several types of mortar containing only one aggregate size from the dam concrete are cast. The aggregates are crushed in order to obtain a same aggregate size distribution in each mortar. Moreover, a sufficient amount of alkali is added to the mortar cement paste to be sure that all the residual reactive silica contained in the crushed aggregates will be consumed during the tests. Consequently, the total swelling measured for each mortar depends only on the residual reactive silica contained in the reactive aggregate. Thus the chemical advancement of the reaction in the aggregate of the dam at the coring date can be deduced from the tests (Figure 4).

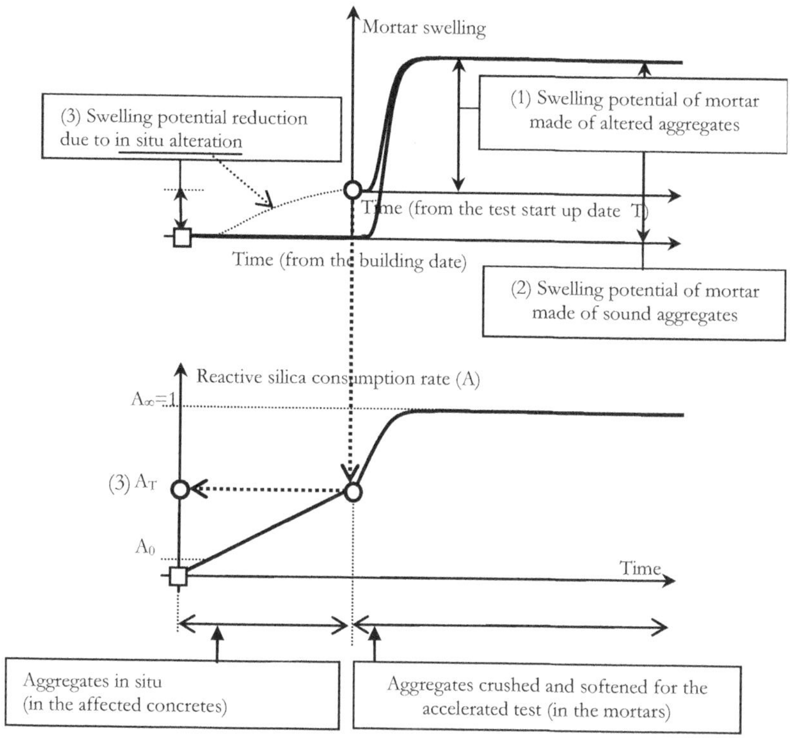

Figure 4. *Principle of the chemical advancement assessment*

The constant representing the kinetics of in-situ chemical advancement (τ^{AAR} in Eq 3) is deduced from the chemical advancements measured for each aggregate size (AT in Fig. 14) and from the environmental conditions taken into account through the integration of Eq 3 (between the beginning of construction and the current date).
3.2.2 Maximal swelling amplitude

The fitting between model and reality requires determination of the maximal volume of gel (V_g in Eq 1). In order to assess this constant, the kinetic parameter τ^{AAR} is taken equal to the one given by laboratory tests (previous stage), so V_g is the last parameter to be fitted. On purpose the model is directly fitted on the observed behaviour of the structure. The constant V_g is then iteratively assessed to adjust the FE model response (in terms of structural displacements) to measurements performed on the structure as illustrated in figure 5.

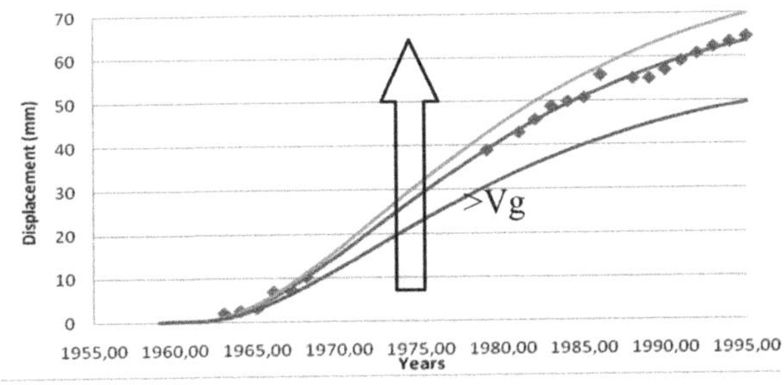

Figure 5. *Volume de gel influence*

After this calibration phase, the model prediction capability is validated through the comparison between the displacement of instrumented points predicted by the calculations and the variations measured on the dam.

3.3. *Dam's Behaviour*

The application of this model to the Chambon dam provides a more comprehensive analysis of dam deformation (figure 6), stresses distributions (figure 7) and cracking pattern evolution (figure 8).

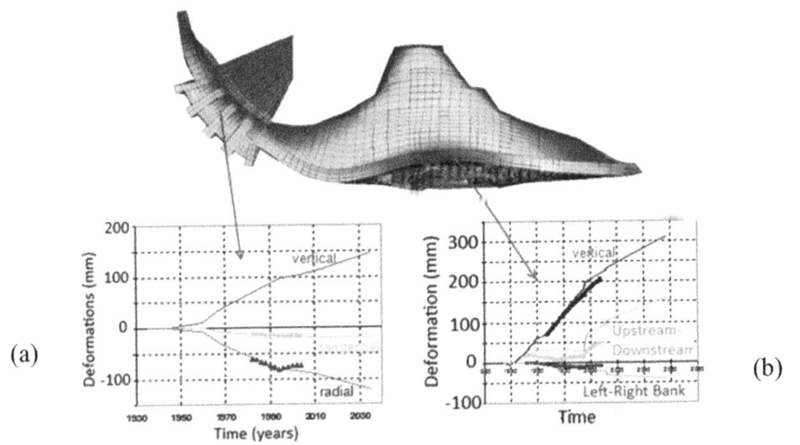

Figure 6. *Comparison of measured and computed deformations in 3D*

As mentioned in the previous paragraph (fitting procedure), the amplitude of the swelling is identified on the vertical displacement of one cantilever (Figure 6b, blue curve). The other measured displacements (other monitoring, in different directions) allow the validation of the model and the hypotheses. Indeed, a good representation of the actual swelling makes it possible to have confidence in our calculation and thus to improve our knowledge of the behaviour of the dam. The other curves (Figs. 6a and 6b) show the right representation of the radial, tangential and vertical displacements at the crest of other cantilever.

In-situ stress measurements with flat jack of over-coring techniques allowed evaluating stress orientation and intensity, providing results in accordance with those provided by the dam modelling (Nadim et al., 2013).

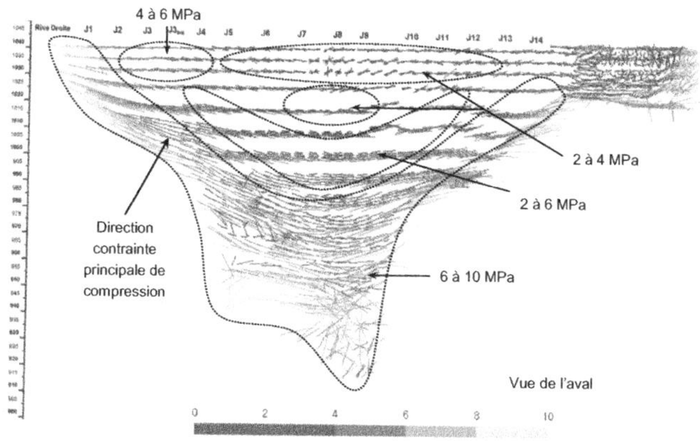

Figure 7. *Vector view of the computed stress state*

The model allows appreciating the evolution of stresses which have been increased since 1940, especially in the contacts with the rock of the foundation. A rotation of the direction of the principal stresses is observed; principal directions of compression directions start vertical in 1940 because a major importance of the weight and become horizontal in 2011 because of swelling effect

Geotechnical investigations including dilatometers confirm the monolithism of the left bank. Inspection of the dam-foundation contact through drainage holes confirm also the absence of discontinuity at the dam-foundation contact excluding the possible damaging through the dam thrust effects.

FEM model allows also estimating the damage zones locations in good accordance with the observed cracking pattern (figure 8).

Figure 8. *Comparison between evaluated damage and observed cracking pattern*

Extensive investigation from the drainage gallery provide a confirmation of the cracks pattern in particular in the curved part, where 3,2 m spaced drains can create a kind of precut zone, location corresponding also to a concrete formulation change.

If the eventuality of an upstream plate or shell could be excluded, the possibility of the existence of isolated block being able in the case of earthquake instability to impair the upstream membrane could not be excluded.

3.4. *Dam's Reinforcements*

Several works were be implemented during two years from 2013 in order to treat each identified pathology in order to extend the life duration of the dam in safety conditions.

Sawing: a new sawing campaign were be performed with large diameters cables (16 mm) on initial slot cutting location with deepening of two of them in order to release the stresses and to prevent the recovering of upstream crest movement in the curved part.

Anchoring and upstream face reinforcement: 410 upstream to downstream faces cables were drilled from the upper part of the dam in order to recover its monolithism (figure 9). They were be dimensioned in order to sustain seismic effects.

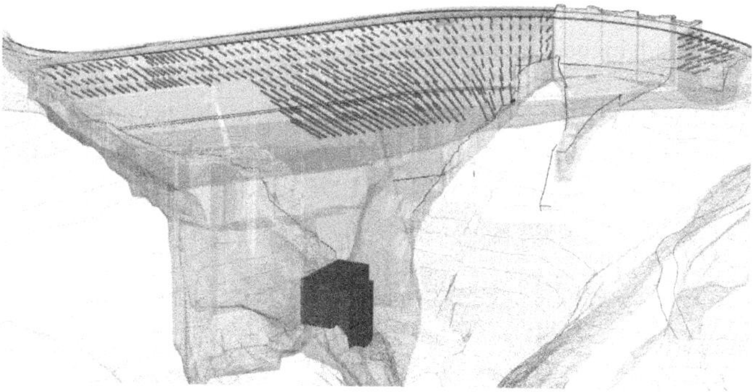

Figure 9. *Dam reinforcement with tendons – Downstream view*

Anchoring density were be completed by a 'nest' formed with carbon fiber strips in order to prevent the fall of possible isolated blocks.

Drillings of the dam have given the opportunity to make an extensive investigation of the dam cracking pattern.

Structural monitoring after remedial works is based on existing devices completed with additional piezometers in order to detect possible piezometric evolution at the dam-foundation contact and inclinometers in order to detect possible abutment movement due to the dam thrust.

4. Conclusions

The physical phenomena involved in the structural effects induced by AAR have been presented and a model proposed for each of them, leading to a global phenomenological modelling. In this modelling, a rheological model is used to control the long-term behaviour of concrete including creep and shrinkage phenomena. An orthotropic damage model based on plasticity and effective stress concepts allows swelling transfer between strain-constrained directions and free-swelling directions to be taken into account. After the calibration phases, the model was used to reproduce the behaviour of Chambon's dam. A comparison with experiments showed the model's capability to reproduce displacements with acceptable accuracy.

The behaviour obtained for all cantilever are similar to those measured by monitoring.

The AAR development produces efforts in the dam. Stresses will increase at the dam-foundation contact and the direction of the main compressive stress will turn from vertical to horizontal. The model represents this evolution.The model also allows us to locate damaged areas of the dam that will need a greater consideration.

Contact with bedrock, hard and stable, resulting in the accumulation of effort in damaged concrete can lead to cracks and endanger the building that must be controlled. Finally the model proposed by EDF allows us to simulate satisfactorily the dam behaviour affected by AAR.

So to follow concrete dams, EDF-CIH has developed a maintenance methodology applicable to all of these structures.

This methodology is based on Structural Monitoring, Laboratory tests and Modeling. It allows to operate safely dams, define long-term consequences of swelling and to evaluate the effectiveness of repairs.

In parallel with this management of concrete structures, CIH continues to perform active research to improve knowledge on AAR, DEF and creep.

It is now necessary to make progress on:

– The definition of acceptance criteria (complex or simplified modeling).

– Repair methods.

– Numerical modeling parameters identification.

– Sharing information between Operators (pathologies, diagnostics and repairs carried out).

5. References

Biot M.A., 1972, "Theory of finite deformation of porous solids", Indiana Univ. Math. J. 21, 579–620.

Coussy O., 1995, *Mechanics of Porous Continua*, J. Wiley & Sons, Chichester, UK.

Gao X. X., Cyr M., Multon S., Sellier A., 2013, "A comparison of methods for chemical assessment of reactive silica in concrete aggregates by selective dissolution", *Cement and Concrete Composites*, 37(null), 82–94.

Gao X. X., Cyr M., Multon S., Sellier A., 2013, "A three-step method for the recovery of aggregates from concrete", *Construction and Building Materials*, 45, 262–269.

Grimal E., 2007, "Caractérisation des effets du gonflement provoqué par la réaction alcali-silice sur le comportement mécanique d'une structure en béton", PhD thesis, Université Paul Sabatier Toulouse, France. (in French)

Grimal E, Sellier A., Le Pape Y., Bourdarot E., May-June 2008, "Creep shrinckage and anisotropic damage in AAR swelling mechanism, part I and part II", *ACI Materials Journal-American Concrete Institute*, 105-m26 and 105-m27.

Grimal E., Sellier A., Multon S., Le Pape Y., Bourdarot E., 2010, "Concrete modelling for expertise of structures affected by alkali aggregate reaction", *Cement and Concrete Research*, Vol. 40 No. 4, pp. 502–507.

Kim T., Olek J., 2014, "Chemical Sequence and Kinetics of Alkali–Silica Reaction Part II. A Thermodynamic Model", *Journal of the American Ceramic Society*, 97(7), pp. 2204-2212.

Morenon P., Multon S., Sellier A., Grimal E., Hamon F., Bourdarot E., (accepted for publication 2017), "Impact of stresses and restraints on ASR expansion", *Construction & Building Materials*.

Nadim C., Grimal E., Lahaie F., Bourdarot E., Bigarré P., February 201,1 "Full 3D stress field diagnosis at the heart of a large gravity dam", *6th International Conference on Dam Engineering*, Lisbon.

Perruchot A., Massard P., Lombardi J., 2003, "Composition et volume molaire apparent", *Comptes Rendus Geoscience*, 335(13), 951-958.

Poyet S., 2003, "Etude de la dégradation des ouvrages en béton atteints par la réaction alcali silice: Approche expérimentale et modélisation numérique multi-échelles des dégradations dans un environnement hydro-chemo-mécanique variable", thèse de doctorat de l'université de Marne la Vallée. (in French).

Poyet S., Sellier A., Capra B., Foray G., Torrenti J.-M., Cognon H., Bourdarot E., 2006, "Chemical modelling of Alkali Silica reaction: Influence of the reactive aggregate size distribution", *Materials and Structures*, 40(2), 229–239.

Salgues M., Sellier A., Multon S., Bourdarot E., Grimal E., 2014, "DEF modelling based on thermodynamic equilibria and ionic transfers for structural analysis", *European Journal of Environmental and Civil Engineering*, 1–26.

Sellier A., Capra B., 1997, "Modélisation physico-chimique de la réaction alcali-granulat: apport au calcul des structures dégradées", *Revue Française de Génie Civil*, 1(3), 445–481.

Sellier A., Casaux-Ginestet G., Buffo-Lacarrière L., Bourbon X., 2013, "Orthotropic damage coupled with localized crack reclosure processing. Part I: Constitutive laws", *Engineering Fracture Mechanics*, 97, 148-167.

Sellier A., Multon S., Buffo-Lacarrière L., Vidal T., Bourbon X., Camps G., 2016, "Concrete creep modelling for structural applications: non-linearity, multi-axiality, hydration, temperature and drying effects", *Cement and Concrete Research*, Vol. 79, pp. 301–315.

Sellier A., 2016, "Model FLUENDO3D Version 20-P", LMDC Internal Report. Toulouse. Retrieved from alain.sellier@insa-toulouse.fr.

Swelling Effects in Fagilde Dam (Portugal)

First Approach to Structural Analysis and Interpretation

JOSÉ PITEIRA GOMES* — ANTONIO LOPEZ BATISTA* — S.P.M. SOUSA**

** LNEC – National Laboratory for Civil Engineering*
Av. do Brasil, 101, 1700-066 Lisbon, Portugal
pgomes@lnec.pt, a.l.batista@lnec.pt

*** FCT/UNL – Faculty of Sciences and Technology, New University of Lisbon*
Campus da Caparica, 2829-516 Caparica, Portugal
sp.sousa@campus.fct.unl.pt

ABSTRACT. *The Fagilde dam is affected by concrete swelling reactions, of the ISR type, which induced serious damage in the dam's body, namely cracking. The structural modeling is supported by the monitoring results and the results of the preliminary tests for the diagnosis of the swelling reactions. A first approach for the structural analysis of dam is presented, done in order to obtain a general overview of the spatial heterogeneous expansion development, that should be considered as a necessary first step for the parameters estimation to be used in the next computations, with damage models to simulate the concrete cracking. However, a reasonable agreement between the results obtained numerically with those measured by the monitoring system was found. A great influence of the swelling effects on the dam's behavior is shown. Nowadays, serviceability and safety conditions of the dam are still acceptable.*

KEYWORDS: *concrete dams, swelling processes, structural modeling, structural effects, safety control.*

1. Introduction

In Portugal there are about 60 large concrete dams that are continuously monitored, of which 19 are affected by swelling reactions. The causes of the expansion phenomena are mainly related with the alkali-silica reaction (ASR) and the internal sulphate reaction (ISR). These dams were built between the decades of 1940 and 1980. The main types of aggregate used in concrete mix were granite (13 dams) and quartzite. In 15 of these dams the swelling magnitude is still moderate, but in the other 4 the phenomenon itself and the related effects are severe.

The ASR type is dominant in three dams, the Santa Luzia, Alto Ceira I and Pracana dams, all having quarzitic aggregates. The remaining expansion potential in the Santa Luzia dam appears to be low and the dam shows very mild structural effects. Alto Ceira I dam was partially demolished and replaced by the new Alto Ceira II dam in 2014. The Pracana dam was submitted to large rehabilitation works in the 1990s, and since then the ASR effects have diminished considerably. ISR is the main degradation phenomena in Fagilde dam, that is, nowadays, the one exhibiting the most serious structural effects due to this pathology (Batista & Piteira Gomes 2012, Batista & Piteira Gomes 2016).

This paper presents the main results of the few studies already done to the swelling phenomenon diagnosis and prognosis of Fagilde dam (Piteira Gomes et al 2016), as well as the first approach for structural analysis and interpretation of the dam behaviour, considering the limited available data from laboratory tests and "in situ" monitoring. In fact, the dam swelling process characterization was based only on the results of a set of physical and chemical tests performed on a few number of samples taken from the dam´s body in 2008, which identified the existence of the ISR type, and on geodetic data from the dam's monitoring.

For the dam's structural behavior simulation was recently developed at LNEC a detailed 3D finite element mesh of the dam and its rock mass foundation (Sousa 2016). The dam's structural response was obtained considering the models and methods developed and available at LNEC (Piteira Gomes 2007). The numerical results were used for the structural behaviour interpretation, considering the monitoring data since the first filling of the reservoir, in 1985, until 2016.

2. Fagilde Dam and its Swelling Process

2.1. *Dam's Characteristics, Objectives, Design and Construction*

2.1.1. *Dam Characteristics and Objectives*

The Fagilde dam, located in the Dão River (north-centre of Portugal), is a concrete structure, 27 m high, composed by three buttresses, between which the spillways are inserted, and two lateral arches (Fig. 1). The structure in founded in a granite rock mass.

Figure 1. *Fagilde dam. Downstream view*

The dam construction lasted from August 1982 to late 1983 and the first filling of the reservoir began in June 1985 and finished in January 1987.

The dam reservoir, with a volume of 2.8 hm^3, is used for water supply to three important municipalities in the neighborhood.

2.1.2. *Design and Construction*

The dam design defined, for the structural elements, a concrete of the B225 class (Hidroprojecto 1979). A ready mixed concrete was used in the construction, with a Portland cement content of 360 kg/m^3, limestone coarse aggregate and siliceous sand (Table 1). Table 2 presents the concrete compressive strength obtained during construction, for concrete quality control, using cubic samples of 20 cm edge, for the ages of 7, 14 and 28 days (LNEC 2001).

Components (size in mm)		Content (%)	Content ($kg.m^{-3}$)
Coarse aggregate	38.1 – 25.4	35	661
	25.4 – 9.5	15	283
	9.5 – 4.8	12	227
Sand	Coarse	27	510
	Fine	11	208
Cement (Portland)		-	360
Water		-	160
Water-cement ratio		0.44	-

Table 1. *Concrete composition*

Age (days)	Number of samples	Average compressive strength (MPa)	Standard deviation (MPa)	Characteristic compressive strength (MPa)
7	24	34.0	4.6	26.7
14	24	38.8	4.1	31.8
28	77	40.5	4.1	33.8

Table 2. *Concrete compressive strength (cubic samples of 20 cm edge)*

The characteristic compressive strength of 33.8 MPa, obtained at the age of 28 days, could be considered high when compared with the design prescribed value of 22.5 MPa. In fact, the concrete maturation was very fast (there was no temperature control during and after casting).

2.2. *Monitoring System*

The dam monitoring system, installed during the construction phase, allows the evaluation of the actions and of the structural and hydraulic responses, by measuring: i) the upstream level, by staff gauges and water level recorders; ii) air temperatures, by a daily maximum and minimum thermometer; iii) uplift pressures, by piezometers; iv) horizontal displacements, by geodetic methods; v) vertical displacements, by precision geometric levelling of the crest; vi) joint movements, by deformeters; and vii) discharged and infiltrated flows, by drains and seepage measuring weirs. Due to the abnormal behaviour observed, in 2008 LNEC proposed the rehabilitation and strengthening of the monitoring system, in order to improve the knowledge about the swelling process itself and its effects (LNEC 2008). The construction of an external seepage measuring weir was done in 2013, the geodetic targets were replaced in 2016 and the drills to insert rod extensometers from the crest were done in 2008, but the installation of the rods was still not possible.

2.3. *Structural Evidences of the Swelling Phenomena*

Evidences of the swelling phenomena were identified just a decade after the construction, by measuring progressive vertical displacements upwards, that reached about 25 mm nowadays (Fig. 2), and axial displacements upstream (Fig. 3). Linear and map cracking was then detected on the surfaces (LNEC, 2006). In Fig. 4 the cracking pattern of the downstream surface, obtained in the survey carried out in 2009, is represented.

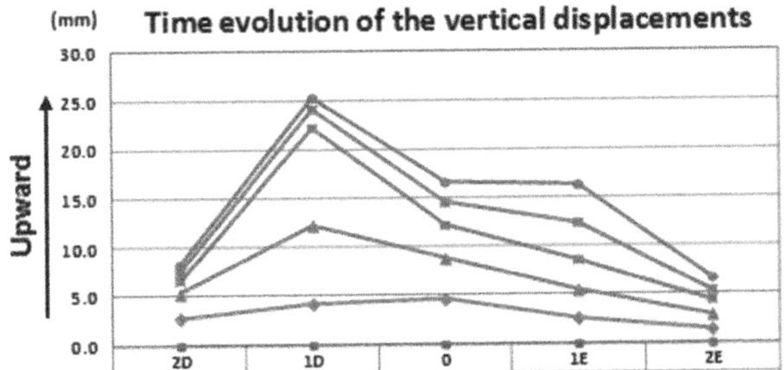

	2D	1D	0	1E	2E
june 1985	0.0	0.0	0.0	0.0	0.0
april 1995	2.8	4.2	4.6	2.6	1.4
se ptember 2000	5.3	12.3	8.8	5.5	2.9
november 2005	6.6	22.2	12.2	8.6	4.4
june 2010	7.5	24.1	14.5	12.4	5.4
october 2015	8.3	25.3	16.6	16.3	6.7

Figure 2. *Accumulated vertical displacements between 1985 and 2015 at the crest, obtained from the results of geodetic levelling*

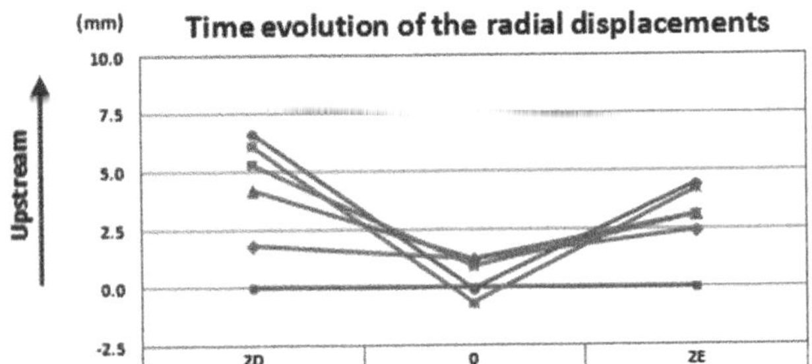

	2D	0	2E
june 1985	0.0	0.0	0.0
april 1995	1.8	1.2	2.4
september 2000	4.2	1.2	3.1
november 2005	5.3	0.9	3.1
june 2010	6.1	-0.7	4.2
october 2015	6.6	-0.1	4.4

Figure 3. *Accumulated horizontal displacements (river axis direction) between 1985 and 2015 near the crest, obtained from the results of geodetic methods*

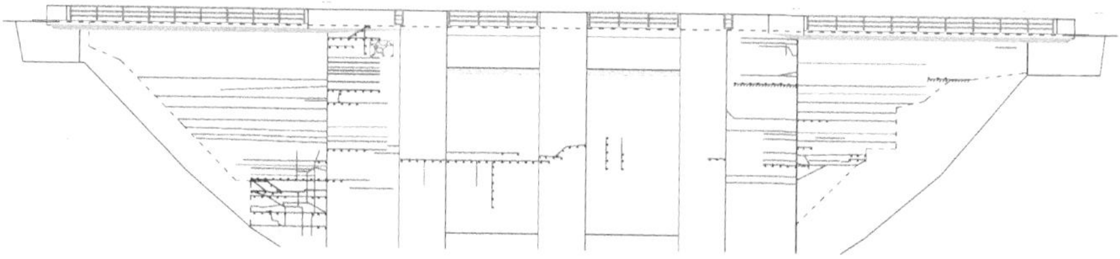

Figure 4. *Cracking pattern of the downstream surface in 2009*

Figs 5 and 6 show photographs, obtained in 2015 and 2016, respectively, of the concrete surfaces cracking of the arches and of the central buttress.

Figure 5. *Downstream views of the bottom zone of the right arch (on left) and of the upper zone of the left arch (on right), in 2015*

Figure 6. *General view (on left) and detailed view (on right) of the central buttress, in 2016*

2.4. *Tests Results of the Preliminary Diagnosis of the Swelling Process*

As mentioned before, the concrete maturation was very fast, which is not adequate for mass concrete structures such as dams. Additionally, the concrete casting was done without temperature control. Due to the high cement content and to the thickness of the structural elements, it is probable that the concrete reached high temperatures that are favorable to the development of the ISR.

Following previous studies (LNEC 2003, LNEC 2008), to do a preliminary diagnosis of the swelling process and strengthening the monitoring system, 7 deep drills 76 mm in diameter were done in 2008 from the crest to the rock mass foundation, to obtain concrete cores and install rod extensometers in these holes. With the extracted cores laboratory tests were performed, namely for: i) petrographic and mineralogical analysis; ii) chemical analysis of the concrete (determination of the cement, sulphate, alkali and silica contents) and the aggregate (determination of the silica and alkali contents); iii) evaluation of the aggregate potential alkali-silica reactivity; and iv) evaluation of the potential for further expansion due to ASR and ISR.

The tests results indicated that the coarse aggregate was mainly composed of limestone, but also granite, schist and siliceous aggregate was identified. Although the soluble alkali content was large, it was concluded that no expansive potential was expected by the ASR type. However, the expansive potential due to the ISR type was greater and the sulphatic content was enough for the continuation of the swelling process (LNEC 2010).

For the mechanical testing of the concrete (compressive and tensile strengths, modulus of elasticity and creep function), 4 short vertical cores 150 mm in diameter were extracted from the crest of the arches and from the lateral buttresses.

The strength tests, done with 14 samples 150 mm in diameter and 150 mm high, showed an average compressive resistance, referred to cubes of 20 cm edge, of 33.7 MPa, and a minimum strength of the 21.1 MPa. The tensile strength, obtained in diametral compression tests, was an average value of 3.0 MPa and a minimum value of 2.25 MPa. By these values, it was considered that the concrete had, at that time, 2008, a moderate properties depreciation.

The modulus of elasticity, obtained from the tests results of 4 samples 150 mm in diameter and 60 cm high, had values between 19.1 GPa and 24.5 GPa, which indicated a considerable decrease of this property, probably due to the microcraking induced by the swelling process.

3. Models and Methods

3.1. *Swelling Evaluation*

The main purpose of the modelling of swelling reactions, for solving engineering problems, consists of integrating in a macroscopic model, the data about the physical and chemical phenomena developed at a microscopic level. It is considered that the development of swelling reactions depends mainly on the variables as follows

(Capra 1997): Q, which represents, in overall terms, the chemistry of the process and which depends on the concrete composition (type of cement and reactive aggregates, their size, shape and space distribution, existence and nature of admixtures, W/C ratio, porosity, water content, etc.); temperature T; relative humidity Hr; stress field σ ; and time t. This dependency can be written in a generic form,

$$\varepsilon^{ASR}(t) = f\,(Q, T, H_r, \sigma, t) \qquad [1]$$

By assuming the hypothesis that the chemical reactions occur on an isotropic material with elastic behaviour, a relation between chemistry and mechanics can be established using the reactive porous media theory (Coussy 1995). This makes it possible to analyse the behaviour of a closed system, in which a chemical reaction in the interstitial solution is produced and a differential equation can be obtained to evaluate the free swelling increment at the time interval dt (Larive 1998),

$$t_c(T,\xi)\frac{d\xi}{dt} + \xi = 1 \qquad [2]$$

where,

$$t_c = \tau_c(T)\frac{1+\exp[-\tau_L(T)/\tau_c(T)]}{\xi+\exp[-\tau_L(T)/\tau_C(T)]} \qquad [3]$$

where τ_L is the latency time, τ_C is the characteristic time and $\xi = \varepsilon_{Free}^{ASR} / \varepsilon_{Free}^{ASR}(\infty)$, being $\varepsilon_{Free}^{ASR}(\infty)$ the maximum value that the swelling process is likely to generate.

This equation originates free swelling curves with an S configuration, which are characterised by an initial period of latency, in which the gel produced by chemical reactions fills the pores in concrete causing neither stresses nor swelling (Figure 7).

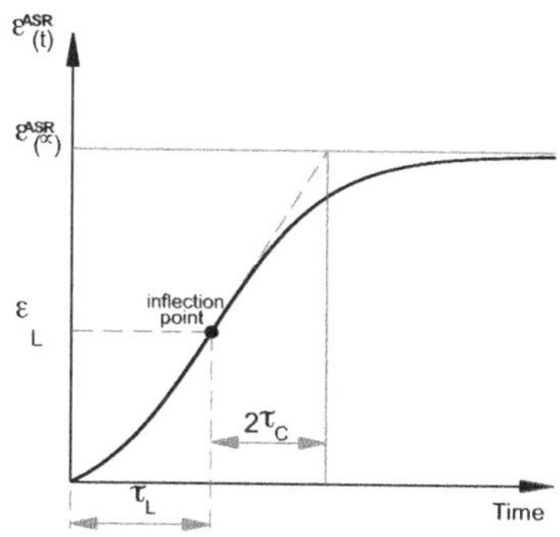

Figure 7. *Free swelling curve obtained in isothermal tests*

The swelling parameters τ_L, τ_C and $\varepsilon_{Free}^{ASR}(\infty)$ can be directly obtained from free swelling tests, carried out at standard humidity and temperature conditions.

To evaluate the influence of the temperature variation during the swelling process, the following equations are used,

$$\tau_C(T) = \tau_C(T_0)\, e^{[U_C(1/T - 1/T_0)]} \quad [4]$$

$$\tau_L(T) = \tau_L(T_0)\, e^{[U_L(1/T - 1/T_0)]} \quad [5]$$

where U_C is the activation energy associated to the characteristic time, U_L the activation energy corresponding to the latency time, T_0 the temperature of the reference free swelling test and T is the temperature in the time interval Δt considered in the discretisation established for the analysis in the time domain.

Therefore, the free swelling in a structure depends on thermal and hygrometric fields, being a function of environmental conditions resulting from their exposure and, in the case of dams, of their operation regimen. The thermal analysis is a problem of heat conduction, of which the solution implies the knowledge of the thermal conductibility of concrete and of the annual thermal waves of air and water, as well as of exposure conditions to solar radiation. The hygrometric computation is similarly done, on the basis of the representative curve of the average local humidity and of the variations in the reservoir water level. These two independent computations make it possible to know the temperature and the relative humidity histories, on each point of the structure, in each time interval Δt of the period considered.

The influence of the stress field on the swelling process is based on the experimental data from tests performed in France, which is considered by the following equation,

$$\psi_x(\sigma, t) = \frac{\varepsilon_x^{ASR}(\sigma, t)}{\varepsilon_{x,free}^{ASR}(t)} = e^{-(a\sigma(t))^b} \quad [6]$$

where $\psi(\sigma, t)$ is the ratio between restrained swelling by the effect of unidirectional stress σ and the free swelling in a certain time interval. The free swelling is assumed as isotropic.

3.2. *Structural Model*

The solution of the time dependent structural problem, considering the swelling and cracking development, is solved by the finite element method, taking into account an incremental procedure in the application of loads and in the evaluation of the delayed behaviour of the concrete and the effects of the non-linearity due to possible damage occurrence.

A displacement formulation is adopted. Using the principle of virtual works, the following equilibrium equation is obtained,

$$\underline{K}(t)\,\Delta\underline{u} = \Delta\underline{f}^{a} + \Delta\underline{f}^{0} + \Delta\underline{f}^{*} \qquad [7]$$

where $\underline{K}(t)$ is the global stiffness matrix, $\Delta\underline{u}$ the increment in displacements, $\Delta\underline{f}^{a}$ the incremental forces due to the applied loads, $\Delta\underline{f}^{0}$ the incremental forces due to the imposed deformations and $\Delta\underline{f}^{*}$ the incremental forces due to the load history.

For each time step this equation is solved in order to evaluate the incremental displacements, the stiffness matrix being updated in each time step (Batista 1998). If considered, the cracking progress is computed using a scalar damage model with two independent variables, d^{+} for the tension damage, and d^{-} for the compressive damage. The effective stress tensor $\tilde{\sigma}$ is decomposed into the stress tensors $\tilde{\sigma}^{+}$ and $\tilde{\sigma}^{-}$, in order to obtain the following constitutive equation (Faria et al 1998),

$$\sigma = \left(1-d^{+}\right)\tilde{\sigma}^{+} + \left(1-d^{-}\right)\tilde{\sigma}^{-} \qquad [8]$$

Tension softening is simulated using an exponential law (Oliver et al 1990),

$$d^{+} = 1 - \frac{r_0^{+}}{\tilde{\tau}^{+}}\, e^{A^{+}\left(1-\frac{\tilde{\tau}^{+}}{r_0^{+}}\right)}, \qquad \tilde{\tau}^{+} \geq r_0^{+} \qquad [9]$$

where $\tilde{\tau}^{+}$ is an equivalent stress for tension, r_0^{+} is a damage threshold for tension and

$$A^{+} = \left(\frac{G_f\,E}{w^{*}\left(f_0^{+}\right)^{2}} - \frac{1}{2}\right)^{-1} \qquad [10]$$

G_f being the fracture energy, w^{*} a characteristic dimension associated to the spatial discretisation so as to guarantee the objectivity of numerical solutions (Oliveira 2000, Oliveira & Faria 2006).

For each time step, the damage is evaluated using an iterative stress-transfer procedure, in order to pursuit the global equilibrium, by updating damage, stresses, strains and displacements. After achieving the equilibrium, the analysis proceeds to the next step (Piteira Gomes 2007).

4. Fagilde Dam Structural Behaviour Analysis and Interpretation

4.1. *Finite Element Mesh*

A 3D finite element representation of the dam and its foundations is used for structural simulation. The structure and the foundation block are discretized by 830 and 3950 hexahedral elements, respectively, each one having 20 nodal points. The mesh has a total of 4780 finite elements and 23227 nodal points. A vertical row of small elements was considered to represent the hinge in the connection between the lateral buttresses and the arches (Fig. 8).

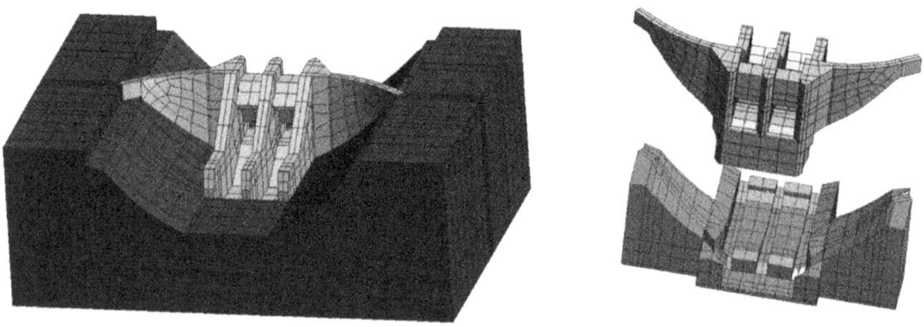

Figure 8. *General view of the finite element mesh (on left) and mesh details at the insertion surface (on right)*

4.2. *Materials Properties*

Based on laboratory tests results (LNEC, 2007) and on the monitoring results, it is assumed that the concrete is homogeneous and isotropic with a rheological behaviour defined by the following Bazant and Panula creep law,

$$J(t,t_0) = \frac{1}{30.0}\left(1 + 3.0\left(t_0^{-0.34} + 0.04\right)\left(t - t_0\right)^{0.18}\right) \quad GPa^{-1} \qquad [11]$$

As regards the foundation bulk the hypothesis of an elastic and isotropic medium is assumed, with a 14 GPa modulus of elasticity, a 0.2 Poisson ratio and with a non-time dependent behaviour. No cracking effects in concrete were considered in this first approach for structural analysis of Fagilde dam.

4.3. *Actions Evaluation and Discretization*

The dead weight of concrete induces very low stresses, so it is not considered in the computations. The water load is represented by the hydrostatic pressure on the upstream surface of the dam (γ_w=10 kN/m^3). Loadings due to water pressure were discretized monthly, considering the water level data in the reservoir represented in Fig. 9.

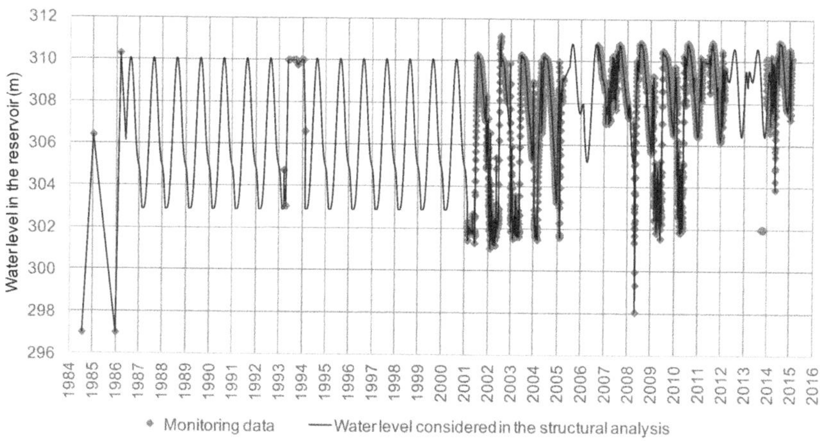

Figure 9. *Water load discretization used in structural analysis*

The thermal actions on the surfaces of the dam are represented by sinusoidal waves of annual period to the air and to the water, of which the parameters were numerically determined by the least square method on the basis of temperatures observed on air and water. A linear thermal dilation coefficient for the concrete α=0.7x10^{-5} /ºC is considered.

The environmental conditions of the dam, namely the thermal and hygrometric actions, are considered for swelling evaluation. The evolution of the vertical crest displacements, recorded by geodetic levelling carried out since 1985 to 2015, are also taken into account to adjust the swelling evolution estimation.

The swelling action is defined considering two zones of different properties. In each one a sigmoid function is considered to represent the swelling evolution over time, characterized by the parameters shown on Table 3. In fact, the only difference between the two zones is related with the reaction speed, which is considered higher in the right arch. An isotropic free swelling development is assumed in both zones.

Zoning	Latency time (days)	Characteristic time (days)	Latency energy (KJ/MOL)	Characteristic energy (kJ/mol)	Long term free swelling
Left arch and buttresses	145	180	9000	11500	2750 x 10^{-6}
Right arch	65	55	9000	11500	2750 x 10^{-6}

Table 3. *Swelling parameters considered in the dam's zoning*

Fig. 10 shows the results of the free swelling computation for three time steps (1995, 2005 and 2015).

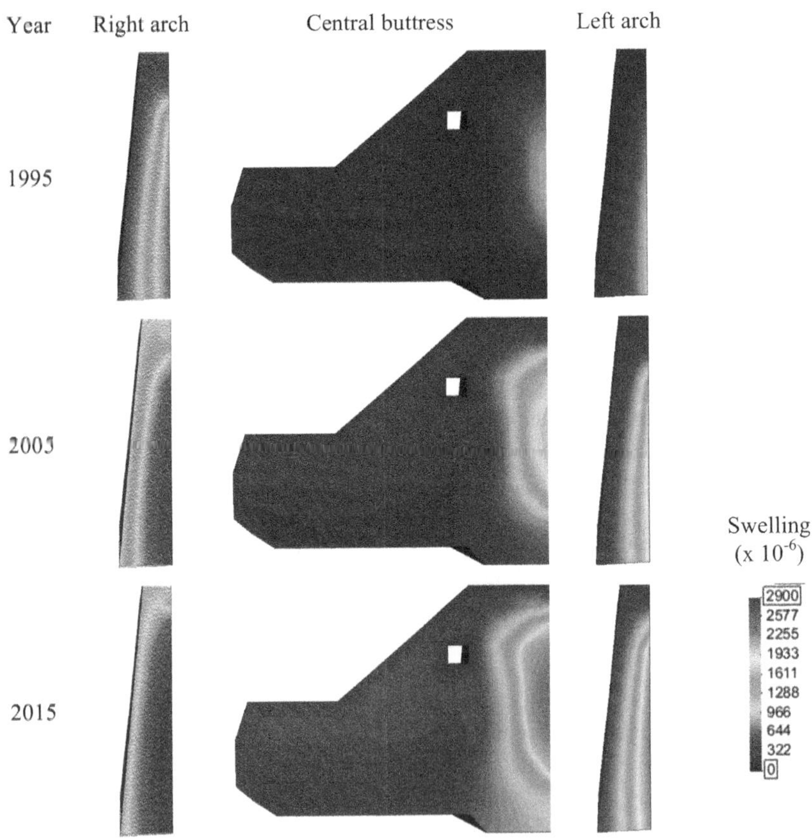

Figure 10. *Swelling patterns computed for the arches and for the central buttress in 1995, 2005 and 2015*

4.4. *Main Results*

The main results, in terms of vertical displacements and stress fields computed on dam surfaces in 2005 and 2015, are showed below.

In Fig. 11 the observed and computed vertical displacements of the dam, since 1985 to 2015, are compared, and a reasonable agreement is achieved.

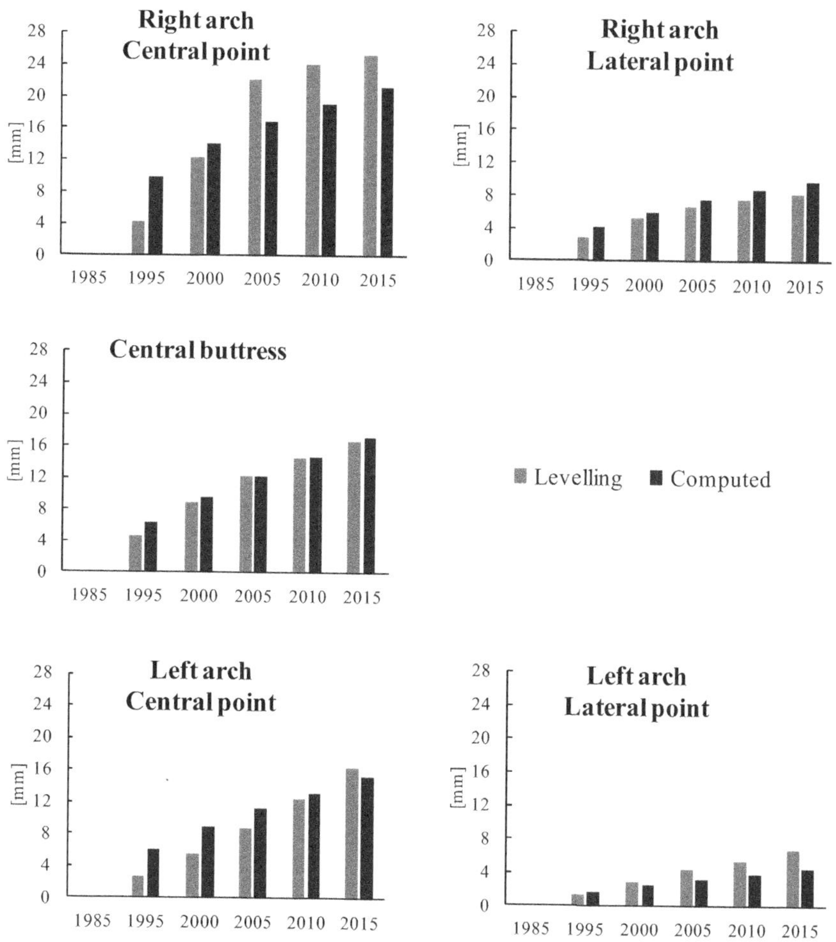

Figure 11. *Vertical displacements, observed and computed, since 1985 to 2015*

In Fig. 12 the principal stresses computed on dam surfaces in 2015 are presented. The maximum compressive stresses, of about 14.0 MPa, are located in the right arch, on the downstream surface, near the abutment basis, and the maximum tensile stresses, of about 8.5 MPa, are also located in the right arch, on its central zone. There are extensive areas of the downstream surface of the left arch with high tensile stresses, but the upstream surface is totally compressed. In fact, the downstream surface of the right arch is more cracked then the downstream surface of the left arch and the water passages through the concrete, from the reservoir, are greater in the right arch.

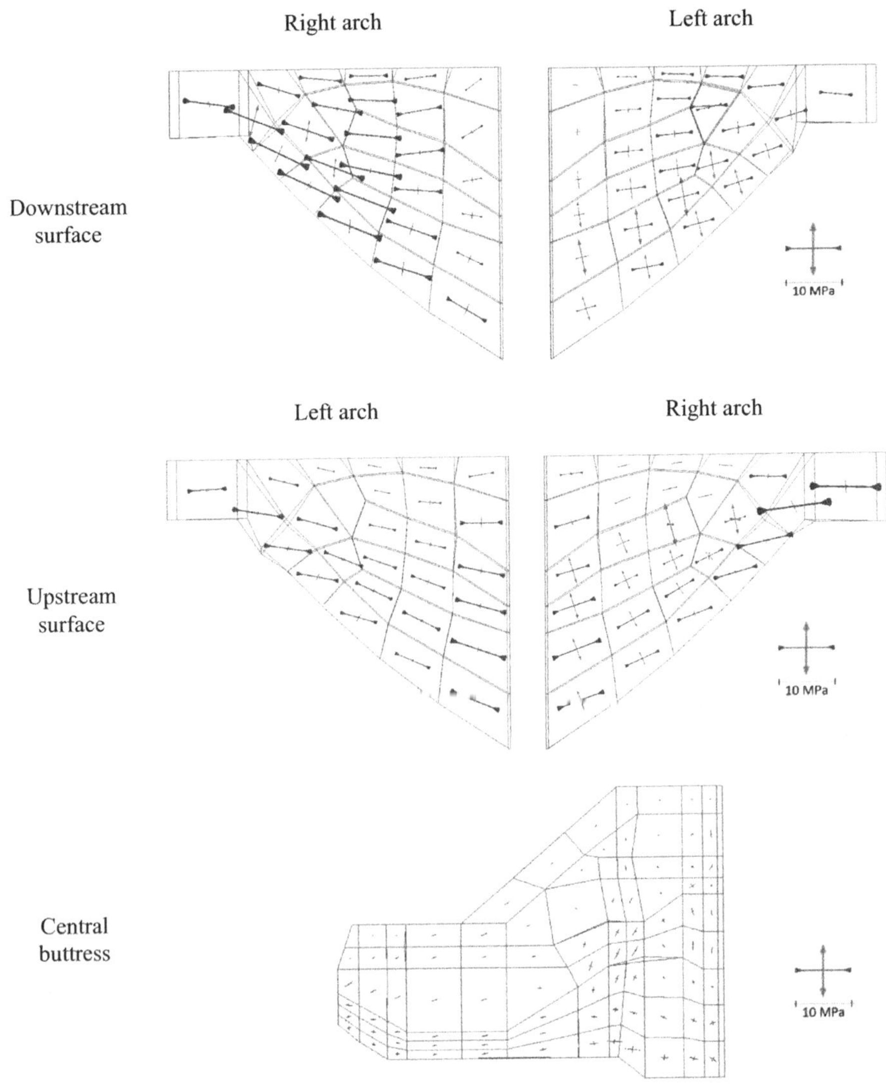

Figure 12. *Principal stresses computed on dam surfaces in 2015*

5. Conclusions

The paper presents the first approach for the structural analysis of Fagilde dam, done in order to obtain a general overview of the heterogeneous expansion development over time that is taking place in the dam's body. The structural modeling is supported by the monitoring results and the results from the preliminary tests for the diagnosis of the swelling reactions. The analysis presented should be considered as a necessary first step for the parameters estimation to be used in the next computations, with damage models to simulate the concrete cracking, in the near future. However, a reasonable agreement between the results obtained numerically with those measured by the monitoring system was found. A great influence of the swelling effects on the dam´s behavior is shown.

There are large areas of the downstream surface, in both arches, with intensive cracking. These preliminary numerical results, obtained in a viscoelastic regime, justify the structural evidences of the dam.

Nowadays, serviceability and safety conditions of the dam are still acceptable. However, the owner is studying a set of different solutions to rehabilitate the dam in the near future.

6. Acknowledgements

Thanks are due to the owner of the Fagilde dam (APA - Agência Portuguesa do Ambiente) for the permission to present the results that were included in this paper.

7. References

Batista, A.L., "Análise do comportamento ao longo do tempo de barragens abóbada". PhD thesis, Technical University of Lisbon, 1998.

Batista, A.L., Piteira Gomes, J., "Practical assessment of the structural effects of swelling processes and updated inventory of the affected Portuguese concrete dams", in *First International Dam World Conference*, Maceió, Brazil, October 2012, 10p.

Batista, A.L., Piteira Gomes J., "Characteristic behaviour of the Portuguese large concrete dams built with granite aggregates and affected by ASR", in *15th International Conference on Alkali-Aggregate Reaction – ICAAR 2016*. São Paulo, Brazil, July 2016.

Capra, B., "Modélisation des effects mécaniques induits par les réactions alcalis-granulats". PhD thesis, École Normale Supérieure de Cachan, France, 1997.

Coussy, O., *Mechanics of porous continua*. J. Wiley and Sons, 1995.

Faria, R., Oliver, J., Cervera, M., "A strain-based plastic visco-damage model for massive concrete structures". *Int. Journal Solids Structures*, Vol. 35, N°14, 1998, pp. 1533-1558.

FCUP, "Barragem de Fagilde. Análise petrográfica do betão". Report, Department of Geology, Faculty of Sciences of University of Oporto, 2006.

Hidroprojecto, *Barragem de Fagilde. Projecto de execução.* Lisboa, 1979.

Larive, C., "Apports combinés de l'expérimentation et de la modélisation à la compréhension de l'alcali-reaction et de ses effects mécaniques". PhD thesis. Laboratoire Central des Ponts et Chaussés, LCPC, Paris, 1998.

LNEC, "Estudo do comportamento da barragem de Fagilde entre 1985 e 2001". Report 289/2001 – DBB/NO, 2001.

LNEC, "Barragem de Fagilde. Metodologia para a caracterização do processo expansivo do betão". Report 12/2003 – DBB/NO, 2003.

LNEC, "Barragem de Fagilde. Levantamento do estado de fendilhação efetuado em 2004". Technical note 58/2006 – DBB/NO, 2006.

LNEC, "Barragem de Fagilde. Estudo da fluência do betão". Report 15/2007 – DM/NB, 2007.

LNEC, "Barragem de Fagilde. Plano da instrumentação complementar ao sistema de observação". Report 111/2008 – DBB/NO, 2008.

LNEC, "Caracterização da acção expansiva do betão da barragem de Fagilde". Report 119/2010 – DM/NB/NMM, 2010.

Oliveira, S.B.M., "Modelos para análise do comportamento de barragens de betão considerando a fissuração e os efeitos do tempo. Formulações de dano". PhD thesis, Faculty of Engineering of University of Oporto, 2000.

Oliver, J., Cervera, M., Oller, S., Lubliner, J., "Isotropic damage models and smeared crack analysis of concrete". *Journal of Engineering Materials and Technologies, Trans. of the ASME*, Vol. 105, 1990, pp. 99-105.

Piteira Gomes, J., "Modelação do comportamento estrutural de barragens de betão sujeitas a reacções expansivas". PhD thesis, Faculty of Sciences and Technology, New University of Lisbon, 2007.

Piteira Gomes, J., Batista, A.L., Silva, J.C., "Avaliação dos efeitos do processo de deterioração do betão da barragem de Fagilde – Parte 1 – Caracterização do estado da obra". II Encontro Luso-Brasileiro de Degradação de Estruturas de Betão, LNEC, Lisbon, September 2016.

Piteira Gomes, J., Batista, A.L., Silva, J.C., "Avaliação dos efeitos do processo de deterioração do betão da barragem de Fagilde – Parte 2 – Análise do comportamento observado e recomendações de mitigação". II Encontro Luso-Brasileiro de Degradação de Estruturas de Betão, LNEC, Lisbon, September 2016.

Sousa, S.P.M., "Analysis of the structural behaviour of dams affected by concrete swelling. The Fagilde dam case study". Master thesis, Faculty of Sciences and Technology, New University of Lisbon, 2016.

Expansions in a Concrete Dam with Bridge over Spillway in South America

Case Study with Expansions of Different Origin

ANA BLANCO* — SERGIO H.P. CAVALARO* — IGNACIO SEGURA** — LUIS SEGURA-CASTILLO*** — ANTONIO AGUADO*

** Universitat Politècnica de Catalunya*
Jordi Girona 1-3, Barcelona 08034, Spain
ana.blanco@upc.edu, sergio.pialarissi@upc.edu, antonio.aguado@upc.edu

*** Smart Engineering SL*
Jordi Girona 1-3, ParcUPC K2M, Barcelona 08034, Spain
ignacio.segura@smartengineeringbcn.com

**** Universidad de la República*
Julio Herrera y Reissig 565, Montevideo 11300, Uruguay
lsegura@fing.edu.uy

ABSTRACT. *A gravity concrete dam built 61 years ago in South America is diagnosed due to the cracking and signs of deterioration detected in the spillway and other elements. Besides the interest to identify the causes, the authorities are concerned by the structural effects on the concrete slab road bridge that is supported on the spillway. This road is one of the main routes for transportation of the country. A comprehensive diagnosis on the state of the dam is presented, including research of the historical documentation, visual inspection of the dam, performance of tests and numerical simulations. The case study is particularly interesting since several phenomena occur, including expansive reactions of different origin.*

KEYWORDS: *gravity dam, expansions, alkali-silica reaction, DEF, finite element method.*

1. Introduction

Expansions in concrete dams may be generated by different causes such as chemical reactions or physical phenomena. The most frequent cause is the alkali-aggregate reaction (AAR) (Stanton 1942; Hobbs 1988; Mehta *et al.*, 2006). Therefore, most of the expansive reactions in dams reported in the literature correspond to AAR (Shayan 1988; Regamey *et al.*, 1995; Ulm *et al.*, 2000, Saouma *et al.*, 2007), even though several cases were reported on internal sulfate attack (ISA) (Chinchón *et al.*, 1995; Ayora et al., 1998; Oliveira et al., 2013a; Oliveira *et al.*, 2013b; Oliveira *et al.*, 2014; Chinchón-Payá *et al.*, 2015) and alkali-carbonate reaction (ACR) (Neville 2004; Blanco *et al.*, 2015). Nevertheless, sometimes the structural response of the dam and the signs of deterioration may not be explained by a single source. In such cases, the diagnosis and evaluation gains special interest due to the scarcity of cases reported in the literature and because they may serve as a valuable example for future diagnosis.

The objective of this paper is to present the comprehensive diagnosis of a concrete dam build 60 years ago in South America with significant cracking and other signs of deterioration that might be attributed to an expansive phenomenon. The outcome of the study reveals that the expansions generated in the dam are caused by a delayed ettringite formation (DEF) and a subsequent ASR.

2. Description of the Dam

The dam is located in South America, less than 100 km of coast and was built in 1955 as an infrastructure for the water supply for the capital of the country. The dam consists of a plain concrete Creager spillway of 100 m of length. The slopes of the spillway are 1:20 in the upstream face and 1:0.63 in the downstream face, being its maximum height from the foundations is 14.0 m. The spillway is divided in concrete blocks of 15.0 m of width, with the exception of the central block that is 7.1 m wide.

The concrete spillway provides support to the piles of a reinforced concrete slab road bridge that crosses a river (see Figure 1a). The thickness of the slab is 0.39 m and it is divided in three parts: a central part of 7.10 m of length and two continuous parts that are supported every 5.0 m. The piles are 0.35 m thick with the exception of the ones supporting the central part which are 0.40 m. The piles were built in two stages with a vertical joint in their axis. The control valves are accessible from the road by means of a concrete stairway located at the central span of the bridge.

Consolidated soil dikes with lengths of 451 m on the left side and 231 m on the right side provide access to the bridge. The width of the dikes is 61.3 m at the base and 9.8 m at the crest. Unreinforced concrete retaining walls contain the dikes.

Figure 1. *View of the downstream face of the dam and the bridge over the spillway*

3. Current State of the Dam

The spillway exhibits horizontal cracks both in the downstream and in the upstream faces that correspond to construction or filling joints (see Figure 1b). This cracking was also reported in previous studies dating from 1989 and 1996, and was described as non-continuous and with no signs of infiltrations. The structure of the stairway is one of the elements of the dam that presents more signs of degradation of the concrete, including loss of concrete cover, deformation of the reinforcement and corrosion. These phenomena are observed in reinforced concrete elements such as columns or walls of the structure. The piles of the bridge that are supported on the spillway also present signs of degradation of different sort, including erosion of the surface, corrosion of the reinforcement and three types of cracking. The first type are cracks that develop perpendicularly from the contact area with the spillway (see Figure 2a). The second type are vertical cracks along the axis of the pile that correspond to a construction joint since the piles were constructed in two stages (see Figure 3b). The third type of crack is induced by the corrosion of the reinforcement and develop vertically according to the position of the rebars in the pile.

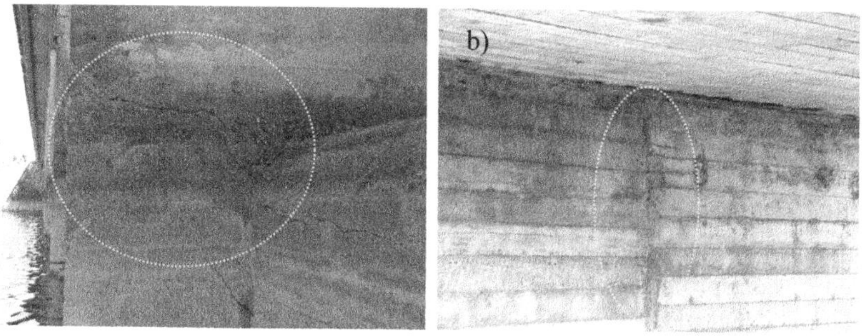

Figure 2. *Cracks in piles: a) in the contact with the spillway and b) in construction joints*

The deck of the bridge does not exhibit significant signs of deterioration. The retaining walls of the dikes exhibit map cracking in the upstream face and both horizontal cracks and map cracking in the downstream face.

4. Historical Documentation

The study of the historical documentation available on the construction and the service life of concrete structure is paramount to provide a diagnosis of their current state. In this particular case, the documentation of the dam is scarce. A descriptive report of the original project and the construction plans dating from 1950-1952 were available. However, no record is available from the construction procedure nor incidents that may have occurred at the time. The information regarding the concrete mix and its components is not accurate.

Regarding the operation and management period, the historical documents and reports reveal that the horizontal cracking in the spillway appeared shortly after the construction since repair and maintenance operations were already conducted in the decade of the 60's. Furthermore, the dam is not instrumented and thus no records are available of the evolution of displacements through the years.

5. Microstructural Analysis

The evaluation of the concrete microstructure was performed by using different techniques on samples (mortar and aggregate) obtained from the dam in 2015, including X-Ray Diffraction (XRD), scanning electron microscope with energy dispersive spectroscopy mode (SEM-EDS) and petrography. The tests were conducted in facilities of the University of Barcelona (Serveis d'Anàlisi del CCiT) and the University of Alicante (Servicios de Análisis) in Spain.

The aggregates samples analyzed correspond to metamorphic rocks, mainly quartzite, quarzitic schists and chalcedonies, which are potentially reactive aggregates and may induce ASR. Even though no signs of amorphous silica were detected, other characteristics were observed in the optical microscope that, according to the literature, suggest their reactivity such as small size of the grain, low crystallinity and presence of undulatory extinctions. The XRD and SEM-EDS analyses of the concrete samples have confirmed the presence of an ASR. Amorphous gel was detected in the samples as shown in Figure 3. The EDS analysis showed it was a product of the ASR in an advanced stage of the reaction due to the ratio between the calcium and potassium peaks.

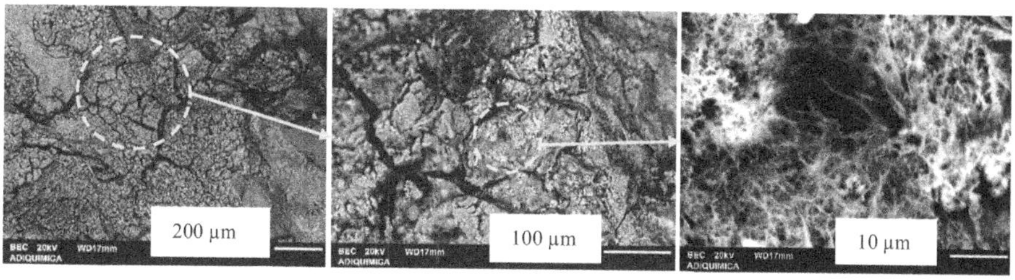

Figure 3. *Alkali-silica gel*

Likewise, massive presence of a sulfoaluminate phase at an incipient stage (monosulfoaluminates phases) and an advance stage (ettringite). In some samples, the ettringite needles were particularly long (see Figure 4a and 4b), which suggest a slow formation process over time. Furthermore, in several samples the sulfoaluminate phases are covered by products of the ASR (see Figure 4c and 4d), which is confirmed by the EDS analysis.

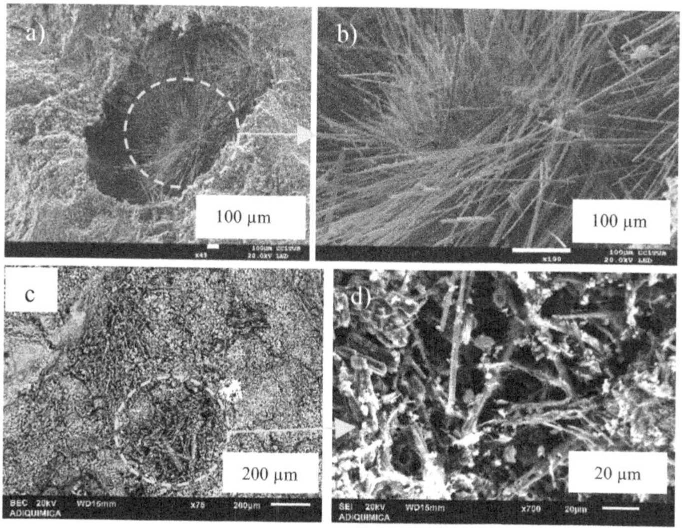

Figure 4. *a) Ettringite needles, b) detail of ettringite needles, c) sulfoaluminate phases covered by product of reaction and d) detail of the ettringite needles covered by the gel*

The presence of ettringite in concrete samples that are more than 60 years old is an evidence of a sulfate attack, either internal or external. Given that the minerals in the aggregates are not susceptible of causing an internal sulfate attach (such as phyrite or pyrrhotite) and the lack of an external source of sulfates in the environment, its presence must be associated with a DEF (Taylor *et al.*, 2001;

Glasser 2002; Brunetaud *et al.*, 2008). Such phenomenon may be justified if the temperature during the hydration and curing of the concrete exceeded the temperature of 70°C. Notice that there are no historical documentation regarding the cement used or the construction of the dam that could have helped to confirm this hypothesis.

The first time that an unusual ettringite content associated to DEF is reported in the literature is in the case of Roxburgh dam built between 1951 and 1956 in New Zealand (Kennerly 1965). Recently, another case was reported in the Vrané nad Vltavou dam in Czech Republic (Šachlová *et al.*, 2014).

6. Thermal Modelling of the Dam

In order to evaluate the hypothesis of a DEF, the possibility of the concrete temperature exceeding 70°C should be verified. For that, a finite difference model that simulates the construction the dam and the different processes of thermal interchange and the heat generation was developed. The result of the simulations allows to predict the historical record of the internal temperature of the concrete and the temperature at the surface.

6.1. *Basic Considerations*

Several phenomena need to be considered to reproduce the thermal behavior of the dam besides the heat due to the cement hydration, namely the heat transfer due to convection, radiation and thermal diffusion. The two former occur in the interface between the concrete and the environment, whereas the latter is internal. The heat transfer due to convection occurs when the atmospheric air warms up due to the higher temperature of the surface of the dam. The warm air reduces its density and its driven away by new air volumes at lower temperatures that contact the concrete surface. This phenomenon takes place successively until thermal equilibrium with the environment is reached. The difference of temperature between the concrete surface and the environment generates a second heat transfer (radiation), which consists in the emission of thermal radiation that contributes to cool the concrete.

Given that the temperature is not uniform inside the dam, other heat transfers must occur among different points in the concrete mass. Due to the heat release from the cement hydration and the heat transfer with the environment, the early age concrete exhibits a non-stationary temperature profile. Assuming such non-stationary condition, the heat transfer in each point is governed by Fick's second law. The total heat variation (dQ) with time in a certain point located at the surface of the dam and the point located inside the dam may be calculated with [1].

$$\begin{cases} \dfrac{\partial Q}{\partial t} = \dfrac{\partial Q_D}{\partial t} + \dfrac{\partial Q_H}{\partial t} & \text{for internal point} \\ \dfrac{\partial Q}{\partial t} = \dfrac{\partial Q_D}{\partial t} + \dfrac{\partial Q_H}{\partial t} - \dfrac{\partial Q_C}{\partial t} - \dfrac{\partial Q_R}{\partial t} & \text{for superficial point} \end{cases} \quad [1]$$

The temperature variation in each point of the dam is calculated by considering the heat content for a certain time in that point. The temperature variation is estimated with equation [2] and depends of the concrete specific heat (C) and the concrete mass (m_H).

$$\partial T = \frac{\partial Q}{m_H \cdot C} \quad [2]$$

6.2. *Description of the Model*

The modelling of dams is usually simplified and performed at a sectional level per length unit (in this case, linear meter). This is possible given that one of its dimensions is significantly larger than the other two. The cross-section of the dam is discretized in square elements of 10 cm. The calculation of the temperature is always conducted taking as a reference the center of gravity of each element.

The heat variation in each element is obtained with the equations described in the previous section. The calculations are performed with finite differences, assuming time intervals sufficiently small to be representative of the real situation. Based on preliminary evaluations, the maximum time interval is set to 5 minutes. Therefore, the simulation of each year in the life of the structure requires approximately 105000 time steps. Table 1 presents the properties of the concrete, defined according the literature, and the input data for the simulations. These values remain constant during the analysis.

Parameter	Value
Cement content ($kg \cdot m^{-3}$)	285
Specific weight ($kg \cdot m^{-3}$)	2300
Specific heat C ($J \cdot kg^{-1} \cdot k^{-1}$)	880
Diffusion coefficient D ($m^2 \cdot s^{-1}$)	7.39E-06
Emissivity ϵ (-)	0.9
Convection coefficient h ($W \cdot m^{-2} \cdot K^{-1}$)	5

Table 1. *Concrete properties and input data for the model*

The model considers the construction procedure of the dam based on casting by layers. For this reason, from the bottom level to the top of the dam, the elements of the same row are activated as the concrete is cast. Such assumption implies a change in the equations used to calculate the temperature in each time step. Before casting a new concrete layer, the horizontal surface of the last layer is unprotected and exchanges with atmospheric air may occur, besides the diffusion with the bottom and lateral layers. When the new concrete layer is cast, transfers with the environment stops, remaining only the diffusion with the other concrete layers.

6.3. Parametric Study

Given that the details of the type of cement and the construction rate of each block are unknown, a parametric study is conducted, where several types of cements and construction rates are considered (see Table 2). Based on the available information, the model assumes that concrete layers are built in layers up to 1.5 m high. Afterwards, the formworks are moved to another block to continue casting. This way, there is a certain time t_{layer} between the finishing of a layer and the casting of the following layer in the same block. Given the dimensions of the mixer used in the construction of the dam, the layer could not be completed with one mixing. In fact, it is assumed that the layer is the sum of several thinner layers of 0.1 m executed in a time t_{cast}. The values considered in the model are presented in Table 2 and correspond to two speeds: fast (V1) and slow (V2).

Parameter	Value
Time between layers t_{layer} (days)	20 (V1) and 30 (V2)
Time to cast a layer of 0,1 m t_{cast} (h)	2.4 (V1) and 3.6 (V2)
Maximum energy released Q_{max} (kJ/kg)	150 (E1), 200 (E2), 250 (E3) and 300 (E4)
Kinetic coefficient (-)	$0.5\text{x}10^{-5}$ (C1), $1.0\text{x}10^{-5}$ (C2), $1.5\text{x}10^{-5}$ (C3) and $2.0\text{x}10^{-5}$ (C4)

Table 2. *Variables of the parametric study*

Regarding the cement hydration, both the maximum energy released (Q_{max}) and the kinetics of the energy release (a) are variables. The first one depends mainly on the composition and the clinker content of the cement and the second is related with the fineness and composition of the clinker. The values assumed for both variables were defined according to heat release curves in the literature for several types of cement. The ambient temperature (T_a) and the initial temperature of the concrete before the casting ($T_{Hinitial}$) remain constant along the simulation with a value of 20°C (T_a is always equal to $T_{Hinitial}$).

6.4. *Results and Discussion*

Figure 5 shows the maximum temperatures reached in different point of the dam assuming a construction rates V1 (the graphic results for V2 are not presented but are commented subsequently). The results for both rates reveal that the highest temperatures occur in the center of the different layers. This is reasonable given that the regions closer to the perimeter are exposed to the heat transfer with adjacent layers or the environment. The results also indicate that the increase of maximum energy released during the hydration and of the hydration kinetics coefficient leads to a temperature rise in the center of each layer. In certain cases, the temperatures reached are consistent with the DEF, even in points close to the surface of the dam.

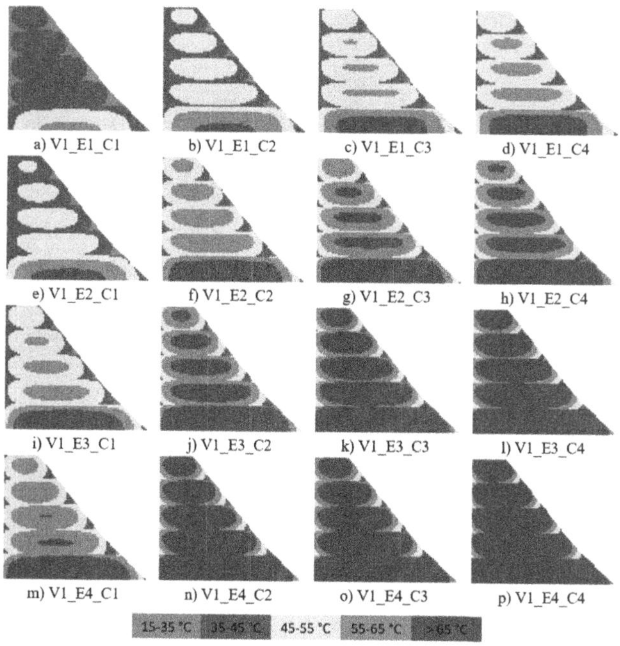

Figure 5. *Maximum temperature reached in each point for construction rate V1*

In the scenarios affected by high temperatures, an expansion in the central area of the layers is expected, which generates compression in the center of each layer and tension in the upstream and downstream faces. If the tension in such locations reaches the tensile strength of concrete, horizontal cracks should appear both upstream and downstream. By comparing equivalent scenarios for the two construction rates, a similar distribution is detected. However, V2 leads to temperatures that are between 1°C and 3°C lower than for V1; nevertheless, the center of the layer still reaches temperatures higher than 65°C.

Figure 6 shows the influence of the maximum hydration energy of the cement in the maximum temperature of points located in the center of the layers, for V1 and different hydration kinetics (C1, C2, C3 and C4) at different heights of the dam (2.25 m, 3.75 m 5.25 m and 6.75 m) for points located at the center (C). The results reveal an approximately linear relation between the maximum energy released in the hydration and the maximum temperature reached. Likewise, the hydration kinetics C2, C3 and C4, with maximum values of energy between 200 kJ/kg and 250 kJ/kg, lead to temperatures above 70°C inside all layers. It should be remarked that the cements currently classified as low heat cement present values of energy release between that range. At the time of construction of the dam, cements with a hydration heat lower than 200 kJ/kg were not common. Considering the above, the DEF may be the cause of the horizontal cracks observed few years after the construction of the dam.

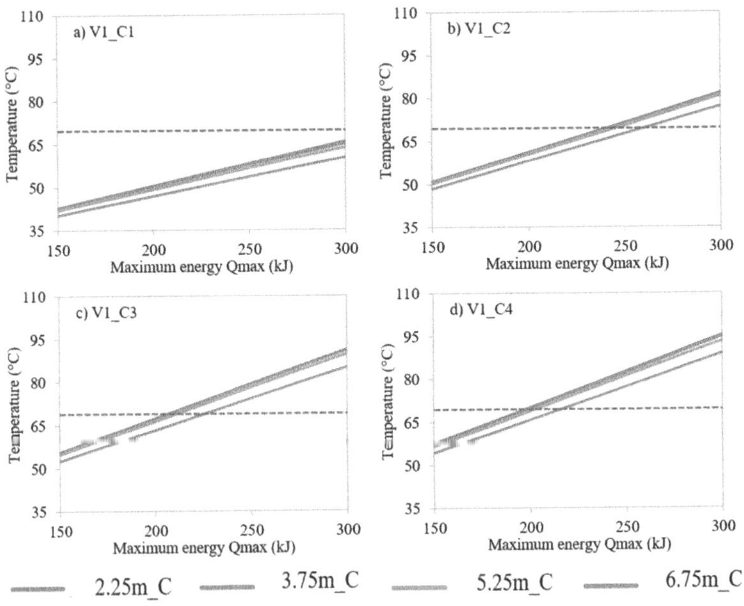

Figure 6. *Influence of the maximum energy released in the hydration in the maximum temperature reached at the center of different layers (at different heights)*

7. Expansion in the Spillway

The ASR and a DEF diagnosed in previous sections may affect the piles of the bridge supported on the spillway. In fact, the expansion generates tension stresses in the piles and, if the tensile strength of the concrete is reached, cracks may appear perpendicularly to the direction of the tension (see Figure 7).

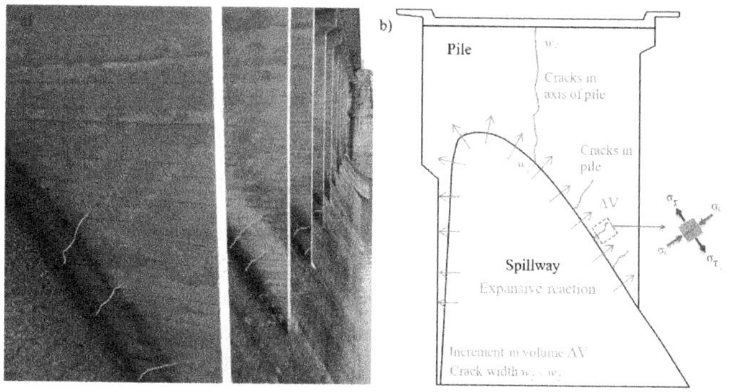

Figure 7. *a) Diagonal cracks in the piles and b) phenomenon causing the cracks*

The vertical crack observed in several piles of the bridge is also due to the expansive reaction, however its development is governed by the presence of a construction joint that is a weak plane and is also more susceptible to cracking.

8. Conclusions

The study presents the comprehensive diagnosis of a concrete dam with significant cracking and other signs of deterioration in different elements, including horizontal cracking in the spillway and map cracking in some of the elements that suggest the presence of expansive phenomena in the concrete.

The DRX and SEM-EDS analyses confirm the presence of an alkali-silica reaction in an advanced stage in the concrete spillway. Massive presence of sulfoaluminate phases both at an incipient stage and at an advanced stage are also detected. Having dismissed the possibility of an internal or external sufate attack and considering the shape of the ettringite needles and how they developed over time, its presence is associated with a delayed ettringite formation (DEF).

This hypothesis is validated through the thermal modelling of the construction procedure of the dam. For that, a specific model based on finite differences and that considers the equations governing the different heat transfer phenomena involved was developed. The outcome of the model reveals that concrete temperatures over 70°C were reached during the construction, which is the cause for the DEF. This phenomenon explains the horizontal cracking in the spillway of the dam. The expansions in the spillway due to the DEF and the alkali-silica reaction generate tension stresses in the piles of the bridge which may lead to the cracking observed.

9. Acknowledgements

The authors acknowledge the economic support provided by the Spanish Ministry of Science and Innovation through the project BIA2013-49106-C2-1-R and the collaboration project with Universidad de la República. The authors thank Professor Servando Chinchón and Dr. Servando Chinchón-Payá for their kindness and willingness to collaborate in the microstructural analyses of this study.

10. References

Ayora, C., Chinchón, J.S., Aguado, A., Guirado, F., "Weathering of Iron Sulfides and Concrete Alteration: Thermodynamic Model and Observation in Dams From Central Pyrenees, Spain". *Cement Concrete Res.*, vol. 8, no. 4, 1998, p. 591-603.

Blanco, A., de la Fuente, A., Cavalaro, S.H.P., Aguado, A., "A century of sand-cement: durability of the concrete in the Camarasa dam". *J. Perform. Constr. Fac.*, vol. 30, no. 4, 2016, p. 04015083-1 - 04015083-10.

Brunetaud, X., Divet, L., Damidot, D., "Impact of Unrestrained Delayed Ettringite Formation-induced expansion on concrete mechanical properties". *Cement and Concrete Res.*, vol. 38, 2008, p. 1343-1348.

Cervenka, V., "Simulating a response". *Concr. Eng. Int.*, vol. 4, no. 4, 2000; p. 45-49.

Chinchó, J.S., Ayora, C., Aguado, A., Guirado, F., "Influence of Weathering of Iron Sulphides Contained in Aggregates on Concrete. Durability". *Cement Concrete Res.*, vol. 25, no. 5, 1995, p. 1264-1272.

Chinchón-Payá, S., Aguado, A., Coloma F., Chinchón, J.S., "Study of aggregate samples with iron sulphides trough micro X-Ray Fluorescence (μXRF) and X-ray Photoelectron Spectroscopy (XPS)". *Mater. Struct.*, vol. 48, no. 5, 2015, p. 1285-1290.

Glasser, F. P., "The stability of ettringite". *International RILEM TC 186-ISA Workshop on Internal Sulfate Attack and Delayed Ettringite Formation*, 4-6 September 2002, Villars, Switzerland, p. 43-64.

Hobbs, D. W., *Alkali-silica reaction in concrete*, Michigan, Telford, 1988.

Kennerly, R. A., "Ettringite Formation in Dam Gallery". *ACI Journal*, vol. 62, 1965, p. 559-576.

Mehta, P. K., Monteiro, P. J., *Concrete Microstructure, properties, and materials*, New York, McGraw-Hill, 2006.

Neville, A., "The confused world of sulfate attack on concrete". *Cement Concrete Res.*, vol. 34, no. 8, 2004, p. 1275-1296.

Oliveira, I., Cavalaro, S.H.P., Aguado, A., "New kinetic model to quantify the internal sulfate attack in concrete. *Cement Concrete Res.*, vol. 43, no. 1, 2013a, p. 95-106.

Oliveira, I., Cavalaro, S.H.P., Aguado, A. "New unreacted-core model to predict pyrrhotite oxidation in concrete dams". *J. Mater. Civil Eng.*, vol. 25, no. 3, 2013b, 372-381.

Oliveira, I., Cavalaro, S.H.P., Aguado, A., "Evolution of pyrrhotite oxidation in aggregates for concrete". *Mater. Construcc.*, vol. 64, no. 316, 2014.

Regamey, J.M., Hammerschlag, J.G. "Barrage d'Illsee—assaintment". Proc., *Research and development in the field of dams*, Crans-Montana, Switzerland, 1995.

Šachlová, S., Pertold, Z., Šťastná, A., Míčka, T. "Factors affecting DEF and ASR in the concrete dam at Vrané nad Vlatavou". *International Journal of Research in Engineering and Technology*, vol. 3, no. 13, 2014, p. 57-62.

Saouma, V., Perotti, L., Shimpo, T., "Stress Analysis of concrete Structures Subjected to Alkali-Aggregate Reactions". *ACI Struct. J.*, vol. 104, no. 5, 2007, p. 532-541.

Shayan, A., "Alkali aggregate reaction in a 60-year-old dam in Australia". International *Journal of Cement Composites and Lightweight Concrete*, vol. 10, no. 4, 1988, p. 259-266.

Stanton, T. E., "Expansion of concrete through reaction between cement and aggregate". *Transactions of the American Society of Civil Engineers*, 107(1), 1942, p. 57-84.

Taylor, H. F. W., Famy, C., Scrivener, K. L., "Delayed ettringite formation", Cement and Concrete Res., vol. 31, no. 5, 2001, p. 683-693.

Ulm, F.-J., Coussy, O., Kefei, L., and Larive, C. "Thermo-Chemo-Mechanics of ASR Expansion in Concrete Structures". *J. Eng. Mech.-ASCE*, vol. 126, no. 3, 2000, p. 233-242.

Modeling Concrete Expansions via Coupled C-M Mesoscale Analysis with Zero-Thickness Interface Elements, and Lab Experiments

IGNACIO CAROL — JOAQIN LIAUDAT — CARLOS M. LÓPEZ

Technical University of Catalonia (UPC)
School of Civil Engineering-Barcelona (ETSECCPB), c/ Jordi Girona, 1, D2, 08034 Barcelona, Spain
ignacio.carol@upc.edu, joaquin.liaudat@upc.edu, carlos.maria.lopez@upc.edu

ABSTRACT. *This paper briefly summarizes ongoing developments on C-M coupled modelling of concrete swelling due to external sulfate attack and due to alkali-silica reaction. Both models are based on a meso-mechanical model for concrete, previously developed for purely mechanical actions, that has been also used as the basis for modelling other coupled phenomena such as drying shrinkage and mechanical effects of high temperatures. Additionally, two new experimental setups are presented. One devoted to the study of ASR expansions at the level of a single aggregate-matrix (cement paste or mortar) interface and the other devoted to study ASR expansions in concrete under triaxial confinement.*

KEYWORDS: *alkali-silica reaction, external sulfate attack, coupled model, FEM.*

1. Introduction

With constantly increasing computer power, sophisticated material degradation models are becoming more popular, including phenomena such as moisture changes, temperature distributions and/or chemical processes coupled with mechanical behavior as well as micro/meso-level analysis of heterogeneous materials. In particular, in the case of concrete and rock, meso-scale multiphysics FE analysis has emerged as a powerful tool for the modelling of the cracking and fracture processes under different loading and environmental situations. During over a decade, the group of Mechanics of Materials at UPC (Barcelona) has been developing such type of combined tools with the distinctive feature of representing cracks via zero-thickness interface elements, for progressively more complex scenarios. The approach was first applied to the purely mechanical behavior in 2D (López et al. 2008a; López et al. 2008b) and 3D (Caballero et al. 2006), then to basic creep (López et al. 2001), drying shrinkage (Idiart et al. 2011b) and external sulfate attack (ESA) in 2D (Idiart et al. 2011a). Current ongoing work is devoted to the application to the mechanical effects of high temperatures (Rodriguez et al. 2015), to the degradation of oil well cement due to the exposure to carbonated brine, to the extension of the ESA model to 3D, and finally to expansion due to Alkali-Silica Reaction (ASR). In this paper, recent work on ESA and ASR expansions in concrete is briefly summarized.

2. Meso-Mechanical Model

The basic meso-mechanical model was originally proposed to represent the behavior of concrete under mechanical loads. In this model, the largest aggregate particles are represented explicitly, and it has the distinctive feature of considering all lines in the FE mesh as potential crack lines, via the systematic use of zero-thickness interface elements equipped with traction-separation constitutive models based on principles of non-linear fracture mechanics (Carol et al. 1997; Carol et al. 2001), while the matrix phase is considered elastic or linear visco-elastic. With this conceptually simple model, the mechanical behavior of concrete was successfully reproduced under a variety of uniaxial, biaxial and triaxial, conditions, creep, and differential temperature expansions (Carol et al. 2001; Caballero et al. 2006; López et al. 2008a; Rodriguez et al. 2015).

3. Transport Model and Coupling Strategy

A standard non-linear diffusion analysis is performed over the same FE mesh as mechanical calculations, in order to obtain concentrations of the relevant ions at each point in the concrete. Interface elements with double nodes are also used to introduce the effects of preferential diffusion of species via interfaces and cracks, as proposed by Segura and Carol (2004). Such interface elements for diffusion incorporate longitudinal as well as transversal diffusivities $\boldsymbol{D}_l$ and $\boldsymbol{D}_t$. In the absence of specific information, the latter may be given a very high value, while $\boldsymbol{D}_l$ is assumed to vary substantially with the crack state, i.e. with a law proportional to the cube or square of the crack opening (Idiart et al. 2011a). The diffusion of aqueous species drives chemical processes that lead to the formation of expansive products, which in general turn out non-uniform and therefore lead to stresses and possibly cracks. These cracks may in turn change the diffusion properties, and are the main vehicle for coupling. The coupling is implemented using staggered strategy for each time step of the time discretization.

4. Expansions Due to External Sulfate Attack

This degradation phenomenon is originated by the ingress of sulfate ions from the environment into the cement pores. These ions then react with solid ingredients of the cement paste leading to volume expansion. It seems accepted that the reaction takes place in two steps: first, sulfate ions react with the portlandite and/or calcium silicate hydrates of the cement paste leading to gypsum, and second, gypsum reacts with various kinds of calcium aluminates leading to formation of secondary ettringite, which is really the expansive product. In the model developed, this mechanism is represented via Eq. [1], which consists of an ionic diffusion term and a sink term which corresponds to the consumption of sulfate ions. Isotropic expansion of the cement paste due to secondary ettringite precipitation is obtained from equation [2], as the product of the total aluminate reacted C_{eq}^{react} by an equivalent molar volume α_s. The second term in the equation accounts for the volume of capillary pores $f\phi_{\text{ini}}$ that has to be filled before any macroscopic expansion is observed.

$$\frac{\partial U}{\partial t} = \frac{\partial}{\partial x}\left(D_U \frac{\partial U}{\partial x}\right) - kUC;\ D_U(H) = D_0 + (D_1 - D_0) f(\beta, \phi_{cap}) \qquad [1]$$

$$\varepsilon_v(t) = max\left|\alpha_s \cdot C_{eq}^{react} - f\phi_{\text{ini}}, 0\right| \qquad [2]$$

Previous work in the group involved the coupled simulation of 2D specimens subject to external sulfate attack, showing clearly a strong C-M coupling effect as open cracks become penetration channels for the sulfate ions and onion peel-type spalling develops (Idiart et al. 2011a). Recent work along this line has been directed to the 3D extension of the analysis, and preliminary results of a simple, uncoupled, calculation are shown in Figure 1. A cubical concrete specimen is subject to sulfate attack from two of its faces (Riera Bayo 2016). As sulfates penetrate the specimen, secondary ettringite is formed, imposing volumetric strains in the mortar matrix.

Consequently, a typical cracking pattern is developed with tendency to spalling at the shared edge of the two exposed faces.

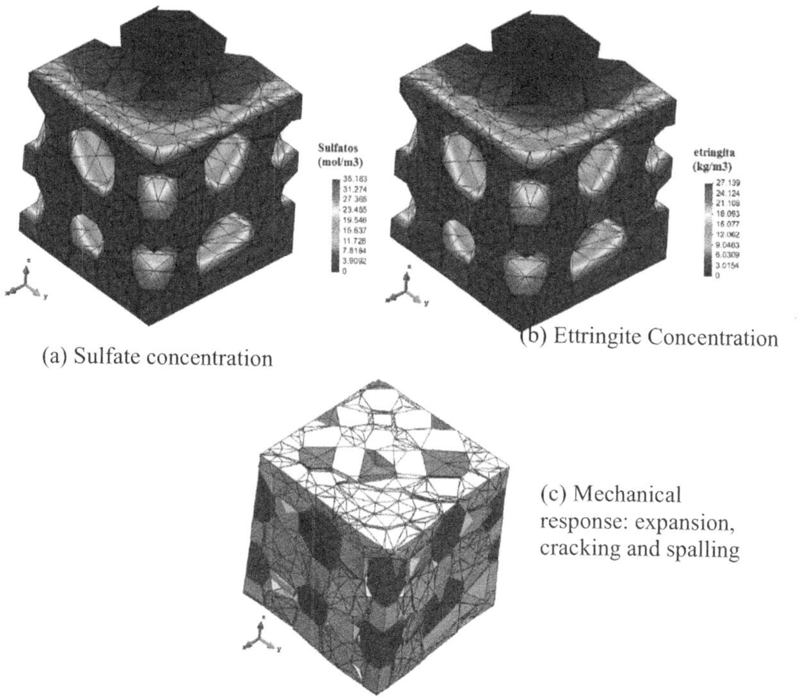

(a) Sulfate concentration

(b) Ettringite Concentration

(c) Mechanical response: expansion, cracking and spalling

Figure 1. *3D simulations at the meso-level of external sulfate attack to concrete after 740 days. In figures (a) and (b) aggregates have been removed to ease the visualization*

5. Expansions Due to Alkali-Silica Reaction

5.1. *Experiments*

Recent experimental work includes the development of two new experimental setups, one devoted to study the ASR expansion mechanisms at the level of a single matrix-aggregate interface and the other to the study of ASR expansions of cubic concrete specimens under triaxial confinement. In the following paragraphs, a brief description of these tests is given. Further details are given in Liaudat et al. (2016).

5.1.1. *Interfacial Expansion Tests*

The aim of this test is to measure the expansions at the level of a single interface aggregate-matrix due to the precipitation of ASR products. For this purpose, two kinds of cylindrical specimens of 33 mm diameter and 66 mm height are elaborated. The first type, named “Active specimens”, involves sandwich – like specimens, with cement paste or mortar on top and bottom of a disc of reactive aggregate in the middle (see Figure 2). The second one, called “Control specimens”, involves purely mortar or cement paste specimens without reactive aggregate. Control specimens are used to assess the deformations of the matrix of cement paste or mortar, solely caused by phenomena such as drying shrinkage or thermal expansions. The alkali content of the cement pore solution is increased by adding NaOH to the mixing water.

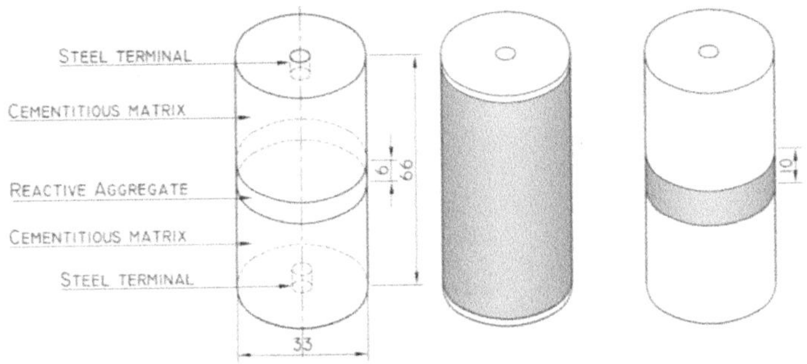

Figure 2. *Configuration of active specimens used in ASR interfacial expansion tests and schematic representation of the three sets of specimens tested with different lateral sealing*

After casting, the specimens are kept in the molds for 24 h in a humid chamber. Once unmolded, they are cured in airtight containers with a solution of sodium hydroxide of 1 mole/liter at 23°C during 27 days. Finally, the containers are introduced in an oven at 60°C until the end of the test. The sodium hydroxide solution keeps the specimens saturated of water and prevents the leakage of alkali ions from the cement paste. During both stages, specimen length changes are measured regularly with a dial indicator until the end of the test.

Once in the oven at 60°C, the active specimens start to expand and to expel a whitish reaction product at the interfaces, until the moment when the amount of reaction product at the interface is high enough to separate the glass from the matrix, which can happen in a matter of days, weeks, or months depending on the dosage (mainly on NaOH added and sand content) of the matrix and the aggregate type. Before the separation, the reaction products at the interface are studied by means of SEM images and EDS analysis of polished sections of the specimens.

From the length change measurements of the active and the control specimens, the expansion curves corresponding to the ASR reaction products precipitated in a single interface matrix-aggregate are obtained using Eq. [3], where L and DL are the length and the length change of the active specimen respectively, e is the width of the glass disc, ε_c is the average deformation of the control specimens, α_{agg} is the thermal expansion coefficient of the aggregate, DT is temperature variation from curing (23°C) to exposure (60°C) and d_I is the expansion at a single interface.

$$d_I = \frac{1}{2}\left[\Delta L - \bar{\varepsilon}_c(L - e) - e \propto_{agg} \Delta T\right] \qquad [3]$$

In Figure 3, interfacial expansion curves obtained with the described methodology for specimens made cement and soda-lime glass as reactive aggregate are presented. Where indicated with a letter D over the last measurement, the test ended because one of the glass faces detached from the HCP. In this case, with the goal of assessing the influence of ionic diffusion through the ITZ, three sets specimens with different lateral sealing have been tested:

– Set #1: Specimens without lateral sealing

– Set #2: Specimens with complete lateral sealing

– Set #3: Specimens with lateral sealing of 10mm width covering aggregate (glass), the ITZ and the immediate HCP.

The results seem to confirm our assumption that the kinetics of the expansion is mainly determined by the flux of external alkalis into the specimens, and that this flux is mainly localized along the ITZ.

5.1.2. *Confined Expansion Tests*

The aim of this test is to measure ASR expansions in cubic concrete specimens under true triaxial confinement, i.e. under different constant stress in each main direction. To do so, an ad-hoc testing machine (AAR-TM) which was originally designed and constructed by Prof. Saouma at the U. of Colorado - Boulder. Later, in 2010, it was transferred to UPC where a number of modifications were introduced to the original set-up. After several preliminary tests, the first experimental campaign with this machine is currently under way. See Figure 4 for a general view of the equipment.

The AAR-TM allows testing cubic specimens of 150 mm side under different pressures from 1 to 9 MPa in each main direction, applied by hydraulic actuators. The temperature of the specimen is controlled by means of temperature sensors and electric heaters attached to the loading plates, allowing rise the temperature up to 60°C. The faces of the loading plates in contact with the specimens are crossed by grooves for a solution to circulate in order to keep the specimen wet and supply reactants (alkalis). Water leakage is prevented by means of an O-ring in between the plate and the specimen. Additionally, the contours of the contact area are covered with silicone sealant. The dimension/shape changes of the specimen during test are measured through nine LVDTs, three in each direction, attached to the loading plates.

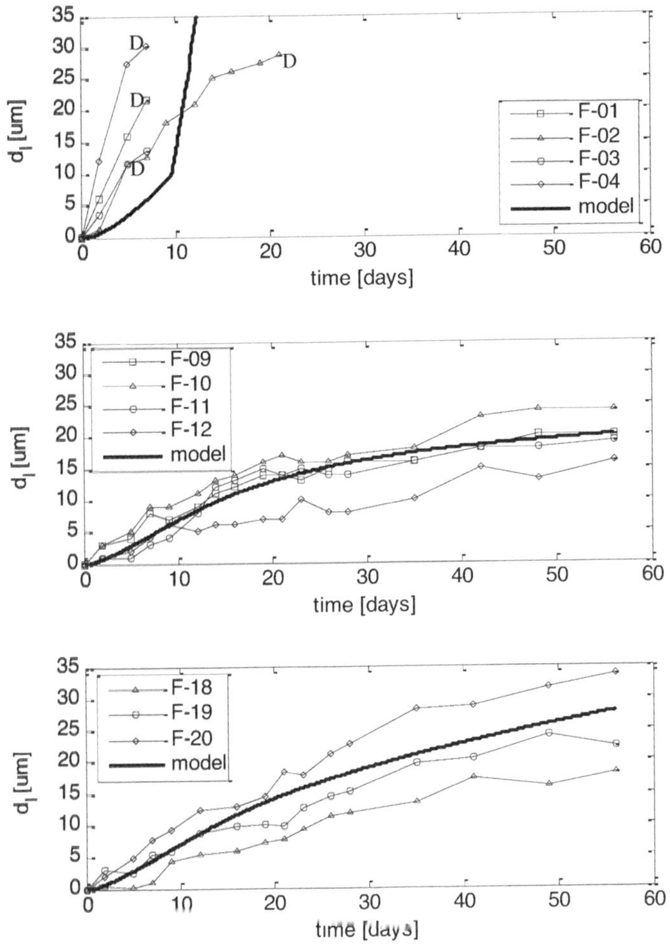

Figure 3. *Interfacial expansion curves: Set #1 - specimens without lateral sealing; set #2 - specimens with complete lateral sealing; set #3 - specimens with lateral sealing only on the HCP-glass interface zone*

Additionally, to assess the effect of the applied confinement on the ASR expansion rate, a number of free expansion tests with specimens of identical characteristics and under the same exposure conditions (alkaline solution at 60ºC) were performed separately. This has to be done with a different testing procedure since the AAR-TM is not appropriated for free-expansion tests ($\sigma_x = \sigma_y = \sigma_z = 0$). This procedure consists in keeping each specimen in a 19L airtight plastic recipient filled with alkaline solution inside an oven at 60ºC. The deformation of the specimen is measured regularly (once a week) in the three main directions by means of a removable 100mm-strain-gage which is posed on stainless-steel datum discs previously glued to the surface of the specimen.

Figure 4. *Machine used for ASR expansion tests under "true" triaxial confinement states*

5.2. *Model*

The proposed reaction mechanism is schematically summarized in Figure 5. The interfacial zone between HCP and aggregate in concrete is represented. Both components are assumed to be porous and fully saturated with water at all times during the reaction. Immediately after casting, the HCP has a certain content of portlandite and its pore water has a certain content of aqueous calcium (Ca) and alkalis (R). Note that sodium (Na) and potassium (K) are treated here indifferently as "alkalis" (R). The aggregates are supposed to have a certain content of reactive silica (SiO_2) and no calcium or alkalis in their initial pore solution.

The concentration gradient of aqueous calcium and alkalis causes a diffusion process from the HCP towards the aggregate, which is followed by an increase of the pH of the pore solution in the aggregate. This increment causes the dissolution of the reactive silica into silicate ions ($SiO_4H_3^-$), which in turn, due again to a concentration gradient, diffuse towards the cement paste. At the same time, these three diffusion processes drive the dissolution or precipitation of portlandite ($Ca(OH)_2$).

Wherever, in the aggregate or in the HCP, the three reactants (aqueous calcium, alkali and silicate) are available, a reaction occurs forming a Calcium-Alkali-Silicate-Hydrate (C-R-S-H) of variable composition, depending on the concentration of the reactants. The composition, in turn, determines the molar volume of this product. The effect of the remaining chemical species usually present in concrete on the development of ASR is neglected.

The variability of the composition is reached by considering two separate reactions, each one with a fixed stoichiometry and its own kinetic law, one reaction forming C-R-S-H rich in calcium called Reaction Product A (RPA) and the other forming a C-R-S-H poor in calcium called Reaction Product B (RPB). Different densities and mechanical properties are assigned to RPA and RPB in order to reproduce the observed dependence of ASR gel swelling on its calcium content (Leemann et al. 2011).

Finally, at a given location, the volume balance of the solid constituents, namely reactive silica, portlandite, RPA and RPB, determines the transport properties and the volume of pore solution. If the volume of precipitated reaction products is greater than the available space in capillary pores plus the additional space liberated by silica and portlandite dissolution, a localized volumetric expansion may occur.

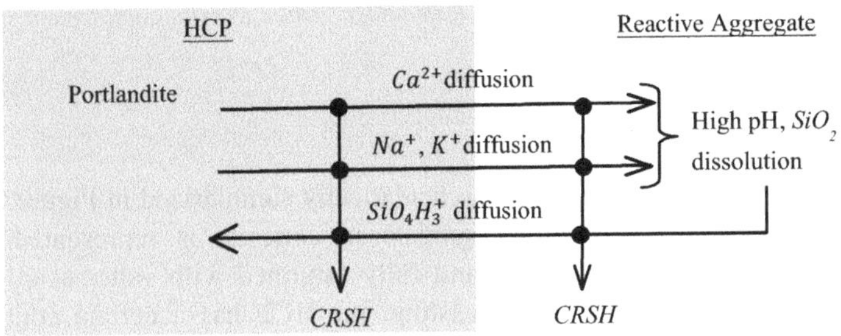

Figure 5. *Schematic representation of ASR mechanism in concrete*

Assuming that the diffusion of aqueous species in the pore solution obeys Fick's diffusion law, averaging it in the saturated porous medium and posing the correspondent mass balance equation, the following diffusion-reaction equations for the continuum porous medium are obtained

$$\begin{cases} \frac{\partial}{\partial t}(\phi c^s) = \boldsymbol{\nabla}^T(D^s \boldsymbol{\nabla} c^s) + q^s \\ \frac{\partial}{\partial t}(\phi c^c) = \boldsymbol{\nabla}^T(D^c \boldsymbol{\nabla} c^c) + q^c \\ \frac{\partial}{\partial t}(\phi c^r) = \boldsymbol{\nabla}^T(D^r \boldsymbol{\nabla} c^r) + q^r \end{cases} \quad [4]$$

where the superscripts s, c and r indicate silicate, calcium and alkali, respectively, ϕ is the total porosity; c^β [mol/m^3] is the concentration of aqueous β-species in the pore solution expressed in moles per unit volume of pore solution, D^β [m^2/s] is the effective diffusivity of aqueous β-species in the porous medium (assumed isotropic), $\boldsymbol{\nabla} = [\partial/\partial x \quad \partial/\partial y]^T$, and q^β [mol/(m^3·s)] is the rate of production /consumption of β-species per unit volume of porous medium, which in turn is a function of the concentration of aqueous silica, calcium and alkalis, i.e. $q^\beta = q^\beta(c^s, c^c, c^r)$. Analogous expressions are obtained for the localized diffusion-reaction processes occurring in a discontinuity, such as a crack or and ITZ.

A general expression of the production rate of β-species, q^β, is given in Eq. [5], where a_α^β is the dimensionless stoichiometric coefficient of β-species in the formation reaction of solid α-species, Γ^α [mol/(m^3·s)] is the reaction rate of solid α-species per unit volume of porous material (positive for solid formation, negative for solid dissolution), which is a function of the vector of local concentration of aqueous species $\mathbb{c} = [c^s \quad c^c \quad c^r]$, expressed in moles per unit volume of pore solution and of the vector of local content of solid species $\boldsymbol{\mathcal{N}} = [\mathcal{N}^S \quad \mathcal{N}^C \quad \mathcal{N}^{RPA} \quad \mathcal{N}^{RPB}]$ which is expressed in moles per unit volume of porous material.

$$q^\beta = -\sum_\alpha a_\alpha^\beta \Gamma^\alpha(\mathbb{c}, \boldsymbol{\mathcal{N}}) \qquad [5]$$

The calculation of the sink/source terms q^β according to Eq. [5] requires stablishing the kinetic laws of the dissolution/precipitation reactions in order to obtain the correspondent reaction rates Γ^α. To do so, it is assumed that the driving force of the dissolution/precipitation reaction of reactive solid α-species is $(\psi^\alpha - 1)$, where ψ^α is the saturation index of the pore solution with respect to solid α-species.

The precipitation/dissolution processes determine the evolution of the material porosity and consequently the effective diffusivity of both the continuum medium and the discontinuities. The mass balance equation of solid α-species is given by Eq. [6a], where $\mathcal{U}$ [m^3] is the total volume of continuum medium or discontinuity. The volume of solid α-species is obtained as function of their molar concentration with Eq. [6b], where η^α is the molar volume and ϕ^α is the intrinsic porosity of solid α-species. Intrinsic porosity is defined as the ratio between gel water volume and total apparent volume of phase α.

$$\frac{\partial(\mathcal{N}^\alpha \mathcal{U})}{\partial t} = \mathcal{U}\Gamma^\alpha; \; \mathcal{U}^\alpha = \left(\frac{\eta^\alpha}{1-\phi^\alpha}\right) \mathcal{N}^\alpha \, \mathcal{U} \qquad [6a,b]$$

Further details of the proposed diffusion-reaction formulation may be found in Liaudat et al. (2016).

Eventually, the volume of Reaction Products (RPs) may grow to a point in which the available space is exhausted. From this point on, any additional mole of RP will exert an internal pressure which will depend on the relative stiffness of the RPs in relation to the surrounding solid matrix. The internal gel pressure is determined in the mechanical model by means of a mixed constitutive law consisting in a parallel Maxwell chain model, in which one branch represents the RPs mechanical behavior and the other the surrounding solid matrix behavior. The compression of the RPs is associated to a reduction of the intrinsic porosity (gel porosity) with consequences in the kinetics of the ASR.

The model has been used to simulate the above described interfacial expansion tests with different lateral sealing. The obtained expansion curves are plotted together with the experimental ones in Figure 3. As it can be seen, the agreement is very good. The model seems capable of capturing the effect of external alkali diffusion in the development of the expansions.

6. Concluding Remarks

The meso-mechanical model for concrete previously developed for purely mechanical actions, which is based on Voronoi-generated geometries of the larger aggregates and cracking via fracture-based zero-thickness interface elements, seems to provide a convenient basis for extensions to coupled durability mechanics calculations.

The formulation proposed for ESA has been proven to be capable of reproducing both the expansion rate and the cracking patterns of affected concrete studied in laboratory tests. The mechanical coupling resulted essential to obtain realistic results in 2D simulations, in particular with regard to the effect of localized diffusivity in concrete cracks. Current work is devoted to extent the model from 2D to 3D problems.

The ongoing research project on the mechanisms of concrete deterioration due to ASR includes both experimental and numerical modelling work. The experimental work consists of an additional series of cubical tests with lime glass aggregates, as well as new tests using real expansive basaltic aggregates.

With regard to the numerical modelling work, the proposed coupled model for ASR expansions seems very promising. From the chemical point of view, the proposed formulation has the advantage that it incorporates the recently identified role of calcium in ASR expansions, i.e. depending on local conditions (including pH and Ca concentration) the model can lead to either generation of RP with low calcium content and high expansive capacity, RP with high calcium content and low expansive capacity, or a mix of both in varying proportion. Preliminary results of coupled simulations indicate that the compressibility of the RPs may be essential to reproduce the effect of the stress state of concrete in the development of ASR expansions.

7. Acknowledgements

This research was supported by grants BIA-2016-76543-R funded by MEC (Madrid), which includes FEDER funds, and 2014SGR-1523 from AGAUR-Generalitat de Catalunya (Barcelona).

8. References

Caballero, A., López, C.M. & Carol, I., 2006. 3D meso-structural analysis of concrete specimens under uniaxial tension. *Computer Methods in Applied Mechanics and Engineering*, 195(52), pp.7182–7195.

Carol, I., López, C.M. & Roa, O., 2001. Micromechanical analysis of quasi brittle materials using fracture based interface elements. *International Journal for Numerical Methods in Engineering*, 52, pp.193–215.

Carol, I., Prat, P.C. & López, C.M., 1997. Normal/Shear Cracking Model: Application to Discrete Crack Analysis. *Journal of Engineering Mechanics*, 123(8), pp.765–773.

Idiart, A.E., López, C.M. & Carol, I., 2011a. Chemo-mechanical analysis of concrete cracking and degradation due to external sulfate attack: A meso-scale model. *Cement and Concrete Composites*, 33(3), pp.411–423.

Idiart, A.E., López, C.M. & Carol, I., 2011b. Modeling of drying shrinkage of concrete specimens at the meso-level. *Materials and Structures*, 44(2), pp.415–435.

Leemann, A. et al., 2011. Alkali–silica reaction: the influence of calcium on silica dissolution and the formation of reaction products. *Journal of the American Ceramic Society*, 94(4), pp.1243–1249.

Liaudat, J., López, C.M. & Carol, I., 2016. Numerical and Experimental study of ASR in concrete at the meso-level. In V. Saouma, J. Bolander, & E. Landis, eds. *The 9th International Conference on Fracture Mechanics of Concrete and Concrete Structures (FraMCoS 9)*.

López, C.M., Carol, I. & Aguado, A., 2008a. Meso-structural study of concrete fracture using interface elements. I: numerical model and tensile behavior. *Materials and Structures*, 41, pp.583–599.

López, C.M., Carol, I. & Aguado, A., 2008b. Meso-structural study of concrete fracture using interface elements.II: compression, biaxial and Brazilian test. *Materials and Structures*, 41, pp.601–620.

López, C.M., Carol, I. & Murcia, J., 2001. Mesostructural modeling of basic creep at various stress levels. In F.-J. Ulm, Z. P. Bazant, & F. H. Wittmann, eds. *Creep, Shrinkage and Durability Mechanics of Concrete and other Quasi-Brittle materials*. pp. 101–106.

Riera Bayo, C., 2016. *Modelització mesomecànica 3D d'atac sulfàtic en formigó*. Master Thesis (in Spanish), Departament d'Enginyeria Civil i Ambiental, Universitat Politècnica de Catalunya.

Rodriguez, M., Lopez, C.M. & Carol, I., 2015. Modeling of Heat and Mass Transfer Induced By High Temperature in Concrete. In E. Oñate et al., eds. *XIII International Conference on Computational Plasticity. Fundamentals and Applications COMPLAS XIII*. pp. 346–353.

Segura, J.M. & Carol, I., 2004. On zero-thickness interface elements for diffusion problems. *International Journal for Numerical and Analytical Methods in Geomechanics*, 28(9), pp.947–962.

CHAPTER 4

Dams and Hydraulic Structures Applications. Remedial Works

Numerical Analysis of AAR Affected Structures with Slot-Cuts

Finite Element Analysis Using Explicit Scheme

VLADIMIR GOCEVSKI* — EMRE YILDIZ**

** Hydro-Quebec Equipment*
855, Sainte-Catherine Est, 10th Floor, Montreal (Quebec) H2L4P5, Canada
Gocevski.vladimir@hydro.qc.ca

*** IDAE s.e.n.c.*
204 rue Saint-Sacrement, Suite #300, Montreal (Quebec) H2Y 1W8, Canada
emre@idae.ca

ABSTRACT. *AAR expansion causes an accumulation of displacements and stresses within the structure. As a remedy, the slot-cutting of the concrete has been used on many structures to reduce the stress and to control the displacements. Although it is an efficient method to reduce accumulated stresses, Hydro-Quebec's experience shows that its effects on long-term behavior should be studied in detail before applying any cuts to the structure. The simulation techniques of slot-cuts and their effects on long-term behavior are discussed in this paper.*

KEYWORDS: *slot-cutting, finite element analysis, explicit analysis, long term effects of slot-cutting, AAR modelling.*

1. Introduction

The Alkali-Aggregate Reaction (AAR) creates an imposed strain on concrete. This strain translates to either displacements of the structure, if the structure is free to move or to stresses if there are restrictions preventing movements.

The displacements, caused by the AAR, may not be necessarily an issue for the structural integrity; however, they may prevent the equipment from functioning properly. Jamming of gates or unbalancing of rotors is common problems in AAR - affected structures. The maintenance of this equipment is required from time to time to increase the displacement tolerances, the straightness of supports and other required interventions to insure their functionality.

When the structure is restrained from moving freely, either by boundary conditions such as presence of rock, or by internal restrictions, such as the geometry of the structure, presence of prestressing or of reinforcement steel, the AAR expansion causes stresses.

For some structures, the geometry of the structure may not allow accumulation of the compressive stress, causing cracking in some areas. In such case, slot-cutting is often used by many owners as a prevention to the releasing of stresses. However, the behaviour of the AAR-affected structure before and after slot-cutting should be studied in detail before taking the decision on applying slot-cuts. Hydro-Quebec's experience shows that this technique may not always yield in a positive long-term behaviour.

2. Hydro-Quebec's Approach to AAR Modelling

Numerical modelling of AAR-affected structures has been a challenge for both academic research and engineering applications. The highly nonlinear behaviour of concrete, combined with the complexity of the AAR swelling – such as anisotropy, humidity and stress state dependence, makes modelling difficult. Many past and current engineering applications are considering the concrete as a linear or nonlinear elastic material subjected to simulated thermal loads that would represent the AAR expansion. While a good match of past displacements can be obtained using this method, it is of limited use for predicting the future behaviour as well as the cracking of the concrete. To overcome the limitations of the linear and nonlinear elastic modelling, much research has been done in recent years to implement both the nonlinearity of the concrete material and the complexity of the AAR expansion. These newly developed numerical models use many parameters based on the chemical properties of the AAR to estimate the expansion. The AAR swelling is then introduced as an implied strain load on the concrete model. This new trend in modelling ASR expansion through in-situ obtained chemical reaction parameters, combined with nonlinear plasticity based material modelling is a very important improvement to the study of the affected structures. It provides more information on the damage and guidance on the future behaviour of the structure. However, the calibration of the various parameters requires a lot of input data and matching procedure to observed displacements and is lengthy and difficult. Moreover, the simulation of nonlinear behaviour of concrete is a significant numerical challenge. Convergence issues arise as the cracking of concrete becomes more important and if ever the convergence is assured through stabilization methods, the iterative nature of the process limits the size of models. The nonlinear modelling of AAR of an average sized hydraulic structure may take many days for each run. By introducing some simplifications for reduction of a number of parameters, the complexity of the execution may drastically be reduced without affecting the accuracy of the results. With the objective of providing a robust nonlinear numerical modelling tool that doesn't require a large number of input parameters and that is easy to use with available data on concrete properties and on instrumentation, the Hydro-Quebec's first AAR model was developed in 1992. This implicit scheme was implemented as a user material within commercially available COSMOS/M finite element software. The earlier versions of the code were able to adequately simulate the nonlinear behaviour of the concrete as well as the AAR strains. Since then, the model is under continuous improvement. The latest versions of the code had most important AAR parameters implemented, such as the anisotropic swelling, anisotropic confinement, relative humidity dependence, time dependence of both ASR and concrete materials, etc. In 2012, the code was moved to Abaqus Explicit [Hibbitt 2016], commercially available finite elements software. The use of explicit analysis provides an significant advantage in dealing with large models with strongly nonlinear behaviour, such as AAR-affected hydraulic structures.

Explicit analyses are widely used in many fields of engineering where large nonlinearities occur in the structure. Metal forming and crush test simulations are examples of the strength of an explicit scheme. The explicit analysis does not require the formation of the stiffness matrix for the whole model and does not use an iterative solver. When the size of the model becomes significantly large, as it is the case of analyzing a hydraulic structure, the applicability of the implicit approach is limited by the available memory to store the large stiffness matrix. Furthermore, as the material, geometrical and contact nonlinearities arise, the iterations become more demanding and time-consuming, often eliminating the practicality of such analysis.

The application of explicit analysis to civil engineering structures for quasi-static loading cases requires attention. As the explicit analysis is dynamic by its nature, the inertial forces caused static loads should be monitored and reduced to ensure that the obtained results are equivalent to a static loading. Another important aspect of the explicit analysis is the critical time step increment required for a stable solution.

This increment is calculated based wave propagation speed within the elements and is usually done automatically by the software. Methods of proper use of the explicit analysis for quasi-static loading cases are described in detail by many references available in the literature [Hibbitt et al. 2016 and Ben Ftima 2014].

The mathematical formulation of the current model is explained elsewhere [Huang et al. 1996 and Pietruszczak 1996]. Recent developments in the mathematical formulation and validation process can also be found in the literature [Moallemi 2017].

The Hydro-Quebec's approach to AAR modelling involves many steps of data collecting, simulation and validation of results with in-situ observations. The following flow chart resumes the methodology developed over the years.

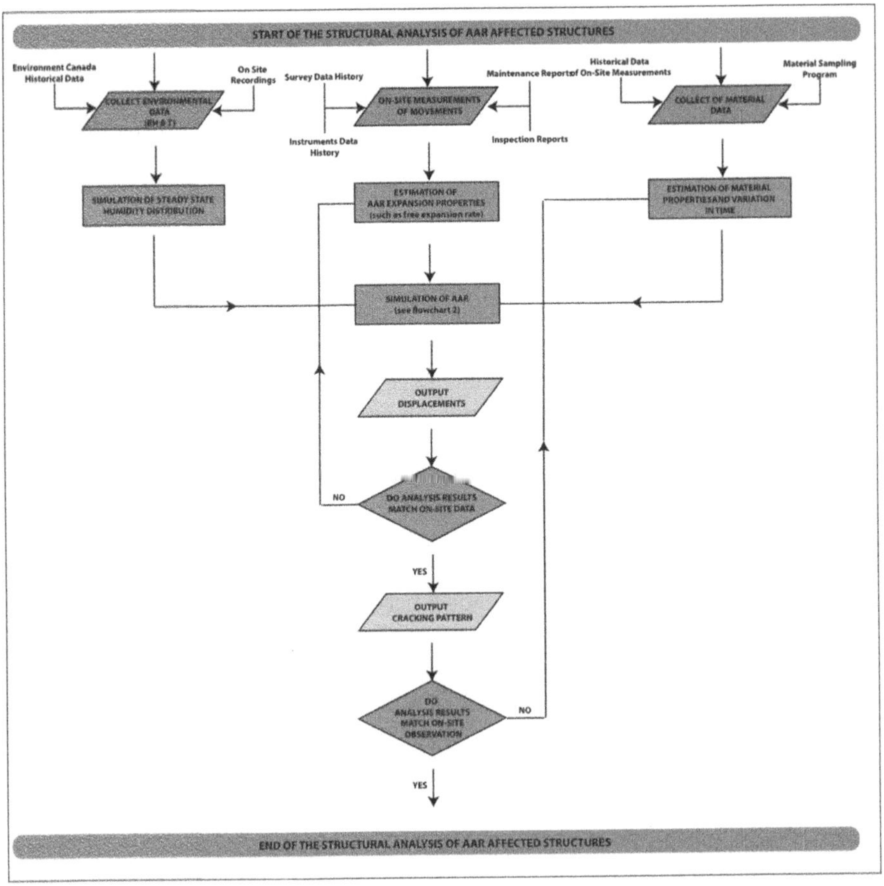

Figure 1. *Flow-Chart of the Hydro-Quebec's AAR Modeling Methodology*

3. Modelling of Slot-Cut in Explicit Analysis

The numerical studies of the long-term effects of slot-cutting require the simulation of the loss of volume of concrete within the model. This is done relatively easily in an implicit approach, as most of the advanced finite element software provide the possibility to change the model geometry during the analysis by removing elements. Since in the implicit approach, the inertial forces are absent from the analysis unless requested by the user, the sudden release of stresses do not introduce instabilities to the model as it would in the case of an explicit analysis. In case of the explicit analysis, a sudden removal of a boundary condition of a stressed material would cause an effect similar to an impact load, inducing large kinetic energy and vibrations to the structure, as shown in figure 2 for a simple case of a concrete bloc.

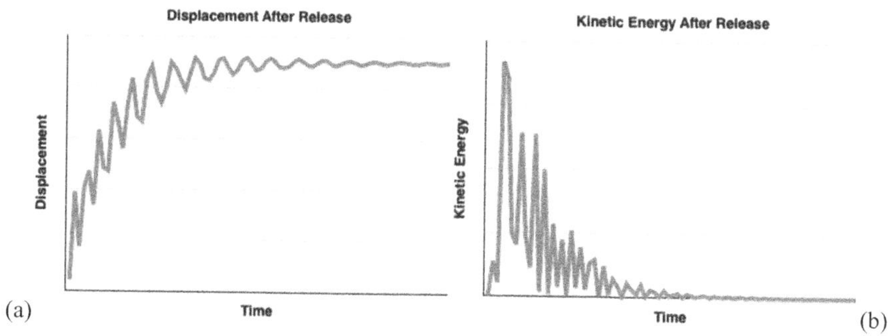

Figure 2. *Effect of sudden release of boundary condition in explicit analysis: a-Displacement after release and b – Kinetic Energy*

To prevent such dynamic effects, Hydro-Quebec has developed a new user material to simulate the slot-cutting in an explicit analysis. This material's stiffness is reduced gradually following a smooth step function as presented in figure 3, before actually removing the elements from the model. The use of smooth step function to reduce the dynamic effects of changes in loading is a common approach in explicit analysis. The duration of gradual reduction of stiffness is an important parameter to ensure that the dynamic effects are negligible. It should be noted by now that the simulation of slot-cutting introduces some artificial duration to the analysis. The time frame introduced by using the smooth step can be significant, comparing to the duration of the AAR. This is especially true if the structure would undergo multiple slot-cuttings over many years, as each cut would require an additional duration.

The AAR is a time-dependent behavior and Hydro-Quebec's AAR model has many parameters that vary with time, such as material properties, AAR rate, etc. It is, therefore, important not to include the artificial cutting time in the calculation of

AAR effects and time-dependent material properties. Hydro-Quebec's AAR model has the ability to interrupt the time dependency during the cutting periods. As an example, Figure 4 shows how the expansion is interrupted during the cutting.

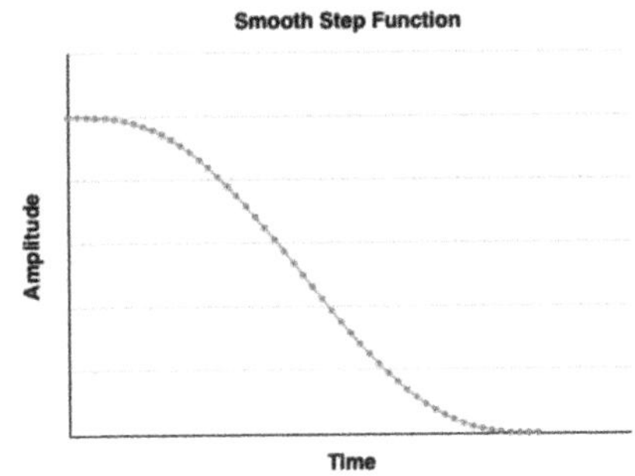

Figure 3. *Smooth Step function used for gradual reduction of stiffness before element removal*

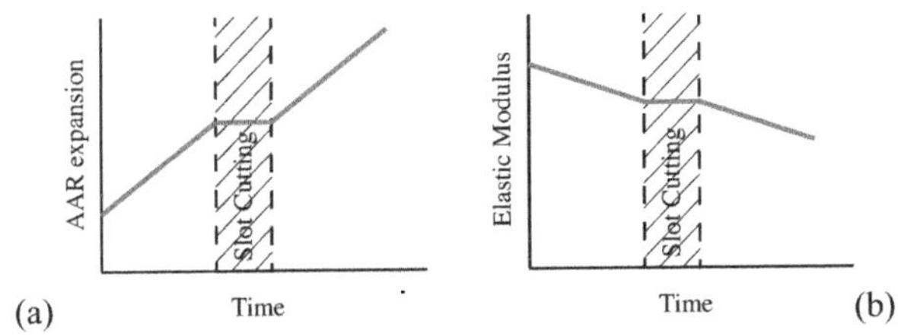

Figure 4. *Interruption of AAR's time-dependent properties for slot-cutting duration: a- AAR expansion and b- Elastic modulus degradation*

4. Validation of the Slot-Cutting Analysis

The validation process for the development of the user material starts with a simple model containing a concrete bloc where a slot-cut is applied at the center, as presented in figure 5.

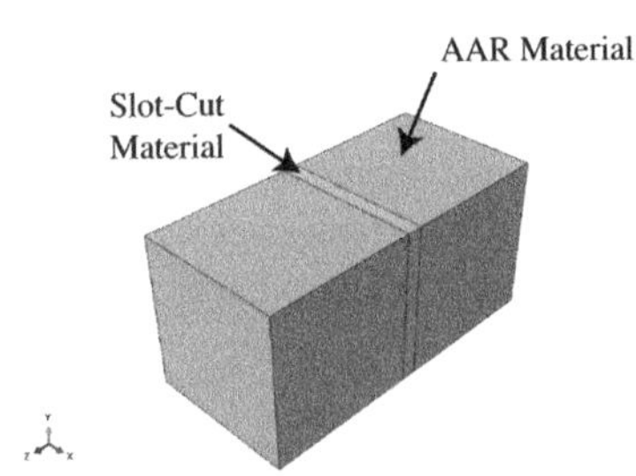

Figure 5. *Validation of slot-cutting material with a concrete bloc – Model*

The gradual reduction of stress in the slot-cutting material is shown in figure 6. Finally, the horizontal displacements at the edge of cuts during slot-cutting is presented in figure 7.

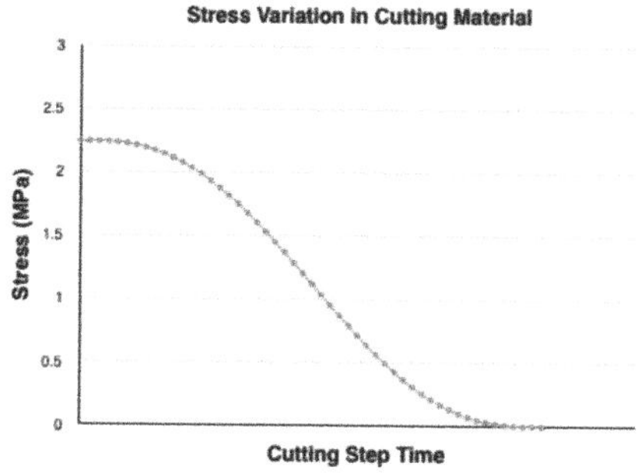

Figure 6. *Validation of slot-cutting material with a concrete bloc – Stress in slot-cut material*

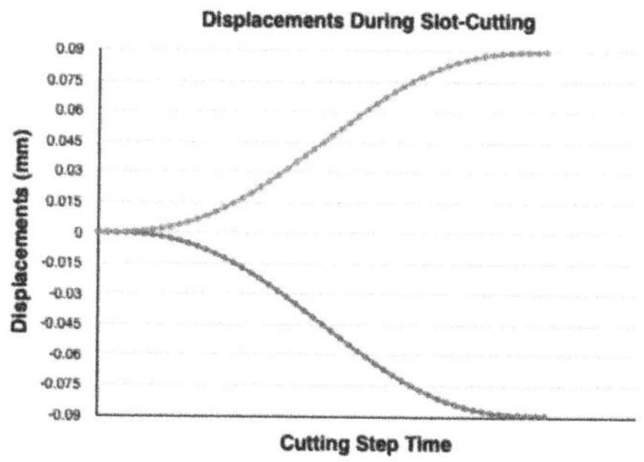

Figure 7. *Validation of horizontal displacements during slot-cutting with a concrete bloc*

The second validation is realized using the geometry of a typical intake structure containing 3 units. This theoretical application is used to qualitatively evaluate the behavior of the slot-cutting material as well as its effects on the rest of the structure. The geometry of the model is presented in figure 8.

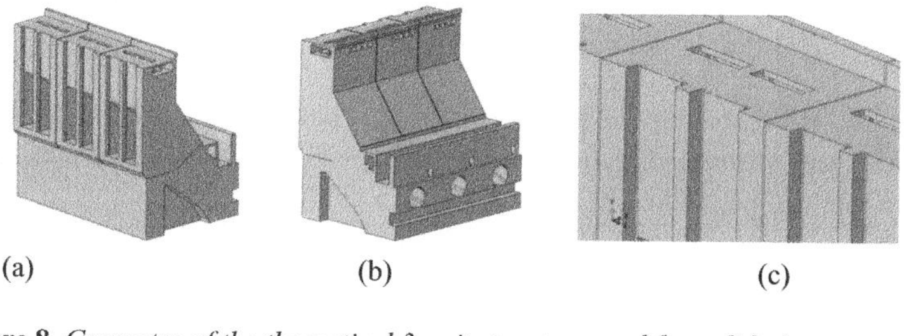

(a) (b) (c)

Figure 8. *Geometry of the theoretical 3-unit structure used for validation purposes: a- Upstream view; b- Downstream view; c- close-up on slot-cut zone*

Two slot-cuts is applied to the structure between each unit. The kinetic energy and horizontal displacements at the edge of slot-cutting after slot-cutting is presented in figure 9. It is noted that the structure exhibits a quasi-static behavior.

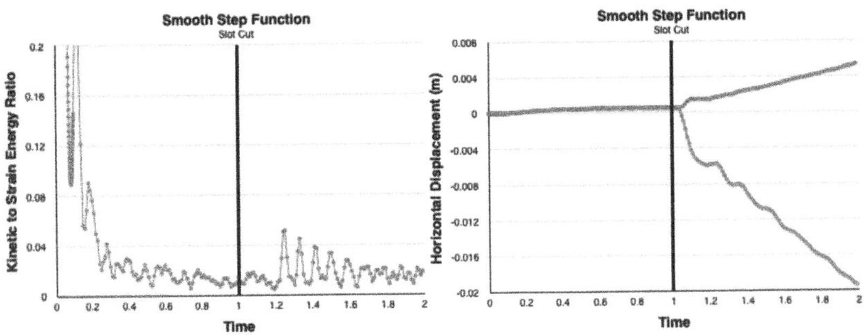

Figure 9. *Kinetic energy and horizontal displacement for the 3-unit model*

The AAR-affected concrete's behavior is then studied. The stresses in the longitudinal direction of the structure are shown in figure 10, before and after the cutting. It is noted that the slot-cutting is successfully removing the accumulated stress from the concrete.

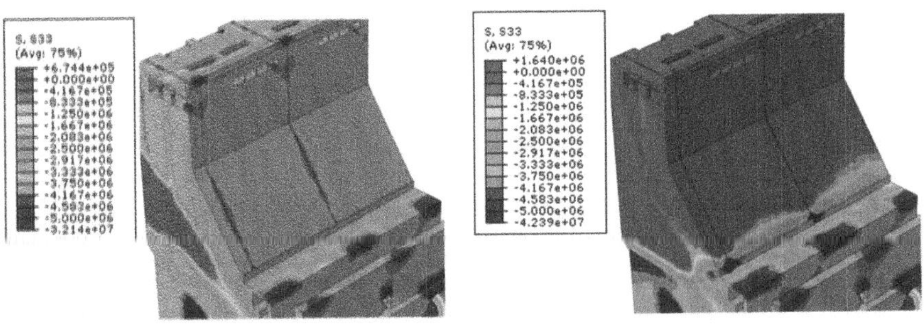

(a) (b)

Figure 10. *Stresses in the longitudinal direction for the 3-unit model – (a) before and (b) after slot-cutting*

Figure 11 presents the AAR expansion rate before and after slot-cutting, for longitudinal and vertical directions, respectively. The effect of slot-cutting on the expansion rate can be studied in detail with such analysis.

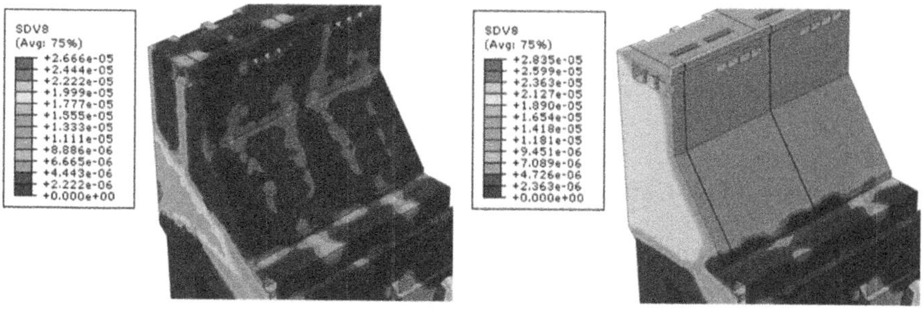

(a) before cuts after cuts

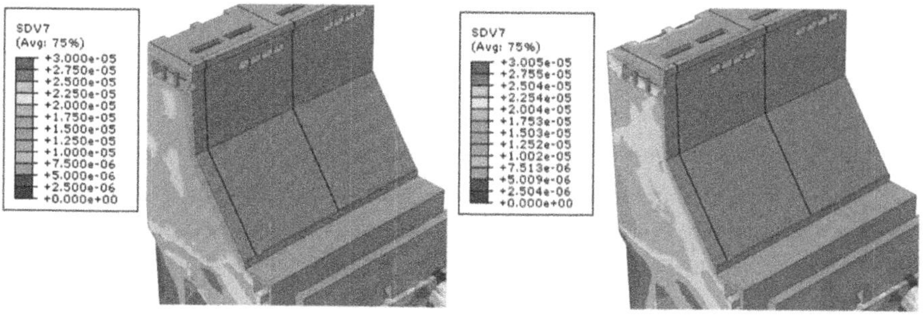

(b) before cuts after cuts

Figure 11. *AAR expansion rate variation for the 3-unit model – (a) longitudinal and (b) vertical direction. In the longitudinal direction, the average expansion rate increases 430% after the cuts. In the vertical direction, the increase is 35%*

The validation test presented above concludes that the slot-cutting user material developed by Hydro-Quebec for explicit analysis simulates appropriately the effects of such interventions. The AAR model of Hydro-Quebec helps to evaluate the effectiveness of slot-cutting while providing information on the expected displacements, stresses and cracking that may occur after the intervention.

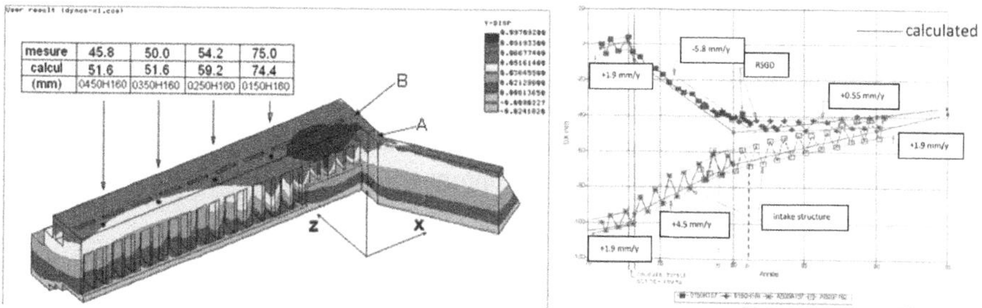

Figure 12. *An example of a real structure analysis with slot-cutting and validation with in-situ measurements*

Moreover, Hydro-Quebec has extensive experience on modelling slot-cutting on real structures. As an example, the structure presented in figure 12 is simulated. Good correlation with in-situ measurements is obtained.

5. Application of Slot-Cutting to structures

The concrete swelling of the AAR-affected hydroelectric power plant structures, is causing distortion of the turbine throat rings, misalignment of turbo generators. Deformation of wicket gates and stay wanes, obstruction of intake and spillway gates, closure of the expansion joints, concrete cracking and occasional infiltration are common problems observed in such structures. Thirty-six (36) Hydro-Quebec structures are affected by AAR. The rate of concrete swelling of these structures varies from moderate to high.

Common measures employed to remedy these problems can be classified as:

– major: the structural behavior is altered. (e.g., extensive post-tensioning to stabilize the cracked concrete masses, slot-cutting of the concrete to prevent displacements in unwanted direction and to release the accumulated concrete stresses)

– minor: the structural behaviour is not altered. (e.g., local corrective measures of the concrete around the affected equipment)

The slot-cutting as a remedial measure was practiced by Hydro-Quebec for more than 20 years between 1970 and 1992. The results of the elaborate studies, realized later starting 1990, of many of the structures that underwent slot-cutting, indicated not only that there were no benefits of slot-cutting but in many cases, the structures were damaged. Results of some of the early studies are discussed in Hydro-Quebec's internal reports and mentioned in some publications (Gocevski, 1995, CDSA/CANCOLD, 1997; Gocevski and Pietruszczak, 2000). Similar conclusions were obtained during the expertise realized by Hydro-Quebec on structures not owned by Hydro-Quebec.

A rational prediction of long-term effects of slot-cutting and justification of any other remedial work based on analytical results requires the use of specialized software with an adequate constitutive law and the time integration scheme. Decisions taken in the absence of such appropriate analysis almost always leads to costly and unnecessary or inefficient methods of maintenance.

As an example, a slot-cutting application on a Hydro-Quebec's structure is discussed here. The accelerated damage of the water passages of an intake structure and the sheared stay wanes of the units close to the slot-cut is shown in figure 12.

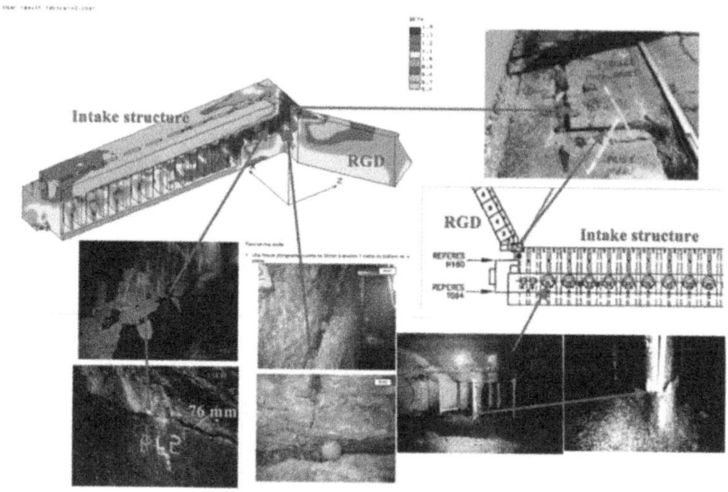

(a) (b)

Figure 13. *The location of the 1972-73 slot-cutting and the concrete damage as a consequence of this intervention: (a) the increased macro cracking in the water passage of group A of the intake structure and (b) sheared stay wanes of units A and B following the slot-cutting at the junction of the intake structure and the right Gravity Dam.*

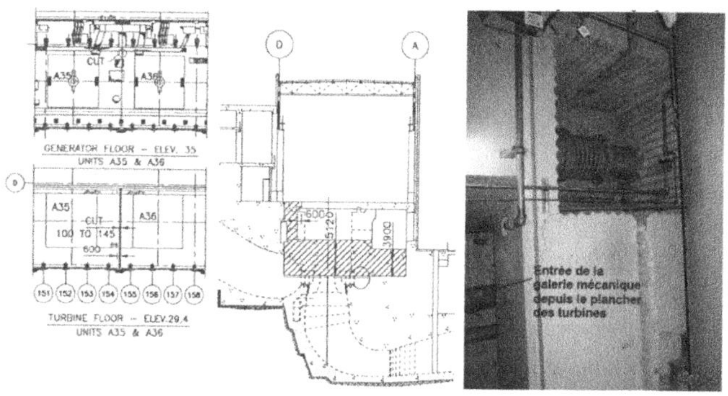

Figure 14. *Slot-Cutting – between the units 35 and 36 (1980–1981)*

The powerhouse presented in figure 13 was commissioned in 1960. The cracking due to the AAR concrete swelling before the slot-cutting in 1981 is shown in figure 15a and 15c. The significant increase in cracking five years after the slot-cutting is presented in figure 15b. The analytical results obtained for the same period had no slot cutting been performed is presented in figure 15d. The uneven expansion of unit 34 and especially the units 35 and 36 caused interruption of production of unit 36 for 3 years causing a production loss of 60 000$ a day. Gates of unit 37 remains until now stuck and they cannot be operated.

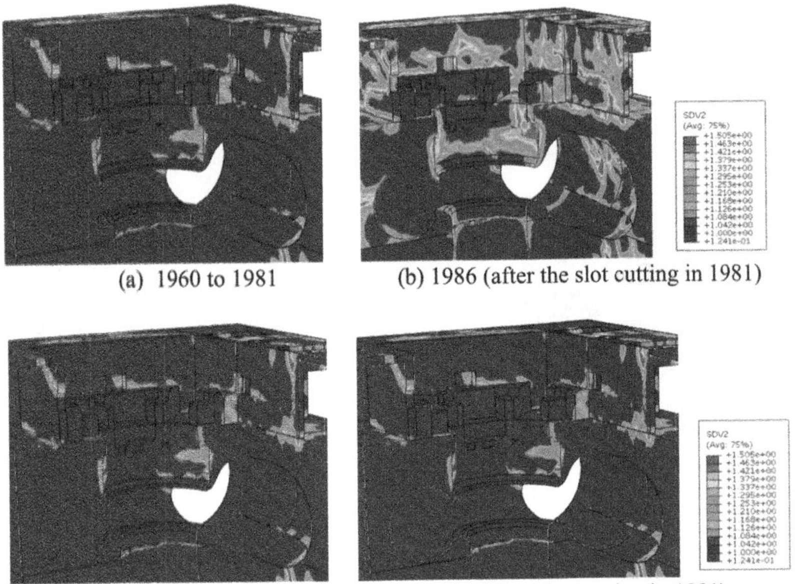

(a) 1960 to 1981 (b) 1986 (after the slot cutting in 1981)

(c) 1960 to 1981 (d)1986 (if no slot-cutting in 1981)

Figure 15. *Slot-Cutting – between the units 35 and 36 (1980–1981); development of extensive concrete cracking in unit 36 fife (5) years after the slot-cutting in 1981*

It is noticeable that the cracking observed in the concrete of Unit 36 for the first 21 years, before the slot-cuts, is relatively minor. There is only minor cracking around the spiral case. On the other hand, during and after slot cutting, the cracking expanded noticeably over a short period of time. The higher level of cracking is more noticeable in the concrete of the upper part of the structure, over the depth of the slot-cut between the units, especially around the spiral case.

From in-situ observations and numerical analysis results, it is reasonable to conclude that the considerable level of cracking is a consequence of these interventions.

6. Lessons Learned on the Effectiveness of Slot-Cutting

The lessons learned from the studies and the interventions undertaken in the past and the gained experience from these interventions from 1947 to 1992 lead to the following main recommendations:

– The in-situ observations and numerical models combined with the laboratory tests, can help the engineers to understand the long-term behavior of AAR affected structures. In many cases, the in-situ observations, instrumentation and survey data available for such structures may document how the structure behaves over a long period of time.

– In most cases, it is more appropriate to not interfere with the established structural behavior by using major interventions. Furthermore, the major interventions may damage the structure, making it more accessible to water, and even may cause stress concentrations. Such interventions should only be applied if justified by the appropriate numerical models and on-site observations.

Therefore, Hydro-Quebec favours the work that would consist of:

– interventions to accommodate the concrete swelling, allowing equipment's safety and functionality;

– strengthening or relaxation of deformed superstructures;

– pinning or anchoring the loose concrete blocks;

– slot-cutting of structural elements without modifying the structural behavior.

7. Conclusion

Hydro-Quebec has many years of experience dealing with AAR-affected structures. During these years, in collaboration with universities and industry, many tools have been developed to appropriately simulate the effects of AAR. The modelling technique used for slot-cuts was described in this paper.

The use of explicit analysis presents many advantages for dealing with large structures with nonlinear behavior. However, introducing the slot-cuts in an explicit analysis requires the development of a special method. Hydro-Quebec's development described in this paper proved to be efficient and accurate. It has been used in the evaluation of the real structures where the results were compared to instrumentation data. Very good correlation was found between the results obtained from numerical analysis to those acquired from the in-situ observations.

Adequate numerical analysis in general and the introduction of modeling slot-cutting using the procedure explained in this article will prevent acceptance of non-adequate refurbishing interventions.

8. References

Hibbitt, H.D., Karlson, B.I. and Sorensen, E.P., *ABAQUS version 2016, finite element program*, Providence, R.I., USA, 2016.

Ben Ftima, M., "Utilisation de la méthode des éléments finis non-linéaires pour la conception des structures en béton armé: application aux structures massives", PhD thesis, École Polytechnique de Montréal, Canada, 2014.

Gocevksi, V., "Monitoring, testing and remedial work at Beauharnois power plant", *Proc. 2nd Int. Conf. On AAR in hydroelectric plants and dams*, Chattanooga, USA, 1995.

Huang M. and Pietruszczak, S., "Numerical analysis of concrete subjected to alkali-aggregate reaction", *Mechanics of cohesive-frictional materials*, Vol 1, pp. 305-3019, 1996.

Pietruszczak, S., "On the mechanical behaviour of concrete subjected to alkali-aggregate reaction", *Int. J. Computers & Structures*, Vol. 58, pp. 1093-1099, 1996.

Gocevski, V., and Biner, A., "Swelling of Concrete Dams", *Proc. CDSA/CANCOLD*, Montreal, Quebec, Canada, 1997.

Gocevski, V. Pietruszczak, S., "The effects of slot-cutting in concrete dams affected by alkali-aggregate reaction", *11th Intern. Symp. on Alkali-Aggregate Reaction*, Quebec, 2000.

Moallemi, S. Pietruszczak, S., "Analysis of localized fracture in 3D reinforced concrete structures using volume averaging technique", *Finite Element Analysis and Design*, Vol 125, pp. 41-52, 2016.

Swelling Arch Dams with Thrust Blocks

Grégory Coubard* — Jérôme Sausse**

* *EDF/CIH*
Savoie - Technolac. 73370 Le Bourget du Lac, France
gregory.coubard@edf.fr

** *EDF/DTG*
18 Avenue Poincaré, 19100 Brive, France
jerome.sausse@edf.fr

ABSTRACT. *The safety reassessment of arch dams with thrust blocks subject to swelling phenomena raises the issue of the stability of these thrust blocks. Classical analysis methods conduct to low stability coefficients that could result in expensive and unnecessary reinforcement works. EDF feedback on a quite important hydraulic park is the continuation of investigations with the help of monitoring, site exploration and development of non-linear model able to estimate irreversible displacements that could occur at the concrete-to-rock interface. This article describes in a more detailed way these different items.*

KEYWORDS: *arch dams, thrust blocks, swelling, monitoring, stability analysis, non-linear model.*

1. Introduction

Whether for geological or topographical reasons, an arch dam site is rarely perfectly symmetrical. This can lead to the designers implementing particular construction specifications in order to retrieve this symmetry. Thus, thrust blocks are very useful for valleys with flat abutments on the upper levels, this disposition permits a significant increase in storage capacity.

A good definition of a thrust block is given by the Engineer Manual 1110-2-2201, (USACE):

> Thrust Blocks. Thrust blocks are another type of restitution concrete. These components are constructed of mass concrete on foundation rock and form an extension of the arch dam crest. They are particularly useful in sites with steep side slopes extending about three-fourths the distance to the top and then rapidly flattening. In such a site, significantly additional water storage can be achieved by thrust blocks without a proportional increase in costs. For small and short extensions beyond the neat line, the cross-sectional shape can simply be a continuation of arch dam geometry; thus, for some distance past the neat line, arch action will resist some of the applied load. Beyond that distance, cantilever action resists the water load and must be stable as with a gravity dam. The extension may be a straight tangent or curved as dictated by the topography

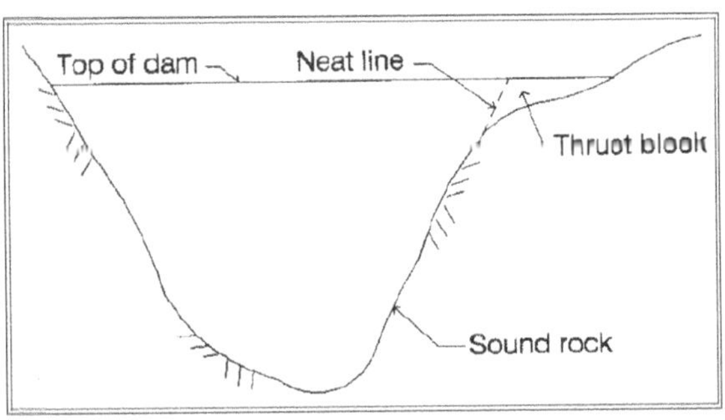

Figure 1. *Schematic Elevation of a simple thrust block (USACE)*

As an initial approach, the stability of a thrust block is assessed in a similar manner to a gravity dam (particularly concerning the sliding stability). Depending on its design, the principal loadings applied to the thrust block are self weight, upstream hydrostatic pressure, uplift pressure and of course the thrust resulting from the arch action of the dam.

In the case of arch dams affected by swelling reactions, these efforts increase with time due to the arc elongation, soliciting more and more the thrust blocks. Thus, the stability coefficient of these thrust blocks is reduced, especially considering additional thermal loads (high temperatures).

EDF is concerned by this problematic: of the 55 arch dams of 20 m height or more operated by EDF, 15 exhibit swelling phenomena and 10 have been identified as arch dams with thrust blocks. These thrust blocks have variable shapes and in particular variable heights: between 20% and 60% of the maximal arch dam height (cf. figure 2).

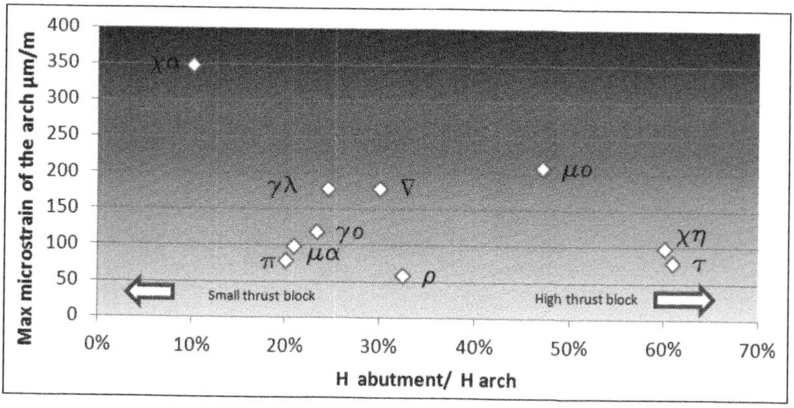

Figure 2. *Horizontal arch strains in function of thrust block heights, for 10 studied dams*

The aim of this article is to share EDF's feedback concerning the monitoring, site investigations and stability studies.

2. Initial Conception and Issues Concerning Arch Dams with Thrust Blocks

2.1. *Initial Conception*

Firstly, it is interesting to consider the design of the structures and in particular the manner that designers have historically studied thrust block stability.

The analysis of the initial conception of the 10 EDF arch dams with thrust blocks affected by a swelling phenomenon indicates that:

– Uplift pressures were rarely taken into account (only 1 case). For 4 cases, this was justified by specific constructive dispositions: "L" shaped thrust blocks with an upstream wall receiving the hydrostatic pressure and a downstream part receiving the thrust from the arch.

– The thrust resulting from the arch action is calculated with an independent arch method. In most cases, only the hydrostatic component was taken into account, the thermal effects have not been retained (only 1 case considers these effects).

– In a conservative manner, thrust block self weight has sometimes been neglected (2 cases).

– One of the thrust blocks was designed with active anchors in order to reduce the mass of concrete.

– In one case, stability analysis was not undertaken at all.

– In two cases, designers incorporated a lateral abutment of the thrust block but without considering it in the stability calculation.

Figure 3. *Example of an "L" thrust block*

2.2. *Reassessment of Thrust Block Stability*

For arch dams affected by a swelling phenomenon, numerical models (finite element models) are implemented in order to update behaviour studies including irreversible evolutions.

These models take into account the geometry of the arch but also of the thrust blocks. The analysis of monitoring measurements permits the calibration of the numerical model under the different loads: self weight, hydrostatic pressure, thermal and irreversible (concrete swelling).

The resultant forces, including thrust transmitted by the arch but also self weight of the thrust block and loadings directly applied on it, are calculated with the finite element model. Depending on the thrust block height and its configuration, the action of uplift pressure can be integrated to the resultant forces obtained. Therefore, a classical stability calculation of the thrust block can be performed. The most critical case for thrust block stability corresponds to a high water level, in summer (high temperatures), cumulated with concrete swelling.

The sliding stability factor "Fs" is classically calculated as:

$$Fs = \frac{c \cdot A + N \cdot \tan(\varphi)}{T}$$

with c, the cohesion of concrete-to-rock contact; φ, the friction angle; A, the surface area of concrete-to-rock contact; N and T the normal and tangent resultant forces given by the FEM.

This methodology of calculation often concludes to low stability factors, due to different conservatisms:

– The mechanical parameters taken into account are generally low for an initial approach (theoretical values being: cohesion neglected and friction angle of 45°),

– Thrust forces due to concrete swelling reaction is overestimated (simplified swelling model with thermal analogy, creep effects are neglected or underestimated),

– In the absence of precise data, the lateral abutment of the thrust block in the foundation is neglected,

– A conservative profile for uplift pressure is often considered.

This stability reassessment approach for thrust blocks is the first analysis step, at the end of which the engineer has to question the plausibility of the obtained results and the incidence on arch dam global behaviour.

Currently, in this simple initial approach, the thrust block is considered as an independent structure on which external forces are applied including the thrust of the arch. However, in reality the intensity of this thrust depends on the thrust block behaviour and more globally on the interaction between the arch and the thrust block. This observation has already been made by different authors and in particular by Yussof Ghanaat [GHANAAT, 2016]:

> Abutment thrust blocks and kinematically moveable rock wedges in the foundation and abutments of an arch dam are two features of potential stability concern. Stability analysis of these features is commonly evaluated separately from the dam by applying dam thrusts additional external loads.
>
> A thrust block supported by the foundation rock with inadequate shear strength could be susceptible to sliding along the foundation contact [...]. Any movements of an abutment thrust block could significantly affect stress distributions with the dam, thus affecting the

magnitude and orientation of dam thrusts exerted of the thrust block. This dynamic interaction, which may also involve opening and closing of the joint between the dam and thrust block, is nonlinear in nature and is best handled using a coupled nonlinear analysis that include both the dam and thrust block in the same model

2.3. *Interaction Between Arch and Thrust Block*

In a second approach, the thrust block is considered as a part of a global system composed of the arch and the thrust block. Non-linear local phenomena, such as joint opening or moderate sliding along the concrete-to-rock contact, are modelled and lead to a slight modification of the stresses, resultant intensities and orientations.

A good example of this non-linear behaviour is illustrated by the study of non-swelling arch dams in a winter situation. Under this winter condition, a linear model (precedent analysis) can conclude in a low stability coefficient for the thrust blocks (cf. figure 4) with the resultant forces oriented in the valley direction: this result does not seem realistic. More realistic results can be obtained via the introduction of a more physical behaviour to the contraction joints, which allows the relaxation of tensile stresses through a non-linear model. This shows that the lateral blocks are not under high loading for this situation (reorientation of resultant forces down, cf. figure 4).

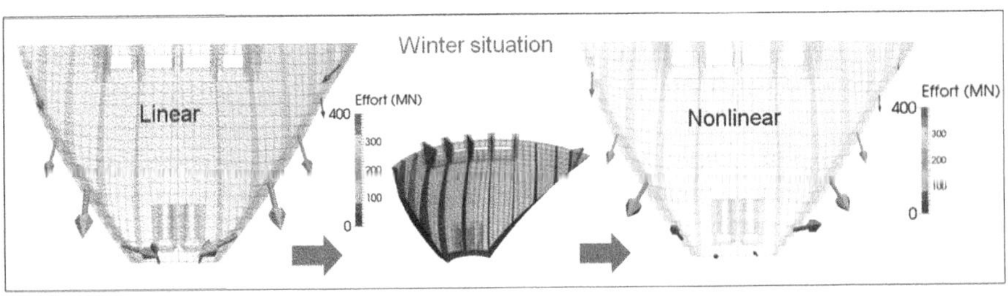

Figure 4. *Linear model and non-linear model with contraction joints in winter situation*

In a similar manner, a slight sliding of a thrust block of a swelling arch dam in a summer condition can lead to a new equilibrium state of the entire system of arch and thrust block. This approach has been implemented several times by EDF in non-linear models, allowing the displacement of the thrust block on its foundation with a constitutive Mohr-Coulomb law. This leads to irreversible displacements not exceeding a few millimeters inducing reorientation and a diminution of stress intensity.

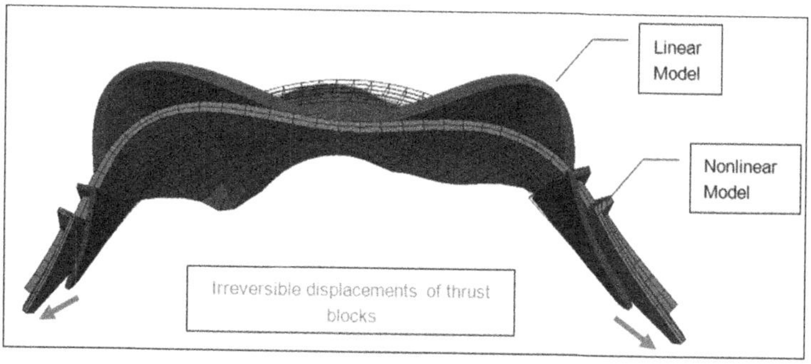

Figure 5. *Linear model and non-linear model with concrete-to-rock joint on a swelling arch dam with thrust blocks*

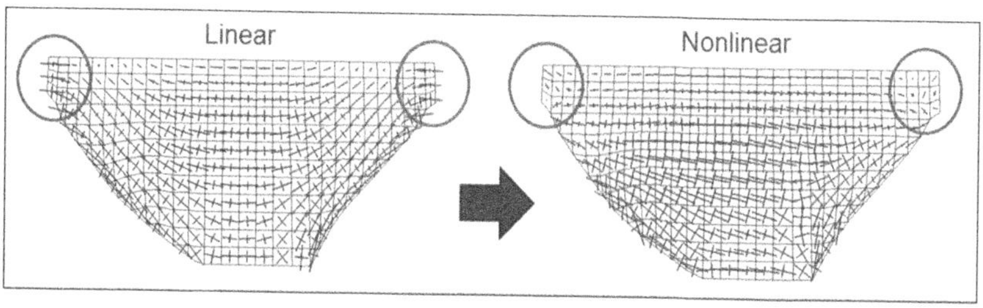

Figure 6. *Linear to non-linear model with concrete-to-rock joint. Stress reorientation and modification on the upstream face*

Therefore, the engineer has to analyse obtained results and decide if the displacements are allowable, referring also to monitoring measurements. One of the difficulties is the capability to measure these small adaptation displacements, cf. § 3.1.

It is interesting to note that the use of non-linear models with joint-elements at concrete-to-rock contact is already proposed by FERC [FERC, 1999]:

> Shear stresses in excess of the shear capacity of the concrete/foundation interface can also be of concern. Massive shear failures [...] are rare, however local shear over stress is more common. [...] If stress can be relieved by re-distribution without overstressing other areas, local shear failure can be tolerated. Elements may have to be allowed to slide by reducing their elastic modulus or disconnecting them. Many FE programs also have non-linear gap friction elements which can be used.

The fact that small irreversible displacements (millimetric) are sufficient to relax forces transmitted to thrust blocks can be explained on one hand by the hyperstatic specificity of arch dams, and on the other hand by the nature of transmitted forces.

Actually, among the forces transmitted to the thrust block, a distinction has to be made:

– Forces directly applied on the thrust block (hydrostatic pressure, self weight, uplift pressure) with no modification in case of thrust block displacement.

– Hydrostatic forces coming from the arch action. These forces can be slightly decreased in case of thrust block displacement.

– Thermal or swelling resulting forces. These forces come from an initial imposed strain and can be totally annulled in case of thrust block displacement.

3. EDF Feedback

3.1. *Monitoring*

The issue of arch dam thrust block stability under swelling phenomena has conducted EDF to better monitor these structures. Thrust block monitoring should permit better understanding of their behaviour and interaction with the arch, in order to validate or adjust numerical models. For few years, a certain number of pendulums have been installed on the thrust blocks of swelling arch dams. Currently, it is difficult to assess the reality of the very low adaptation displacements obtained by non-linear models (cf. § 2.3). The measurements made on recent pendulums do not benefit from feedback over a sufficient length of time. Also, initial monitoring devices (topography when it has been implemented) are not always precise enough (e.g. about 1 mm over a 30 year period of swelling). Despite this, recent monitoring devices already allow the appreciation of the thermal behaviour of arch dam thrust blocks, which can help to analyse swelling behaviour.

The interaction between arch, thrust block and foundation can be very different from one example to another. The following examples show two main situations in terms of thermal displacements for a seasonal heating load (cf. figure 7):

– irreversible displacement oriented towards the bank and downstream;

– irreversible displacement oriented, in the opposite direction, parallel to the valley and upstream.

This illustrates, on the one hand, the relative flexibility of the thrust block, which allows low reversible displacements under thermal loading. Thus, this justifies the possibility for the structure to be adaptable under irreversible loading. On the other hand, it gives information on the interaction between the thrust block and the arch and more particularly on the thermal displacements that are not only imposed by the arch's thermal expansion (displacements oriented towards the bank) but also by the expansion of the thrust block itself (displacements oriented parallel to the valley). Finally, it is important to keep in mind that the pendulum implantation zone (downstream face or in the center, near the bank or near the arch) and their anchoring zone (in the foundation or on the thrust block) are significant for observed displacements.

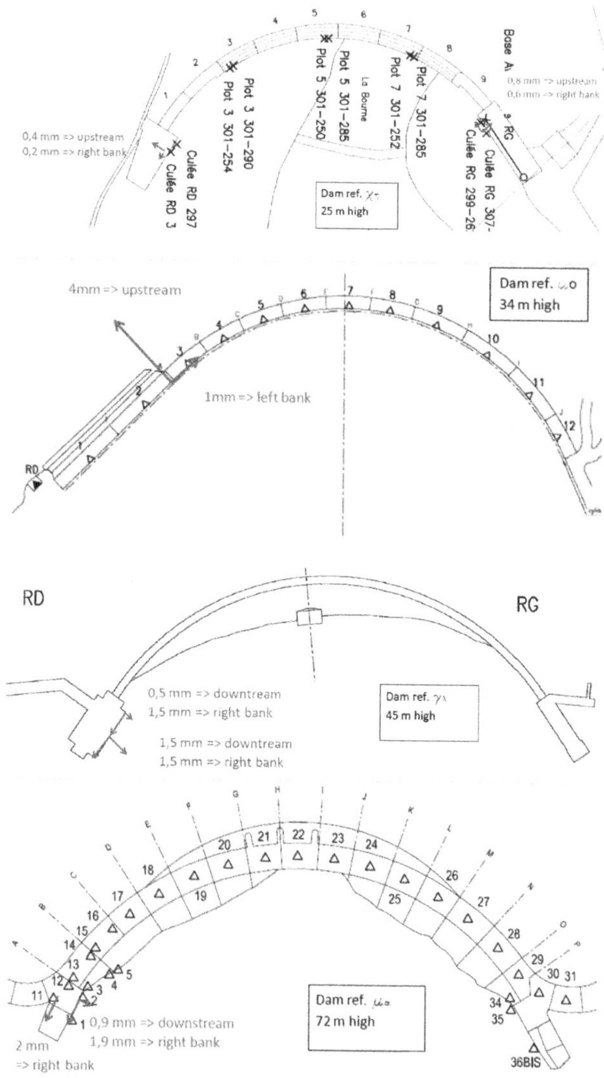

Figure 7. *Seasonal reversible displacements (seasonal heating) measured on thrust blocks of 4 arch dams*

The joint-meters implemented at the interface between arch and thrust blocks do not show significant displacements (neither reversible nor irreversible opening). The concrete swelling can only close low residual openings that may exist at the joint, just after joint grouting. Thrust block monitoring also includes vertical displacement measurements (vertical pendulums or levelling) which allow the appraisal of local swelling rates.

Piezometers implanted at the concrete-to-rock interface, just beneath the thrust blocks, allow the assessment of the uplift pressure hypothesis taken into account in the stability calculations and in particular, the zero value at the downstream end considered in case of "L" thrust blocks, cf. §. 2.1. In the case of active anchors, the implementation of load cells allows the measurement of the residual tensions, which have a tendency to increase under concrete swelling effects.

Finally, all these monitoring devices also allow the close surveillance of all potentially critical situations such as a hot summer (heat wave) with a full reservoir. It is important to keep in mind that this configuration, which creates the highest loading of the thrust blocks, does not correspond to extreme global displacements (hydrostatic pressure involving a downstream displacement of the arch dam/heating and swelling involving upstream displacement).

3.2. *Site Investigations*

After an initial analysis of the swelling arch dam behaviour and more particularly of the thrust block stability, site investigations can be useful to reduce some conservatisms and uncertainties.

Coring undertaken from the thrust block crest to the foundation, associated with borehole imaging, allows the:

– acquirement of information on the concrete quality and characterize its swelling phenomenon;

– retrieval of intact concrete-to-rock core samples in order to test their shear strength;

– estimation, with the help of borehole imaging, the quality of the bonding between concrete and rock;

– characterisation of the rock foundation and its joints;

– confirmation that the geometry of the concrete-to-rock interface is compliant with drawings;

– definition of bank-to-bank or upstream-downstream abutments of the thrust blocks, where they exist.

In-situ stress tests can also be implemented in order to compare stresses obtained by finite element model analysis with real in-situ stresses. This allows the estimation of the intensity of forces transmitted to the thrust block.

The most classical devices used to measure these stresses are flat jacks. Nevertheless, these flat jacks measure the stress on the structure skin and EDF's experience shows that results can be misleading or incorrect, due to the swelling reaction being more active in the heart of the structure than on the skin. Paradoxically, significant tensile stresses have even been measured on the skin of structures affected by swelling phenomena!

Other devices permit the measurement of internal stresses but are not proven and do not currently allow representative feedback.

3.3. *Step-by-Step Calculation Method Applied by EDF*

EDF feedback on swelling arch dams with thrust blocks call for a step-by-step methodology permitting an optimised safety reassessment. This methodology of calculation is made up of the following steps:

– Analysis of the initial design and associated loadings.

– Linear finite-element model taking into account the arch and the thrust block. Simulation of concrete swelling by thermal analogy in a primary approach, taking into account a slight concrete creep due to swelling itself.

– Classical stability analysis of the thrust block with resultant forces determined using the linear model and conservative parameters (low shear strength, no lateral abutment).

– If a low stability coefficient is obtained, implementation of a non-linear model with joint-elements at the concrete-to-rock interface with a Mohr-Coulomb law. Evaluation of the irreversible displacements obtained.

– Comparison of these irreversible displacements with monitoring measures.

– If necessary, monitoring device reinforcement of thrust block.

– If necessary, realisation of site investigations in order to estimate the effectiveness of lateral abutment, shear strength and stress levels.

– If necessary, development of a more sophisticated non-linear model able to describe the behaviour of swelling concrete.

This step-by-step method enables a conclusion concerning the global behaviour of the entire arch and thrust block system, with the possibility to make forecasts introducing long term swelling effects.

4. Conclusions

Behaviour analyses of arch dams rely on monitoring, site investigations and numerical modelling. For arch dams with thrust blocks submitted to swelling phenomena, the sum of actions highlights an increase in thrust block loading with time. Classical stability analyses can lead to conclude that sliding stability coefficients are low. This primary analysis can be seen as conservative, mostly because it neglects the non-linear phenomena that occur within hyperstatic structures such as arch dams. Indeed, non-linear models demonstrate that millimetric displacements of thrust blocks are sufficient to reach a new acceptable equilibrium state.

An adapted reinforcement of monitoring devices and specific site investigations permit a better understanding of the global behaviour of arches and thrust blocks, feeding the finite element models. All these tools are fundamental to the better assessment of this specific structural behaviour. The holistic step-by-step methodology allows adapted surveillance of the structure, and where necessary, optimised remedial works to be proposed.

5. Acknowledgements

Philippe Bourgey, Eric Bourdarot who worked on the subject between 2008 and 2010. Damien Ghenassia, who collected precious information during his internship. Samuel Bale, who gave a precious help for the English redaction of this article.

6. References

Bourdarot, E. (2016). Discussion about arch-dam failure mechanisms. Informations provided by accidents and incidents analysis. Chambery: Colloque CFBR: “Sûreté des barrages et enjeux”, 23–24 novembre 2016.

U.S. Army Corps of Engineers (1994). *Engineering and Design – Arch Dam Design*. EM 1110-2-2201, 31 May 1994.

Ghanaat, Y. (2016). Nonlinear seismic evaluation and acceptance criteria for arch dams. USSD 2016 annual conference, 2016.

Federal Energy Regulatory Commission (FERC) Division of Dam Safety and Inspections (1999). *Engineering Guidelines for the Evaluation of Hydropower Projects, Chapter 11 - Arch Dams*. 1999.

Bimont Dam Case

Studies and Investigations Inside the Dam Body

CHRISTINE NORET* — KATIA LALICHE**

** Tractebel Engineering*
5, rue du 19 mars 1962, 92622 Gennevilliers Cedex, France
christine.noret@tractebel.engie.com

*** Société du Canal de Provence et d'Aménagement de la Région Provençale*
Le Tholonet – CS 70064, 13182 Aix-en-Provence Cedex 5, France
katia.laliche@canal-de-provence.com

ABSTRACT. *The Bimont dam developed a network of cracks in some cantilevers soon after construction. Their existence was initially put down to geology, but subsequent investigations showed that specific areas of concrete were affected by Delayed Ettringite Formation. This phenomenon brought about changes in the dam's equilibrium, resulting in the formation of superficial and internal cracks. A numerical model of the dam with elastoplastic features ran in parallel with two special investigation campaigns. These measures allowed for a more in-depth understanding of how the network of cracks was formed, its spatial extent, and its probable future evolution, information which proved invaluable for the design of the dam renovation programme. The project involves mainly the treatment by cement grouting of the cracks and joints and the waterproofing of the dam upstream face.*

KEYWORDS: *arch dam, concrete, cracking, investigation, rehabilitation, swelling.*

1. The Bimont Arch Dam: Description and History of Dam Performance

Located near Aix-en-Provence in the south of France, the Bimont Dam was first brought into service in 1952, and became a part of the *Société du Canal de Provence* concession in 1963. It ensures the water supply to the cities of Aix-en-Provence and Marseille, as well as answering the needs of industry and agriculture in the region. Given its large capacity, the dam also serves to reduce the impact of flooding. Moreover, it features two hydroelectric micro power stations, and if necessary, the upstream turbine can be operated in pumping mode to restock the upstream network.

The Bimont dam is a concrete, double-curved, arch-type structure measuring 86.50 m in height and 180 m in length (along the crest). It consists of 15 cantilevers between two abutments, and its thickness ranges from 4 m at its crest to 13 m at its foot. The reservoir behind the dam is 25 hm^3 under the normal operational water level of 341 m NGF (10 m under the crest). Since 1982, the water level in the reservoir has been kept 10 m lower.

On the occasion of the geological study of dam supports, carried out following the Malpasset disaster (end of 1959), cracks were discovered on the lower section of the upstream face of cantilever 2, on the right bank. In the following years, a network of cracks gradually appeared at the base of the cantilevers on the right ban. Several groups of cracks developed in horizontal, vertical and diagonal directions, including a clearly visible one on the downstream face in front of cantilever 3, which is now 13 mm wide. The development of these cracks was accompanied by irreversible movements on the right bank of the dam. The cantilevers moved upwards and tilted towards the upstream side and left bank of the crest, and some construction joints opened (Fig. 1).

Figure 1. *Bimont Arch Dam*

Various investigations were conducted in order to assess and understand changes in dam performance, and tests performed on concrete samples concluded that the concrete was sufficiently strong and of good quality.

Different monitoring techniques were used: topographical surveys, crackmeters and plumblines were installed to map the cracks network and monitor its evolution.

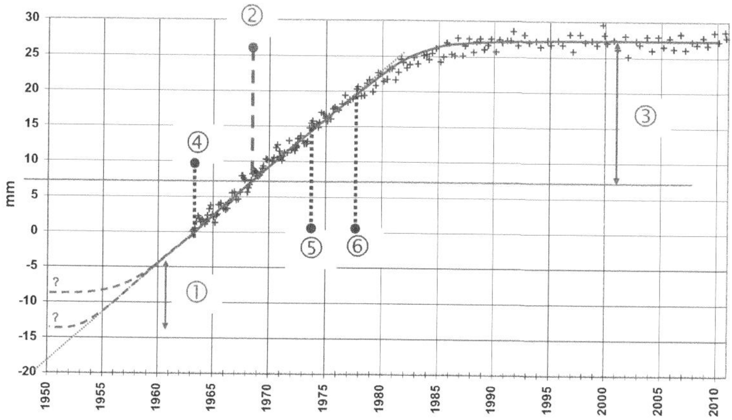

Figure 2. *Cantilever 2 crest levelling showing the swelling evolution (1: Unknown previous swelling; 2: Joint B re-grouting – March 1969; 3: Swelling after March 1969; 4: Beginning of levelling survey; 5: P2 crest plumbline installation; 6: P7 crest plumbline installation)*

Figure 3. *Drilling and swelling zones location (in white)*

The series of studies performed on the structure resulted in a rejection of the hypothesis that the cracks were caused by geology.

In fact, a comparison of a picture taken during the dam's construction and the precise location of the damage helped to establish that the cracks were the result of defects in the cement supplied at a certain point in the construction period.

2. Investigations into Concrete Swelling

2.1. *Early Investigations on Concrete*

2.1.1. *Laboratory Tests on Concrete (1965–1979)*

For a period of fifteen years between 1965 and 1979, a number of different laboratory tests were performed on concrete samples from cantilevers 2 and 14. These included mineralogical, mechanical and chemical characterisation tests, as well as seismic and sound-velocity readings. The tests were performed at different points in time following extraction of the samples and thus at different stages of deterioration. Some were subjected to thermal or wetting/drying cycles.

These tests showed relatively significant concrete swelling in cantilever 2 (2.10^{-3} mm), while swelling in plot 14 reached only half this level. A clay vein was identified in the foundation rock for cantilever 2, making it all the more susceptible to swelling.

The tests also showed that the pressure exerted by the dam's weight was not sufficient to contain the effects of swelling, and that the test samples suffered from mechanical deterioration over time.

The samples analysed demonstrated general carbonation, and at the time it was thought that this may have played a role by first accelerating the shrinkage of fresh concrete, then gradually blocking the play of the cracks through crystallisation.

For the first time, the lab teams studied the possibility of alkali reaction affecting calcareous aggregates on which magnesia and clay were dispersed. However, no alkali presence could be detected in the binding agent.

Although it demonstrates the shortcomings of studies based on laboratory sample tests, an order of magnitude for the maximum foreseeable relative expansion was given: $1.2.10^{-3}$ mm/m, which translates now as 48 vertical millimeters for cantilever 2. This later proved to be correct, as shown in the graph in Fig. 1. Furthermore, it was predicted that the movements would stop at the beginning of the 1990s.

2.1.2. *New Laboratory Tests (2003)*

The prediction turned out to be correct, and by the time these tests were performed on samples of swelling concrete from cantilever 2, the concrete was and had been stable since the beginning of the 1990s.

Apart from a superficial, partially-carbonate fringe of between 5 and 10 mm, tests revealed a largely compact micro-structure with normal sulphate content. The capillary absorption coefficient is weak, which is typical of high-performance concrete.

These tests point to relatively large quantities of massive ettringite with a tendency to expand; consequently, the cause of concrete swelling in the Bimont dam was attributed to the presence of massive ettringite.

The dimensional stability of the concrete samples was also analysed by way of a test designed to measure residual swelling in concretes affected by alkali-aggregate-reaction phenomena. The samples were kept for one year at a temperature of 38°C and with 100% of relative humidity. The results showed that residual expansion was less than 100 mm/m, and therefore negligible.

2.2. *Preliminary Studies of the Renovation Project*

2.2.1. *Two Borehole Campaigns (2009, 2012)*

These campaigns were conducted in 2009 and 2012, and their primary objective was to examine the area of contact between the dam and its foundation under the downstream base of the cantilevers on the right bank, which were likely to have been loosened by the dam's tilt (observed and calculated) towards the upstream.

Although the evolution of the cracks observed on the faces had been monitored for several years, the thickness and variations in depth of these openings was not well known, and there was no solid evidence of internal cracking.

During the first surveys, piezometers were installed in the foundation in order to validate the uplift introduced into the numerical model.

The objective of the second surveys was to obtain more detailed information about the expansion of the internal crack which was identified during the first surveys. Holes were carefully drilled around each swelling nucleus, based on the results of the numerical model, and the investigation programme was continuously adapted according to the results obtained.

A total of 49 boreholes (and 46 camera inspections) were carried out:

– 5 vertical or sub-vertical boreholes at the downstream base of the structure, of which 3 were equipped with piezometers and 2 with extensometers;

– 6 inclined boreholes in which piezometric cells with vibrating wires were installed at mid-thickness of the dam, at the contact between concrete and rock;

– 38 boreholes in the concrete around the swelling nucleus.

At the end of the surveys, most of the boreholes were left open in order to drain the cracked areas, pending the completion of reinforcement work.

The internal "parallel" crack was identified in 28 of the holes, and openings of up to 20 mm were observed. Video footage taken inside each borehole clearly showed the crack's outline and extension, as well as dip and opening of cracks which were discovered in-situ.

The network of cracks spreads across the seven cantilevers on the right bank (Fig. 12). The total area affected (including the swelling zones) is estimated to be approximately 2,000 m^2. These physical observations fairly accurately reflect the results of the numerical model.

Further investigations below the downstream foot of the dam have not found any signs of decompression, as first suspected. The foundation appears to be sound.

2.2.2. *Recent Laboratory Tests (2013)*

Cores of deteriorated concrete were sent to the LMDC laboratory specialised in concrete pathologies, in order to carry out new tests on swelling concrete (cantilevers 2 and 4) and sound concrete (cantilever 4).

The samples show typical levels of free water, bound water, sulphate and alumina content. Observations made using a scanning electron microscope confirm the presence of massive secondary ettringite resulting from an internal sulphate reaction.

Tests were performed on reconstituted mortar in order to estimate the swelling potential of Bimont concrete vis-à-vis alkali aggregate reaction phenomena. The results were negative.

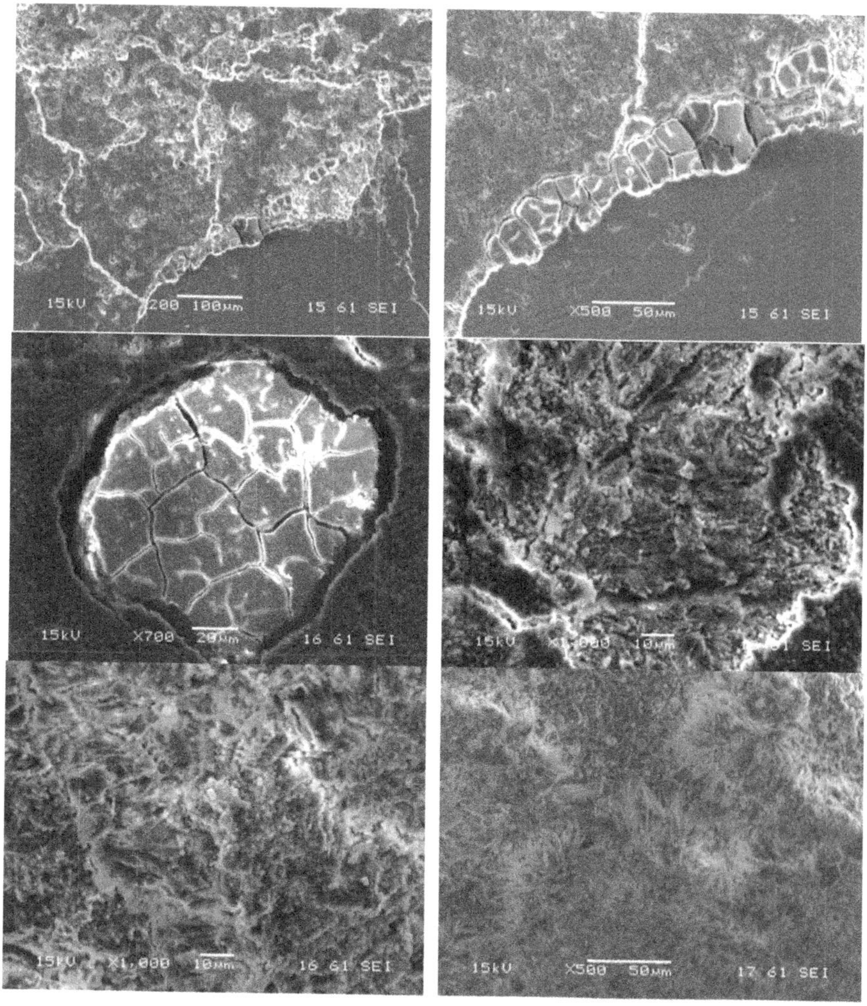

Figure 4. *Concrete sample area S4: Expansive Secondary ettringite, MEB, polish section, SEI, HV, various zooms (extract from LMDC Report)*

In addition, a specific test was performed in order to assess the concrete's capacity for residual expansion. A temperature of 20°C was chosen, so as to reduce the risks of ettringite dissolution as much as possible, and the tests were carried out over a period of 106 weeks. Starting in week 88, the water in the trays was changed regularly in order to accelerate the phenomena. These tests demonstrate that, unlike concrete samples taken from areas in good condition, those extracted from the swelling nuclei show a significant propensity for swelling.

Core name	Localization **Residual expansion**	Localization **Residual expansion**	Localization **Residual expansion**
S1	1,70 à 1,90 m	3,50 à 3,70 m	5,00 à 5,20 m
	60	**98**	**171**
S2	1,40 à 1,60 m	3,60 à 3,80 m	4,80 à 4,95 m
	72	**88**	**94**
S3	1,65 à 1,85 m	3,50 à 3,70 m	5,10 à 5,30 m
	54	**84**	**82**
S4	1,65 à 1,85 m	3,25 à 3,45 m	4,70 à 4,90 m
	306	**129**	**155**
B13	1,50 à 1,70 m	3,50 à 3,70 m	5,15 à 5,35 m
	35	**25**	**31**

Table 1. *Residual expansion between 8 and 106 weeks in µm/m/year*

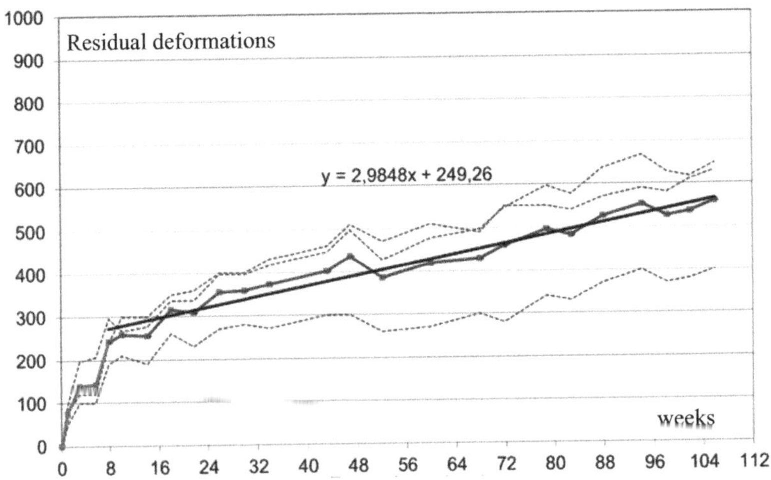

Figure 5. *Expansion test on hardened concrete (extract from LMDC Report)*

3. Numerical Modelling

3.1. *Model Description*

A numerical model was set up in order not only to reproduce the irreversible movements induced by concrete swelling, but also to simulate and understand cracking mechanisms; it also served to calculate the extent of internal cracking parallel to the surface, which was initially assumed but not established and later confirmed by means of surveying campaigns. The model was then used to simulate the effects of a rise in the water level following grouting work on the construction joints and cracks.

The finite element model represented the arch dam and its immediate foundation. The zones of swelling concrete were identified in the model. Joint elements were placed along the line of contact between the foundation and the dam, and along some construction joints and along the main crack between the swelling zones of cantilevers 2 and 4. The Young's rock and concrete moduli, adjusted in order to align calculated displacements with observations, are 10 and 25 GPa respectively.

The swelling phenomenon was reproduced using the nonlinear and anisotropic law SCAS, initially developed by Coyne et Bellier in their COBEF software, in order to simulate the swelling caused by the alkali-aggregate reaction. The swelling's maximum stress limit is 5 MPa and the rate of free swelling is 90 μm/m/year. Although local, the high swelling rate produces tension stresses that are released by construction joint openings or crack formation. Another non-linear module (NOTEN) was used to transform these tension stresses into plastic strains.

Because of the non-linear features of the analysis, the results depend on the load path. Insofar as possible, loads were therefore applied in accordance with the dam's actual history, with obvious limits imposed by the model's relative simplicity.

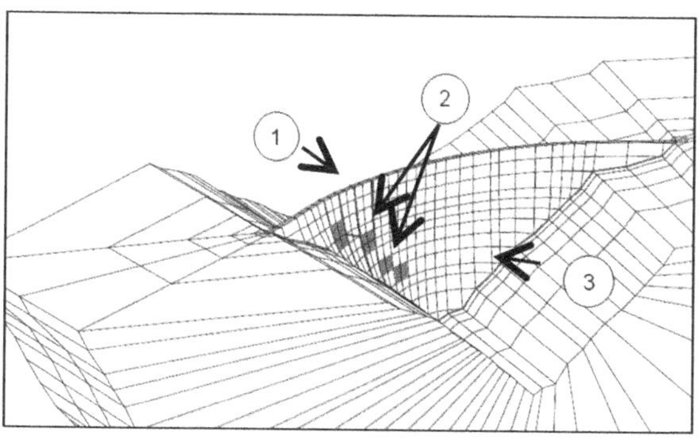

Figure 6. *Dam mesh: (1) Joint between two blocks, (2) Swelling areas and (3) Joint along the dam footprint*

3.2. *Results*

3.2.1. *Crack Mechanism*

The analysis adequately reproduced both irreversible displacements measured by the monitoring the dam, and the mapping of visible cracks, thus validating the process and allowing the crack formation mechanism to be analysed. The cracks are a direct result of swelling, due to the mechanism explained in Figure 6. The zones affected by these cracks are weakened, and there is a major impact on stress distribution with possible consequences on overall dam resistance.

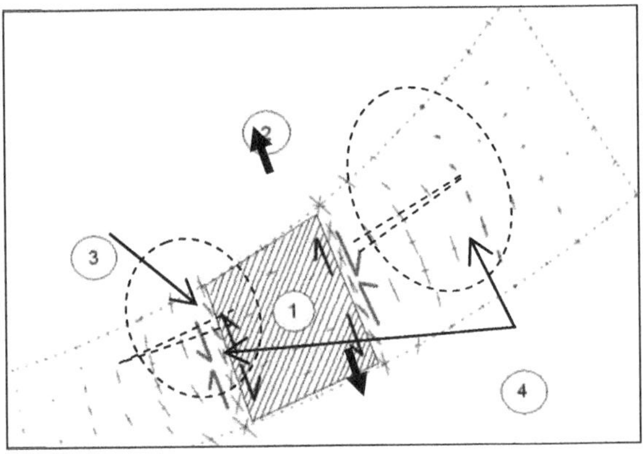

Figure 7. *Cracking mechanism around swelling areas: (1) P4 Swelling area of P4, (2) Increase of the swelling bloc, (3) Shear stress at the origin of cracks and (4) Plastic strain (≈ cracks in the normal plan)*

This cracking mechanism exists all around the swelling zones. The model reproduced the inclined crack, the U/S-D/S vertical cracking plane, the construction joint openings, and the internal cracks parallel to the dam's faces.

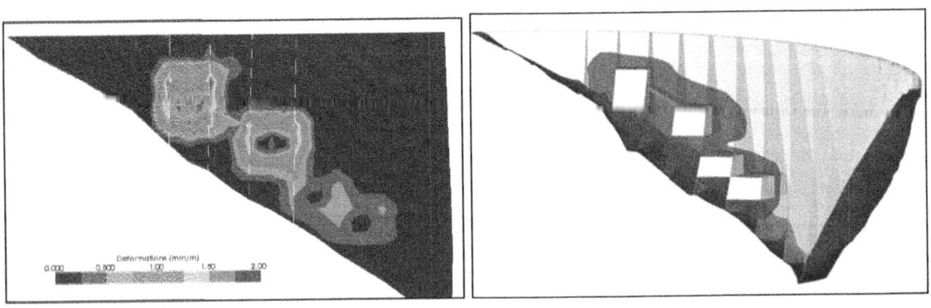

Figure 8. *Developed downstream view of the dam right bank – Left: Plastic strains showing the internal cracks area. Right: Extension of internal crack area as deduced from investigations*

3.2.2. *A Disturbed Stress Distribution*

The swelling phenomenon results in global changes to the dam's performance. Arch stress paths are no longer governed by hydrostatic loading, but by internal pre-stressing caused by the swelling zones. The right-bank area of the dam has suffered considerable disturbance and damage: large cracks have formed, resulting in an increase in compressive stress (10 MPa). However, this stress is mainly concentrated close to the dam's faces, and does not affect its overall capacity to withstand compressible forces: the mean value of compressive stress in the swelling zones is below 4 MPa. Meanwhile, the performance of the dam's left-bank area has improved on account of the increased forces pushing towards the bank. A first analysis without swelling loading showed that some openings could appear at the heel of the left cantilevers, but this disappears when the swelling load is applied.

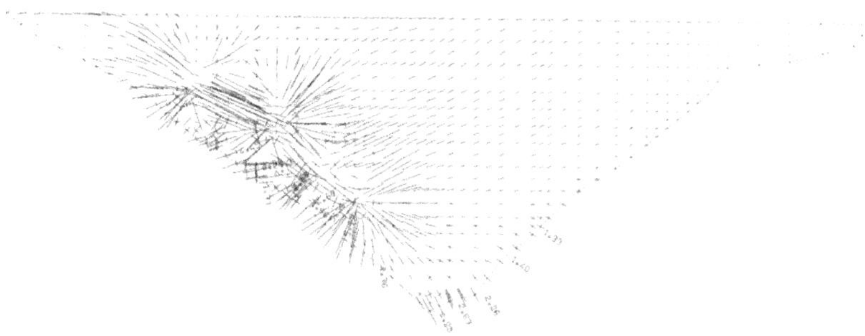

Figure 9. *Principal stress on downstream face after swelling with a water level of 329*

The second stage of the analysis simulated the rise in the water level after grouting work on construction joints and cracks. Since the swelling effect is stronger than the hydrostatic one, the effect of the raised water level is marginal. The response of the (unaffected) left side of the dam remains sound. However, the rise in the water level slightly increased stress distribution on the right bank.

4. Renovation Project

The dam will undergo renovation, which aims to restore its monolithism. It will be able to handle the full reservoir load once more.

The renovation project consists mainly of cement grouting the internal and exposed cracks and joint openings, as well as reinforcing watertightness on the upstream face in order not to trigger the same reaction when raising the water level. It also includes foundation drainage and abutment stability.

Moreover, the dam's monitoring system is upgraded to include short-term monitoring of arch performance during the post-works filling phase, as well as long-term performance over the coming thirty years.

Construction works should last for about two years, and the reservoir shall remain empty for a long period of time. A dam bypass has been set up before works begin this year in order to meet the water demand during this period.

4.1. *Grouting of Internal Cracks*

Internal cracks will be grouted with mortar to consolidate the Bimont Dam by addressing the following issues:

– because the cracks are open and empty, hydrostatic forces are for the most part concentrated in the upstream part of the dam;

– pressure loading due to uncontrolled seepage can disrupt the downstream part of the dam, which would then be subjected to the entire hydrostatic load; and

– by partially cutting arches in two, they cancel shear constraints necessary to the development of active arches, and make the dam less rigid to the detriment of the cantilevers.

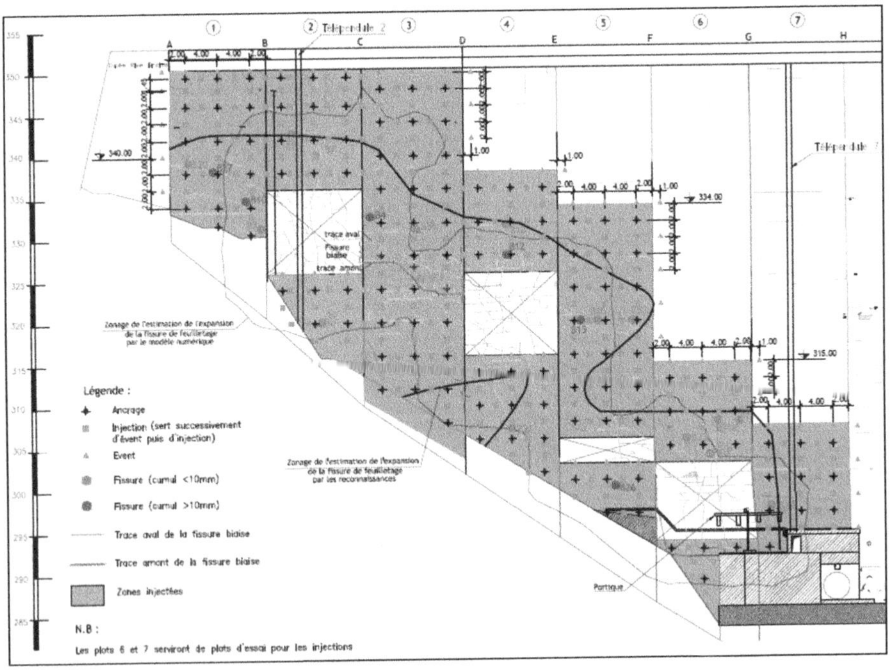

Figure 10. *Anchor drilling (black) and grouting (blue) locations – downstream elevation*

Grouting will be done bottom up, using carefully monitored pressures below 1 bar and by zone of limited height (2 to 4 m), in order to prevent a potential extension of cracks and unbalanced conditions of slivers located between a crack and one of the dam faces through hydro-fracturing caused by grouting pressure.

In order to avoid damaging the dam, affected cantilevers will be reinforced by passive anchors Φ63.5 mm prior to grouting. The anchor bars network suited with a 4mx4m mesh will regain the pressure exerted by the mortar during grouting. Anchor bars will be sealed over their full length.

During grouting, mortar volume and grouting pressure will be continuously monitored and adjusted. Perimeter vents will surround the right bank cantilevers perimeter involved, in order to control the absence of mortar arriving in areas not fitted with anchor bars.

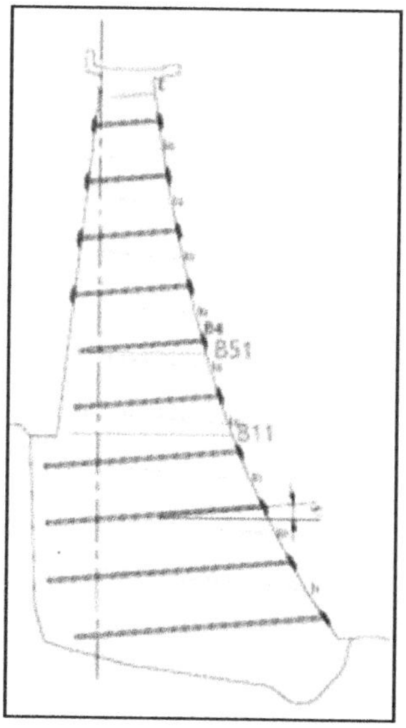

Figure 11. *Anchor drilling (black) and grouting (blue) locations – section of cantilever 3*

4.2. *Grouting of Construction Joints*

Joints grouting will be carried out bottom up in a similar fashion to that of crack grouting. Injection drilling will take into account the higher and lower limits of the existing arch's grouting compartments. Prior to joint grouting, a video inspection will be carried out in each hole to identify the open/closed state of each joint and injecting only if the latter is open.

Joints will be grouted at a low pressure (with a maximum pressure of 4 bar) for a maximum joint opening of 0.2 mm. Movement monitoring will mainly use plaster markers placed across the joints, and 1D fissurometers.

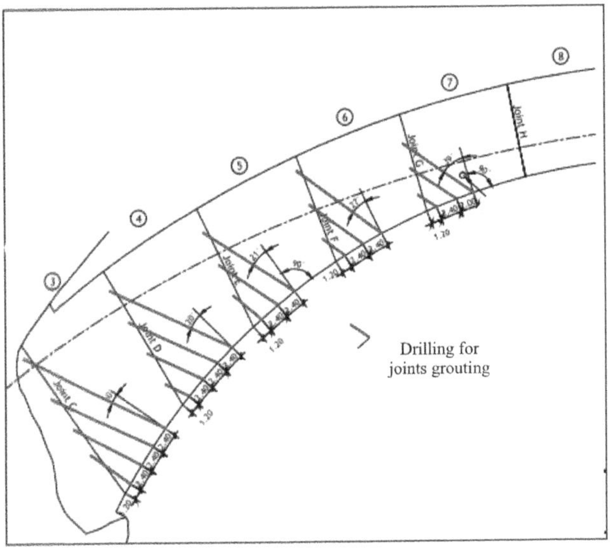

Figure 12. *Joints grouting – Horizontal section between El. 315,00 and 318,00*

4.3. *Watertightness of the Upstream Face*

Watertightness of the upstream face aims to neutralise the moistening of swelling nucleus, which could reactivate the alkali aggregate reaction. Waterproofing works consist in installing a géodrain covered by a geomembrane on the upstream face of the right bank cantilevers, mechanically affixed to its support with metal profiles anchored to the facing.

The geomembrane will extend to the upstream concrete plinth to limit water bypassing the foundation. In the higher part, the geomembrane will stop at the maximum water level. A drainage system will ascertain the effectiveness of the watertightness system.

4.4. *Drainage of the Foundation*

Grouting works in the foundation were carried out at the time of construction. However, there is actually no drainage system of the superficial foundation.

Calculations showed that dam stability was sensitive to pore pressures and recently installed pore pressure cells revealed relatively high levels compared to expectations. A drainage curtain in the foundation from the downstream toe of the dam will thus be set up to limit pore pressure under the dam.

4.5. *Reinforcing the Right Gravity Abutment*

The arch leans against two gravity abutment near the banks. In order to take into account additional forces due to concrete irreversible swelling and extreme thermal gradient in summer, the stability of the abutments was carried out based on the French recommendations for gravity dams.

In the absence of shear strength measurement of the contact surface between abutment and foundation, the foundation's mechanical properties were determined according to Z.T. Bieniawski's classification – the foundation rock is considered a "good rock" of class II, which corresponds to a 0.3 to 0.4 MPa cohesion and a 35 to 45 degrees friction angle.

To guarantee a sufficient safety margin, the right abutment requested strengthening by adding a complementary vertical force to the structure (7.5 MN). Four vertical active anchors will be thus installed in this abutment.

5. Modeling the Renovated Dam

The numerical model presented in chapter 3 was used to check the good behaviour of the structure, after the works for higher reservoir levels. The model could simulate combinations of normal (new NWL +10m), accidental (5,000 year flood – MWL) and rare operating water level (earthquake = 0.35g and 100,000 year flood – MMWL).

Numerical modelling took into account the following residual swelling ratios for swelling areas – 0.2 mm/m for cantilevers P2 and P4, and 0.1 mm/m for cantilevers P5 and P6 in the next 30 years. In order to take into account potential temperature variations caused by climate change, exceptional cases of summer heat waves (+10°C on the average of historical summer temperatures) or of harsh winters (-10°C on the average of historical winter temperatures) under normal operation were also simulated.

Static cases results indicate that constraints are maximised by irreversible swelling of right bank cantilevers and thermal swelling in the summer months, rather than by the reservoir level. High stress areas are mainly located on the right bank, around swelling nucleus. Maximum values reach 15 MPa, an 8% increase compared with known dam values. These values are deemed acceptable since laboratory measures of compression resistance for healthy and damaged concrete range from 23 to 38 MPa. Simple compression tests will be conducted every ten years as part of safety reviews to quantify potential concrete degradation.

In the long term, a further opening of the arch cannot be completely ruled out if concrete swelling continues, as evidenced by the calculated joints opening around cantilever 2 (~0.6 mm). Caution is therefore in order with regard to future joints behaviour.

6. References

Casteigts C., Laliche K., Noret C. Bourgouin T., "Renovation of Bimont arch dam", Q99R50, ICOLD Congress, Stavanger, 2015.

LMDC Report "Caractérisation du béton du barrage de Bimont", December 2013.

Noret C., Molin X., Carrère A., Casteigts C, Grawitz B. Laliche K., "Combined contributions of investigations and modeling in the analysis of Bimont arch dam behavior", Q95R45, ICOLD Congress, Kyoto, 2012.

Chambon Dam

A Struggle against AAR

OLIVIER CHULLIAT — ÉTIENNE GRIMAL — ÉRIC BOURDAROT

EDF-Hydro Engineering Center
Savoie Technolac, 73373 Le Bourget du Lac, France
olivier.chulliat@edf.fr, etienne.grimal@edf.fr, eric.bourdarot@edf.fr

ABSTRACT. *Chambon dam suffers severe alkali-aggregate reaction expansion, causing several pathologies, mainly resulting in shear stresses zones in the structure and significant structural cracking, likely to affect its integrity under earthquake. New reinforcement works were performed in 2013-2014, in the continuity of those already carried out in the 1990's: (i) installation of 415 prestressed upstream to downstream tendons crossing the structure, supplemented with a carbon fiber composite net on the upstream face, (ii) realization of 7 vertical diamond wire slot cuts and (iii) replacement of the existing watertightness geomembrane.*

KEYWORDS: *AAR, tendons, carbon fibers, slot-cuttings, geomembrane, earthquake.*

1. Introduction

The Chambon dam was completed in 1934 and the reservoir filled at the first time in 1935. It is a concrete gravity dam located in the french Alps and operated by EDF at elevation of 1,000 m on the Romanche River with a reservoir capacity of 50 hm^3. The right bank and the central sections are straight, while the left bank section, where the spillway was located, is curved (Figure 1). It is 137 m high above the subglacial deep narrow channel (88 m above river bed). Its crest is 294 m long and 5 m wide. The downstream face is H/V=0.75/1 and the upstream face is vertical. It's made of cyclopean concrete, whit a cement content varying from 150 to 250 kg/m^3.

The Chambon dam suffers Alkali-Aggregate Reaction. Aggregates came from the local quarry of gneiss with numerous layers of black micaschists. The swelling was first noticed in 1958. It causes increasing cracking and deformations in the concrete. The central section moved 1 mm/year downstream, while the curved left wing moved 5 mm/year upstream, with a high risk of spillway gates blocking. A horizontal crack opened between the drainage gallery of the spillway and the downstream face. The crest rose 2.4 to 3.6 mm/y (Figure 2). In the downstream left bank, the yearly increase of leakage flow was 3 l/min at the Q10 measurement point.

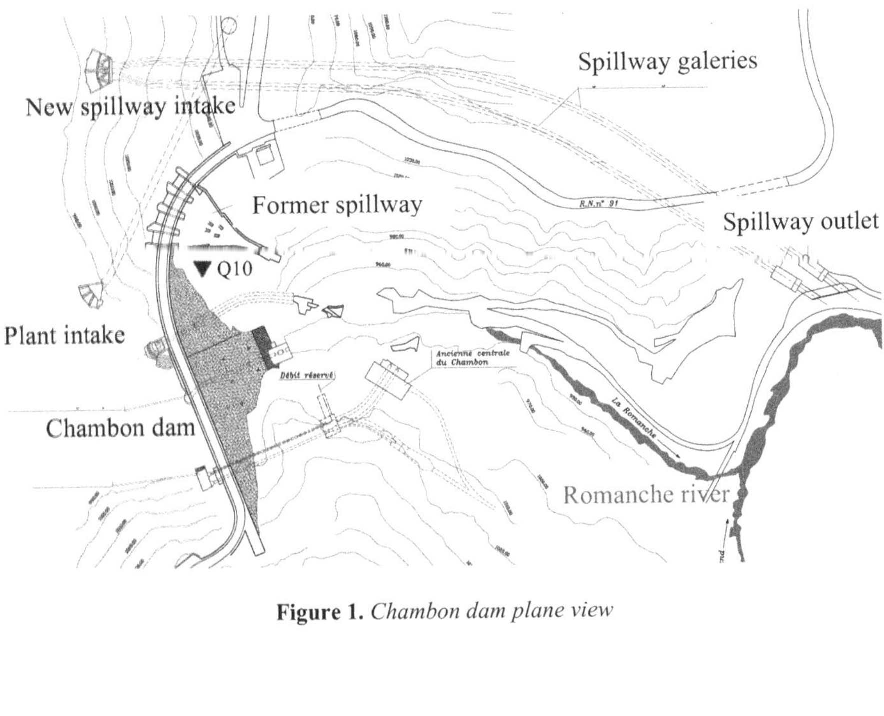

Figure 1. *Chambon dam plane view*

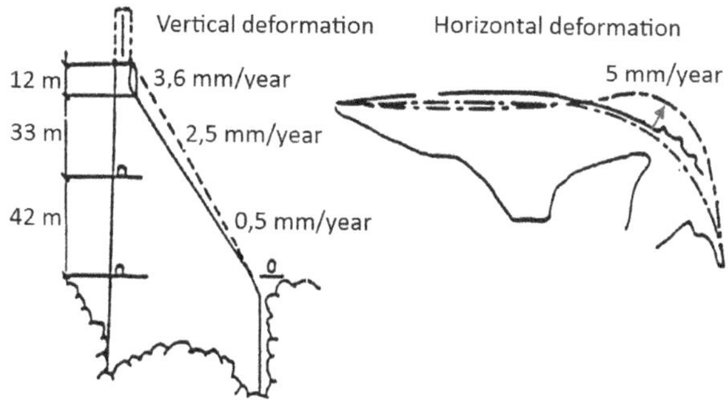

Figure 2. *Main swelling deformation rate at Chambon dam between 1984 and 1994*

2. The First Reinforcement Works Campaign in 1992–1993

In 1991 a working group gathering the owner EDF, the engineering firm Coyne et Bellier and the Public Authorities, proposed an alternative to the construction of a new dam. Giving the measured longitudinal compression stress ranging from 2 to 6 MPa, the first idea was to separate the spillway from the dam, in view of eliminating the risk of gates blocking. The new spillway was built in the left rock abutment (Figure 1) in a view to providing full safety during the reinforcement works.

The second aim was to release the swelling stresses by innovative slot cutting through the crest (Chambon dam was the second dam in the world to experiment this type of works). An isotropic temperature rise-swelling law was implemented in a no-tension model by Coyne et Bellier in an attempt to predict the effect of slot cutting on the dam swelling. The qualitative dam behavior was well described, however longitudinal compressive stresses were two or three times higher than the measured ones. In such conditions, the stresses calculated by the model were unacceptable after 10 to 30 years at the top of the left bank. The slot cutting intended to relieve such inacceptable stresses. However, the model showed that the effects of this stress relief should not last over 20 years and that other works should be required.

Before cutting, the significant opened lift joints, seen in the upper part of the downstream face (5 m to 20 m below the crest, 2 m to 3 m deep over a 200 m length) were grouted (Figure 3a). The grouting was necessary to ensure static and seismic stability of blocks separated by concrete lift joints and vertical saw cutting. All the cracks were filled and sealed with several grouts ($0.5<W/C<2$) at a maximum pressure of 0.2 MPa in two seasons: 1992 and 1993. 20 m^3 of grout were used. Some leaks occurred in the drainage shafts. Quantities grouted in the curved left wing were twice those in the central section.

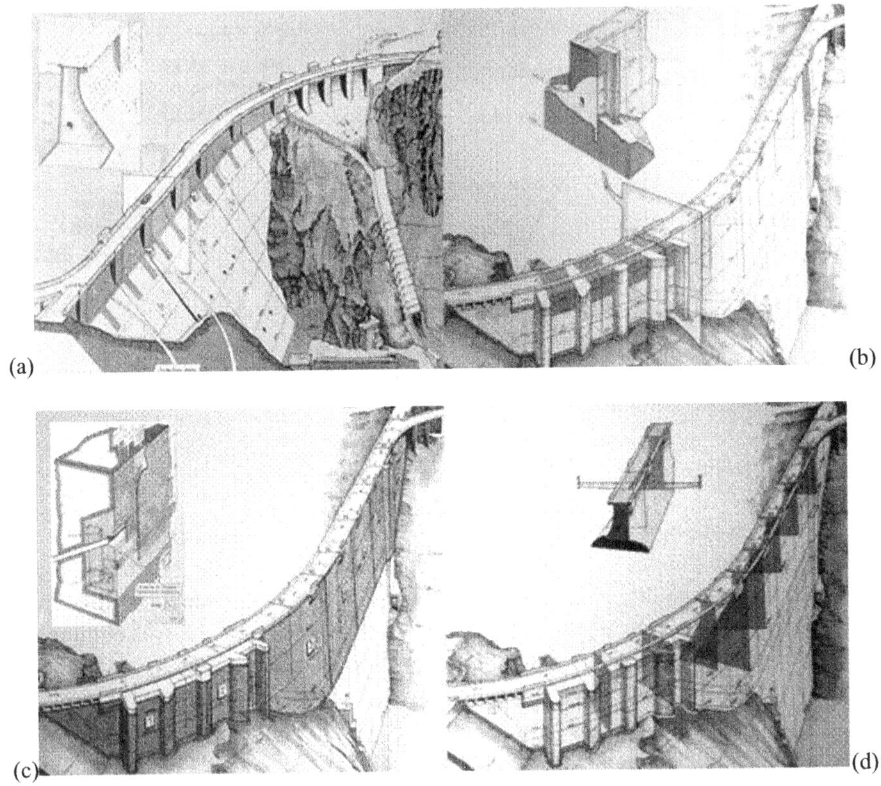

Figure 3. *Repair works conducted in the 1990's in relation with AAR*
a: crack grouting, b: old spillway concrete, c: upstream sealing, d: slot cuts

From 1993 to 1995, 9000 m^2 of Carpi PVC geomembrane was placed on the top 40 m of the upstream face in a view to eliminating the uplift in the upper part of the dam and to solve the question of the slot watertightness (Figure 3c). The upstream face was smoothened with sprayed mortar. A reinforced beam was concreted at the base of the geomembrane. The drainage and puncture protection is ensured by a geogrid. The geomembrane was tightened between vertical profiles 1.85 m apart. The drainage system was separated in 9 isolated blocks with their own drainage monitoring. In such a manner the localisation of the entry point of any leak can be facilitated. Thanks to the complementary foundation grouting works and the concrete shell poured at the spillway toe, the left abutment leakage (Q10) dropped from 50 l/min in 1994 to 22 l/min in 1995 and up to 17 l/min in 1997.

Before starting the slot-cutting works, the initial spillway was decommissioned. The gates were removed and the gate openings were filled in with 4,000 m^3 of concrete (Figure 3b). Three campaigns of slot-cutting were required from 1995 to

1997 (Figure 3d). 2 slots in 1995, 3 in 1996 and 3 in 1997 were completed in depth below the crest from 18 m to 33 m with 11 mm diamond wire. The main displacements of the straight central section are longitudinal, parallel to the dam axis and oriented toward the thalweg, while the curved left wing moved back downstream. During the last campaign, the 3 slots did not close and surrounding slots and joints opened, this was the sign of the effectiveness of the stress relief.

3. The Assessment of the Efficiency of the 1992–1993 Campaign

Fifteen years after the first reinforcement works, the slots closure monitoring showed the slow crest recompression and the pendulums monitoring exhibited the restart of the curved left wing crest movement toward upstream (Figure 4). Following the upstream movement of the crest, the horizontal crack below the spillway slightly reopened.

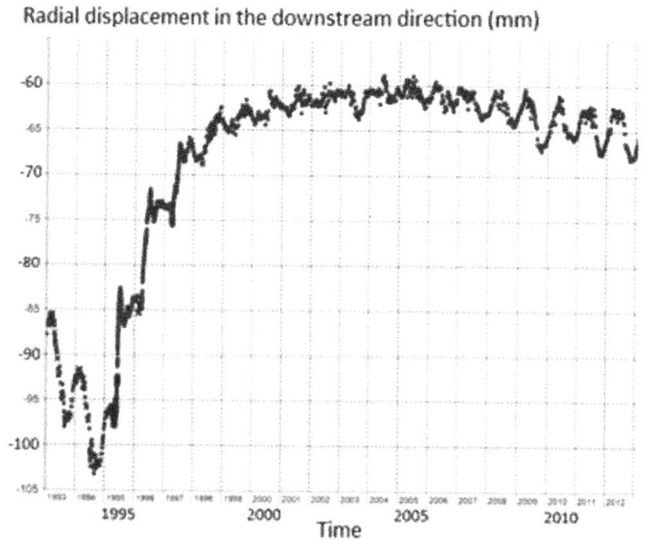

Figure 4. *Displacement of pendulum in the left curved wing before and after the works*

According to those observations, an extensive investigations campaign was launched between 2007 and 2010. The aim was the diagnosis of the dam condition and the definition of the next works to continue the operation of the dam in safe conditions. These investigations included drill holes in the dam body to identify the extent of crack networks, detailed inspection of the concrete-rock interface by digital borehole logging inside drill holes, identification and characterization of swelling laboratory tests on samples representative of the different areas of the dam, on site stress measurements with flat jack devices inside the galleries and from the downstream face. Overcoring in-situ tests were also performed in order to get the three dimensional stress tensors, leading to more relevant results of the compressive stresses. Foundation modulus measurements by dilatometer tests were also carried out.

This extensive campaign led to the conclusion that it was not possible to exclude some upstream concrete blocks instability in earthquake conditions. They might be precut by the upstream elevation drainage curtain and possible extended horizontal cracks. The 800 mm (in red and green on Figure 5) and 300 mm (in blue on Figure 5) diameter vertical drains are separated by an average distance of 3.20 m; they could work as a "precut line" along the axis of the dam. The 3.20 m to 4.80 m distance between upstream face and the drainage curtain might be cracked somewhere between the crest and 40 m below it.

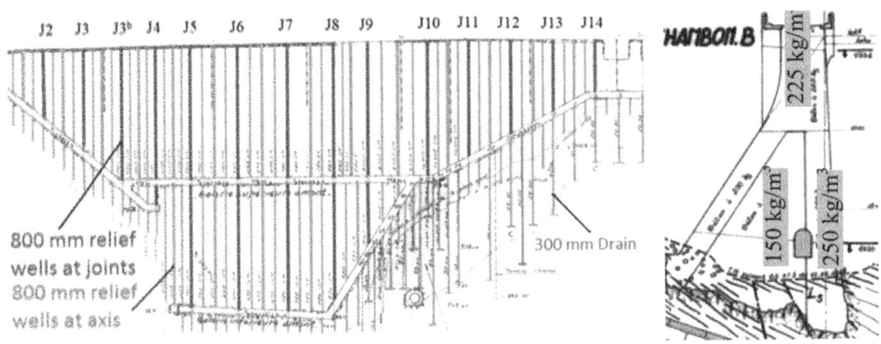

Figure 5. *The vertical drainage curtain seen from downstream and cross section*

The differential expansion rate on the upstream concrete containing 250 kg/m^3 of cement and the downstream one containing 150 or 225 kg/m^3 could explain the downstream irreversible deformation of the straight part. Combined with the drainage curtain effects, it could also lead to the development of a vertical longitudinal crack along the dam axis. Most of these cracks are observed in the drilling holes through the curved zone and are several millimeters to more than 1 centimeter opened. As the different cracks are most of time not linked inside the dam body, the presence of a continuous upstream slab can likely be excluded, but potentially unstable blocks under earthquake loading may exist. These virtual blocks are delimited by the vertical planes in the transversal direction along dilatation joints and saw cuts (5.30 to 16.30 m), and by cracks potentially linking existing drains (singular points), with an average space of about 3 m and horizontal planes by the construction joints (2.40 à 2.90 m). Between structural cracks, the concrete displays good mechanical properties: compressive strength greater than 20 MPa, modulus of instantaneous deformation greater than 20 GPa. Moreover, the contact between limestone (Trias) and igneous rock (gneiss) in the left bank abutment is not disturbed by the thrust of the dam: it is closed and of good quality (with deformation moduli varying between 6 and 14 GPa) excluding any risk of shearing surface development inside the left bank abutment under the stresses developed by the dam swelling.

The new multi scales swelling law for concrete implemented in ASTER computer code developed at EDF was used at this occasion. It provided a very good fitting with the monitored dam behavior (Figure 6).

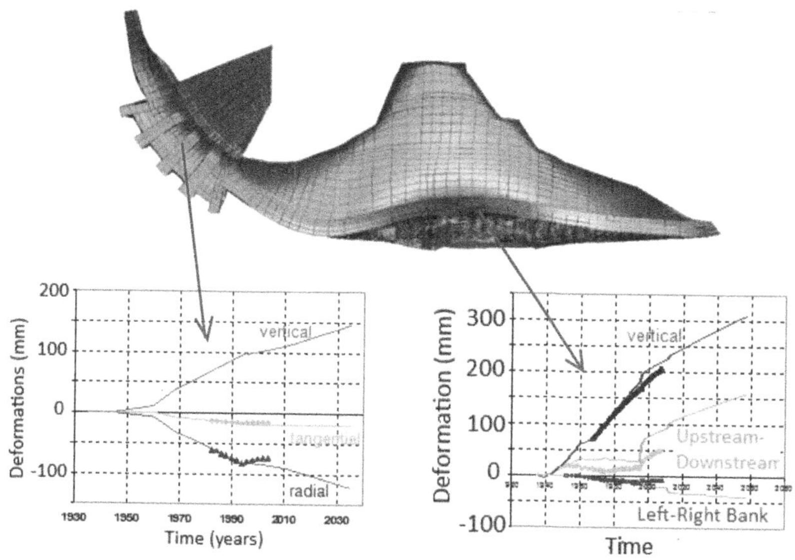

Figure 6. *Comparison of measured and computed deformations in 3D modeling*

The FEM calculations showed that the benefits of the slot cuts done in the 1990's still remain in the upper part of the structure, confirmed by the monitoring of the deformations in the curved right wing. They nevertheless displayed noticeable stresses parallel to the abutments.

The swelling of concrete should continue with a constant rate, even if a gradual slowing is not excluded, for several decades (Figure 7). This swelling rate draw the attention on the shearing conditions of the upper concrete-rock interface in a medium term.

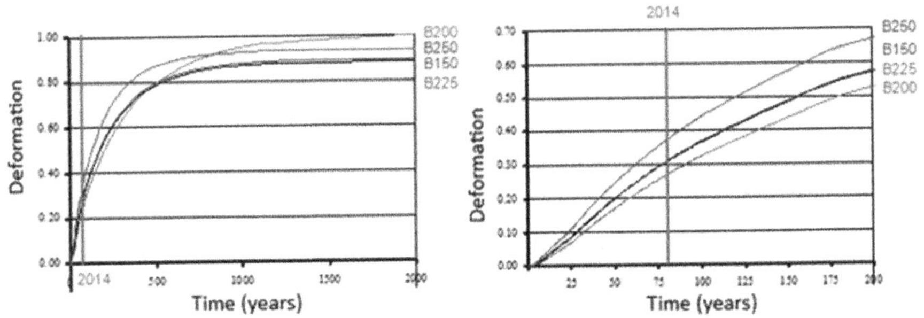

Figure 7. *The predicted swelling in the future at Chambon dam*

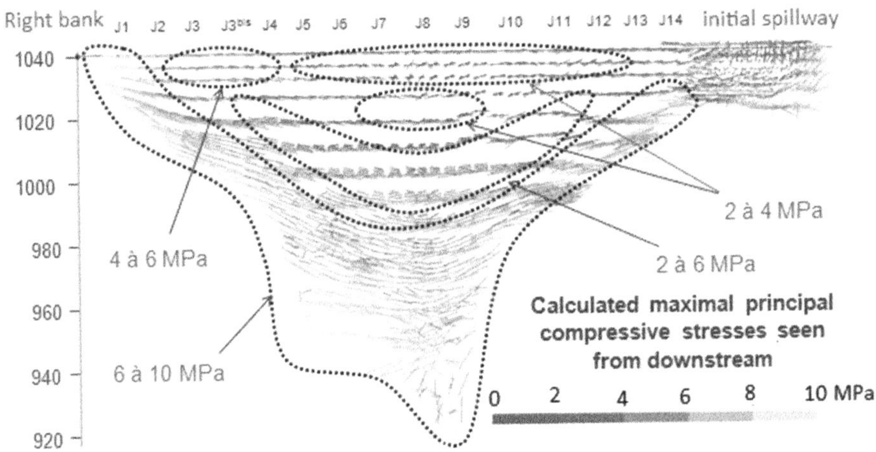

Figure 8. *The principal stresses tensors calculated in the dam*

Based on that diagnosis, a new reinforcement campaign was decided and received in 2010 an agreement from the Permanent Technical Committee on Dams and Hydraulic Structures (commissioned by the French Ministry of Industry).

4. The Second Project of Reinforcement Works

The main objectives of the second campaign were to reinforce the integrity of the upper part of the dam and to prevent any upstream block falling that could lead to membrane damaging. The upper part confinement was obtained by the installation of 415 tendons (Figure 9). Consisting of horizontal cables, with greased sheathed strands type T15, crossing the structure from upstream to downstream, they were pre-tensioned and non-grouted. They are 3.70 m (horizontal) and a 4 m (vertical) spaced, with about one tendon for 15 m^2. The obtained mesh fits the most likely size of the blocks, delimited by vertical cracks parallel to the axis and the upstream face, vertical joints or former slot cuts parallel to the stream direction and the lift joints in the horizontal direction. Upstream and downstream cable heads are fully embedded in reservation drilled in the dam, with diameter ranging between 500 and 700 mm. The sheath outside is not subject to grout injection, as the successful completion of the work could be impaired due the presence of cracks. The corrosion protection is enhanced on the upstream side by concreting the reservations and by the presence of the sealing membrane. A progressive tension increase due to concrete expansion remaining possible, detensioning operations could be necessary in order to fit the required tension.

Figure 9. *The tendons pattern (red) and drainage curtain seen from downstream*

The tension of 66 tendons is monitored during dam operation. They are instrumented with sensors, each containing three vibrating wires, connected to the dam existing remote monitoring system. Punctual in-situ weighing of each tendon is also possible at any time. Downstream cable heads have been designed in order to maintain their tension within specified adequate limits in any circumstances:

– a lower limit equal to the driving force due to earthquake, thus preventing any block movement;

– an upper limit of 80% of the tendons yield stress (including earthquake forces, additional tension due to concrete swelling, differential thermal expansion between steel and concrete, as well as measurement uncertainties).

The reinforcement project is designed for a 50-year duration, with a theoretical tendons de-tensioning every 20 years (representing a theoretical elongation of 24 mm for the longest tendons). Tension design includes the following stresses:

– the pull-out forces due to earthquake: horizontal acceleration (0.18 g peak ground acceleration combined with a maximal amplification at crest about 7) and the thickness of the blocks to confine;

– further swelling concrete, considering isotropic rate of 50 microns/m/year;

– differential thermal expansion between steel and concrete (maximum 10°C);

– uncertainties relating to the tension measurement.

The implemented tendons type varies from 3T15 in the lower part (length: 24 m, 369 kN $\leq$ T $\leq$ 595 kN) and 7T15 in the upper part (length: 5 m, 1060 kN $\leq$ T $\leq$ 1388 kN), locally 10T15. The 4180 m drill holes necessary for the tendons are cored from upstream, with a tolerance of 1% deflection. They are systematically inspected by logging digital in order to map the cracks network and to represent them in a 3D digital model.

In addition to tendons, a carbon fiber composite net is added on the upstream face. It consists in the sticking of 6,000 m carbon fiberstrips. The 20 to 30 cm wide strips link the tendon heads, along vertical, horizontal and diagonal lines (Figure 10). The carbon fibers net serves confinement of small blocks that could escape the tendons action. Strips composite material (carbon strips glued with epoxy resin), glued to the concrete surface previously sand blasted, are designed to form a "chainstitch" and to resist tensile stress due to earthquake. An anchoring device, designed after laboratory tests, allows the connection between the carbon fiber strips and the tendons upstream heads (Figure 11). They are composed of eight rods (type M24 to M40 according to the efforts to be transmitted) inserted into the upstream reservation and a first plate maintaining the diagonal bands. A second plate located upstream of the first one allows the coupling of the horizontal and vertical bands. Bands are continuous in current zone and are re-curved over the anchoring devices located on the periphery of the reinforced zone. The ones which are not subjected to symmetrical efforts are strengthened accordingly.

While carbon fiber composite is a material commonly used in civil engineering, its implementation as an anti-seismic net is relatively innovative and requested qualification tests in laboratory. The test device was composed of ten concrete blocks simulating tendon head hooped recesses at scale 1 and defining the average mesh (3.70 m x 4 m), connected two-by-two by the test composite bands in different configurations encountered. The tests consisted in applying a force perpendicular to the band to simulate a block extraction under earthquake, till break of the band or the anchoring device. The resistance values from the trials were then included in the design and taken into account in a calculation model to determine the number of layers required for each band.

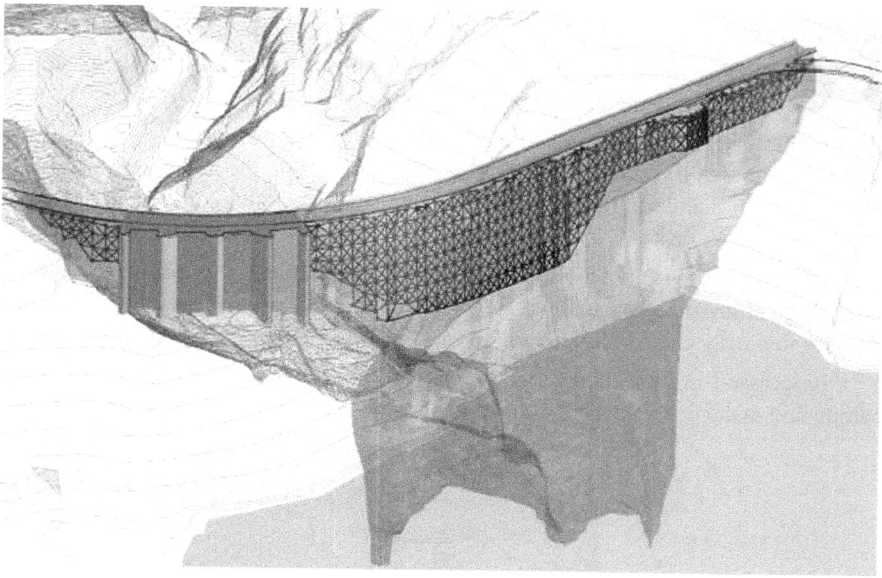

Figure 10. *The carbon fiber net seen from upstream*

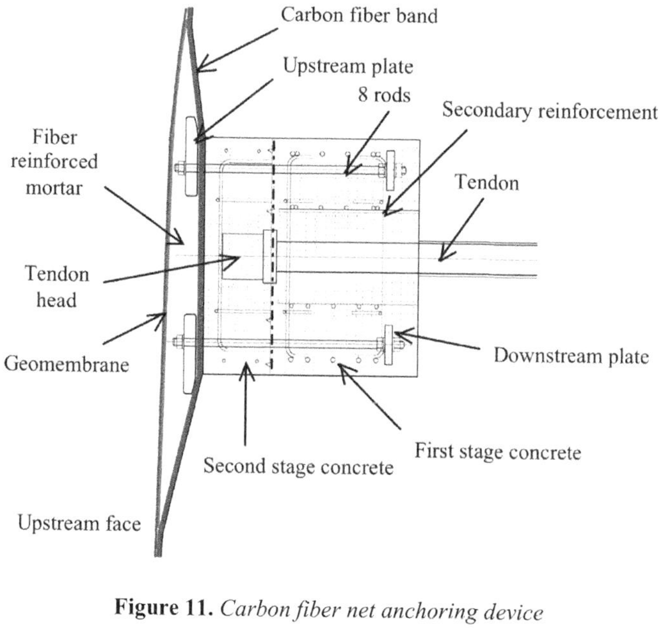

Figure 11. *Carbon fiber net anchoring device*

The new campaign by diamond wire sawing was defined using the numerical model (i) to avoid the re-compression of the upper part of the structure and (ii) to decompress the stress paths parallel to the abutments. Slot cuts mostly follow the same tracks than in the 1990's, with the following exceptions (Figure 12):

– slot cut S3 in the central part of the dam is not sawn again, due to the presence of a cable stuck during the previous campaigns;

– both slot cuts flanking S3 (S2 & S4) are deepened, in order to reach 42 m, leading to an area of 650 m² each, in order to compensate the non-resawing of S3, but mainly to increase the stress reduction efficiency along the concrete-rock interface;

– the diamond wire is 16 mm wide (against 11 previously) to take advantage of technological progress made since the 1990's and avoid any risk of jamming in the event of non fully closed former slot cuts;

– progresses in slot cutting technology allow nowadays far greater rates of cutting and make possible the 2,500 m² total cut in six months, with two equipments.

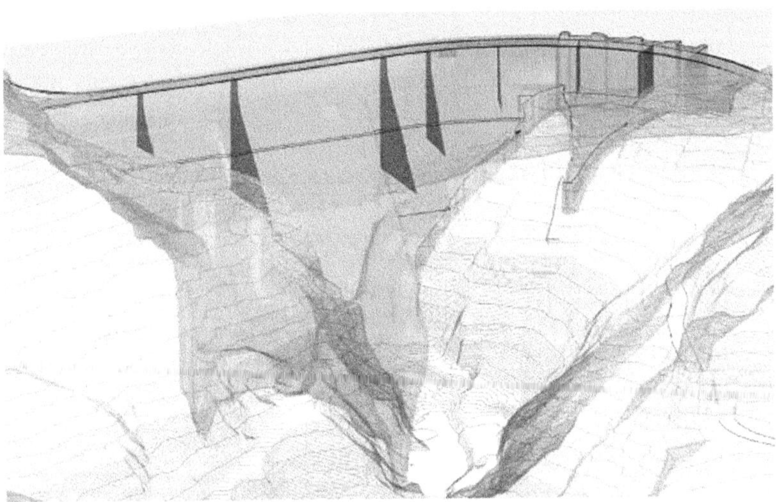

Figure 12. *The design of the new campaign of slot cutting*

5. The Second Campaign of Reinforced Works in 2013–2014

Several main on-site constraints, mentioned below, have conditioned the procedures and the project schedule:

– the presence of a roadway of international importance on the dam's crest connecting Italy and Grenoble and serving major ski resorts without any possible deviation;

– the altitude (over 1,000 m), in the Alps, with cold temperatures and snow;

– the situation in an environmentally sensitive area, close to the Ecrins National Park;

– the ongoing operation of the hydraulic power plant;

– a tight schedule, with many interfaces between different types of works.

Access to the work sites was performed thanks to scaffolding of the entire surface concerned by the works on both upstream and downstream faces of the dam. Scaffolding main platforms were set about 1 m below the tendons heads, with a 4 m vertical step, completed with intermediate levels for the gluing of carbon fiber strips on the upstream face and for the repairing of the coating surface on the downstream face, representing a decking total length of more than 8,000 m. Aerial work platforms were also suspended under these scaffolds, in order to (i) restore the entire downstream coating surface damaged by freeze with time, (ii) replace the sealing membrane PVC layer below the tendons on the upstream face and (iii) permit the due decennial visual inspection of the lower part of the upstream face. Two 75 m jib tower cranes were the main lifting equipments for moving up to 3 tons loads over the entire site. One was placed against the upstream foot of the former surface spillway. The second one was fixed on an old crane pad located in the upper part of the dam downstream face. Pending the finalization of the reinforcement and the requalification of the dam, the reservoir elevation was lowered by 30 m. A preventive volume for storing a 10-year flood was also maintained under the lowest point of the works.

The works lasted from January 2013 to December 2014. The schedule fits the reservoir low level natural periods, situated in winter for reasons of low water supply, as the river's regime depends mainly on the snowmelt in spring. The first winter was devoted to the upstream sealing membrane removal. The second winter was dedicated to treating lower areas, by taking advantage of the natural seasonal lowering of the reservoir, limiting the energy losses. Implementation of tendons and upstream carbon fiber composite net required the removal of almost 8,725 m² of existing waterproofing membrane from elevation 1,041 to elevation 1,003.50. The new waterproofing system has been installed on the same areas, once the upstream face released from scaffolds, from elevation 1,040.40 to elevation 1,003.50, for a total surface of 9,065 m². At the bottom of slots S2 and S4, the system was prolonged down to approximately elevation 1,001.50, fully covering the two slots cut deeper than before. The new waterproofing system was conceptually identical to the one completed in 1995, with a few small modifications and improvements. The new waterproofing liner was Sibelon® CNT 3750, a 2.5 mm thick PVC geomembrane heat-bonded during fabrication to an anti-puncture geotextile. The mass per unit area of the geotextile has been increased from 200 g/m² (installed in the 1990s) to 500 g/m², to provide higher anti-puncture protection to the PVC waterproofing geomembrane. The drainage system behind the PVC geocomposite was the same of 1995, with the addition of a 1 m high longitudinal band of geonet at bottom and of two new drainage discharge holes drilled at the bottom of the system in correspondence of the deeper slots S2 and S4. The drainage system consisted of a drainage geonet installed on the full face of the dam, and doubled at bottom over a height of 1 meter, of the vertical drainage conduits formed by the tensioning anchorage system, and of the stainless steel box drain embedded as bottom collector at the upstream face. The drainage system was divided in 11 compartments by vertical watertight fixations made with an 80 x 8 mm stainless steel batten strip, bolted to the concrete with chemical anchors at 0.15 m spacing. The behavior of the geomembrane remained satisfactory after nearly twenty years of operation, and gave the opportunity during a geomembrane waterproofing replacement to use a large part of the metallic components placed almost 20 years before.

6. Acknowledgements

Works were implemented by Bouygues-VSL company (tendons, carbon fibers, civil engineering and facilities), Marietta spa (slot cutting), Carpi Tech BV (geomembrane).

7. Conclusion

The slot-cutting works, the installation of tendons and carbon bands, the removal and re-installation of the waterproofing system lasted two years, from January 2013 to December 2014. The good planning, the strict coordination of the numerous different tasks, that made these works so unusual, the good cooperation between EDF and the various contractors involved, allowed the works completion successfully and within the deadline. Due to the adopted disposals and site equipments the traffic of the two-lane international road on the crest was never stopped. Reservoir impounding was successfully operated during 2016 spring period.

The extended monitoring system enables to follow precisely dam behaviour and will provide in the next future complementary information relating to the adapted dam behaviour, after these second remedial works campaign, under ongoing concrete swelling.

CHAPTER 5

Long-Term Behavior. Risk Reduction

The Diagnosis and Prognosis of ASR and ISR in Miranda Dam, Portugal

João Custódio* — José Ilídio Ferreira** — Antonio Santos Silva* — António Bettencourt Ribeiro* — António Lopes Batista*

* *LNEC – National Laboratory for Civil Engineering*
Av. do Brasil, 101, 1700-066 Lisboa, Portugal
jcustodio@lnec.pt, ssilva@lnec.pt, bribeiro@lnec.pt, albatista@lnec.pt

** *EDP – Gestão da Produção de Energia*
Rua Ofélia Diogo Costa, 45, 4149-022 Porto, Portugal
ilidio.ferreira@edp.pt

ABSTRACT. *In the last decades, a significant number of problems related to concrete deterioration have been detected in large concrete structures in Portugal, the leading cause being alkali-silica reaction (ASR). Currently, it is still difficult to perform a complete assessment of the actual condition of a structure affected by internal expansive reactions, the accurate prediction of the deterioration of the mechanical properties and, consequently, of the period during which the structure will effectively perform its function, essential for the timely and cost-effective planning of the necessary mitigation/rehabilitation/reconstruction works. This paper aims to contribute to the ongoing discussion of this topic by the scientific community and, therefore, presents the methodology followed to assess the condition of a dam in Portugal that is affected by swelling reactions, the Miranda dam, in which the concrete was produced with granitic aggregates.*

KEYWORDS: *internal expansive reactions, concrete, granitic aggregates, dam, diagnosis, prognosis.*

1. Introduction

Among the various reactions that can cause concrete swelling, the most common in Portugal are the alkali-silica reaction (ASR) and the internal sulfate reaction (ISR). From the 60 large concrete dams that are currently being monitored in Portugal, there are 19 dams in which the concrete swelling phenomena has already been identified. ISR is the main degradation phenomenon in only one of those dams (Fagilde dam). The main types of aggregate used are granite (13 dams) and quartzite. Severe ASR induced concrete swelling has only been observed in three dams, namely Santa Luzia, Alto Ceira I and Pracana, all having quarzitic aggregates. The remaining expansion potential in the Santa Luzia dam appears to be low and the dam shows very mild structural effects. Alto Ceira I dam was demolished and replaced by the Alto Ceira II dam in 2014. The Pracana dam was submitted to large rehabilitation works in the 1990s, and since then the ASR effects have diminished considerably. At present, in Portugal, Fagilde dam is the one exhibiting the most serious structural effects due to concrete swelling. In the other 15 dams the swelling phenomenon is still moderate, except in the case of Covão do Meio dam, in which the vertical strains, computed from the vertical displacements measured, reached a value of about 500×10^{-6} in the last 30 years (Batista & Gomes 2016). Miranda dam, the case treated in this paper, has accumulated average free swelling strains of about 300×10^{-6} during its lifetime of about 55 years (Batista & Gomes 2016).

Currently, it is still difficult to perform a complete assessment of the actual condition of a structure affected by internal expansive reactions (*e.g.* ASR and ISR), the accurate prediction of the mechanical properties deterioration and, consequently, of the period during which the structure will effectively perform its function, essential for the timely and cost-effective planning of the necessary mitigation/rehabilitation/reconstruction works. This paper aims to contribute to the ongoing discussion of this topic by the scientific community and, therefore, presents the methodology followed to assess the condition of a dam in Portugal that is affected by swelling reactions, the Miranda dam. It includes an extensive experimental campaign, performed to aggregates collected locally and to concrete cores extracted from the structure, to allow for a better characterization of the expansive reactions, and also a structural assessment, to understand, in an integrated way, the relations between the different actions and causes, and their major effects.

2. The Miranda Dam Case Study

2.1. *Dam's Characteristics*

Miranda dam is a buttress structure, 80 m high, located in the beginning of the international stretch of the Douro River (which is shared by Spain and Portugal), that is founded on a granite and schist rock mass (Fig. 1). The dam construction works were developed between 1958 and 1961 (HED 1961). The dam monitoring system, installed during the construction, allows the evaluation of the actions and of the thermal, structural and hydraulic responses, by measuring: (i) the upstream and downstream levels, by staff gauges and water level recorders; (ii) air temperatures, by both thermometers and thermograph; (iii) uplift pressures by piezometers; (iv) horizontal displacements by coordimeter bases of the set of plumb lines; (v) horizontal displacements by geodetic methods; (vi) vertical displacements by precision geometric levelling of the crest; (vii) joint movements, using devices for measuring the movement of joints and deformeters; (viii) temperatures in the concrete, by thermometers, as well as by devices for measuring the movement of joints, by strain-meters and stress-meters; (ix) strains, by groups of two, five or nine Carlson strain-meters; (x) stress, by stress- meters; (xi) pressures in concrete pore fluid, by pressure cells; and (xii) discharged and infiltrated flows, by drains and seepage measuring weirs.

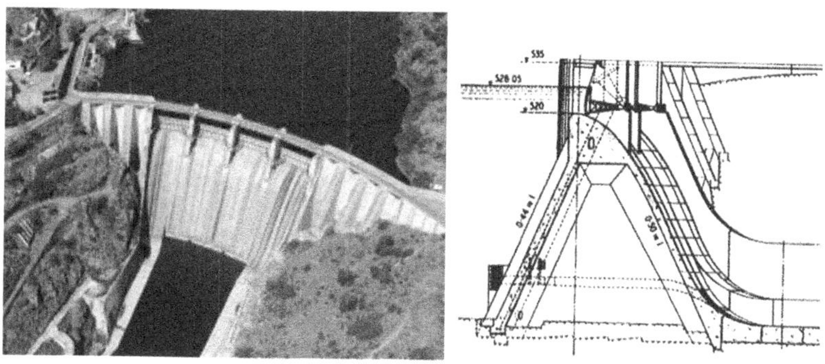

Figure 1. *Miranda dam. Downstream view and buttress profile (Batista & Gomes 2016)*

2.2. *Dam's Concrete*

The aggregates used in the dam concrete were of the granitic type and obtained from a quarry at the construction site. At the time the dam was constructed, the aggregate was considered to be alkali non-reactive (HED 1983) and, therefore, precautionary measures against ASR were not adopted. The dam concretes were produced with four different maximum sizes of aggregate (MSA), namely 150 mm, 75 mm, 38 mm and 19 mm. The concrete placed in the main body of the dam had a MSA of 150 mm, while that used in the facing used a MSA of 75 mm. The physical and chemical characteristics of the portland cement are presented in Tables 1 and 2. Unfortunately, data concerning the cement alkali content was not found in the existing documentation. Two different water-cement ratios were used, 0.53 and

0.58, the lower value was used in concretes having a cement content of 275 kg/m^3, 250 kg/m^3 and 225 kg/m^3; whilst the higher ratio was only used for the concrete with a cement content of 205 kg/m^3. Fig. 2 shows a buttress illustrating where the concretes with different cement contents were applied. The heat of hydration of the cement is presented in Table 3.

Sampling date	Tests		No. of tests	Specified values	Test results		
					Average	Extreme values	
06/09/1957 - 31/01/1959	Setting time	Initial	86	$\geq$ 1 h, $\leq$ 8 h	206 min	130 min	290 min
		Final	86	-	293 min	215 min	415 min
	Soundness (mm)		93	$\leq$ 4	1.3	0.5	3.3
	Flexural strength (MPa)	7 d	86	$\geq$ 3.4	5.0	4.2	5.6
		28 d	86	$\geq$ 4.9	6.3	5.7	6.9
	Compressive strength (MPa)	7 d	86	$\geq$ 17.7	27.3	19.9	32.5
		28 d	86	$\geq$ 30	43.8	34.2	52.5

Table 1. *Cement physical characteristics (HED 1983)*

Sampling date	Tests	No. of tests	Specified values (%)	Test results (%)		
				Average	Extreme values	
07/01/1960 - 19/11/1960	Loss on ignition	10	$\leq$ 4.0	1.39	0.74	2.39
	Insoluble residue	10	$\leq$ 1.5	0.47	0.30	0.62
	MgO	10	$\leq$ 4.0	2.13	1.84	2.37
	SO_3	10	$\leq$ 2.5	1.96	1.70	2.32
	Fe_2O_3	10	-	5.43	5.08	5.84
	$SiO_{2,total}$	10	-	19.29	18.54	20.18
	Al_2O_3	10	-	6.36	5.88	6.91
	CaO	10	-	62.67	61.96	63.75
	CaO_{free}	10	-	3.11	1.33	4.75
	C_3S	10	-	39	35	54
	C_2S	10	-	26	12	30
	C_3A	10	-	8	6	10
	C_4AF	10	-	17	16	18

Table 2. *Cement chemical characteristics (HED 1983)*

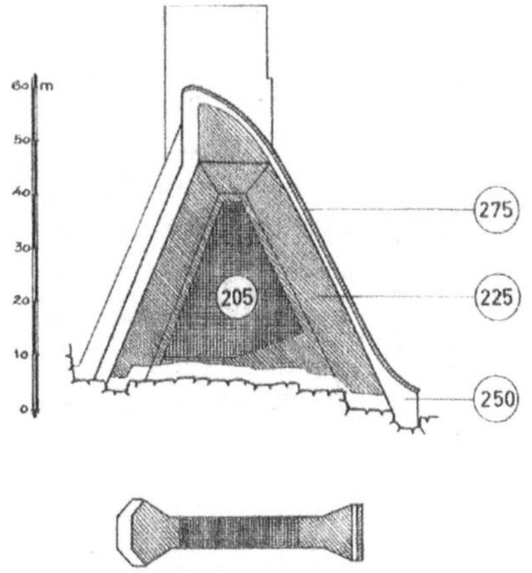

Figure 2. *Identification of the concretes used in the different areas of the buttress, through their cement content (205, 225, 250 and 275 kg/m^3)*

Sampling date	Tests	No. of tests	Age (days)	Specified values (J/g)	Test results (J/g)		
					Average	Extreme values	
07/01/1960 - 19/11/1960	Heat of solution method (LNEC)	9	3	≤ 251	251	301	230
		19	7	≤ 314	289	335	251
		10	28	-	322	351	293
14/02/1959 - 26/01/1961	Heat of solution method (on-site)	84	3	≤ 251	238	293	180
		75	7	≤ 314	272	331	238
		10	28	-	335	381	293

Table 3. *Cement heat of hydration (HED 1983)*

The portland cement used in the dam is coarser and has a lower strength than modern cements (Table 1), being comparable to a CEM I 32.5 N. Table 2 shows that the cement has a low loss on ignition and insoluble residue and, therefore, it was probably constituted only by clinker and calcium sulfate. Considering nowadays cements, the cement used in the dam has higher iron (III) oxide and free calcium oxide contents, and lower sulfate content. The C_3A content, the low C_3S and high C_2S contents are characteristic of the cements produced at that time. The evaluated cement samples exhibited an average heat of hydration between 272 J/g and 289 J/g (Table 3) when determined at 7 days, therefore, it cannot be referred to as a low heat common cement (CEN 2011) or a very low heat special cement (CEN 2015); instead, the obtained values are typical of an ordinary portland cement, type CEM I (Neville 1996).

2.3. *Structural Evidences of the Swelling Phenomenon*

Evidences of the dam's swelling phenomenon were detected 20 years after construction, by the progressive vertical displacements upwards (Fig. 3), as well as vertical sliding on the central contraction joints, near the top, and the increase of strains measured in the stress-free strain-meters. Linear and map cracking was detected on specific zones of the galleries and of the downstream surface.

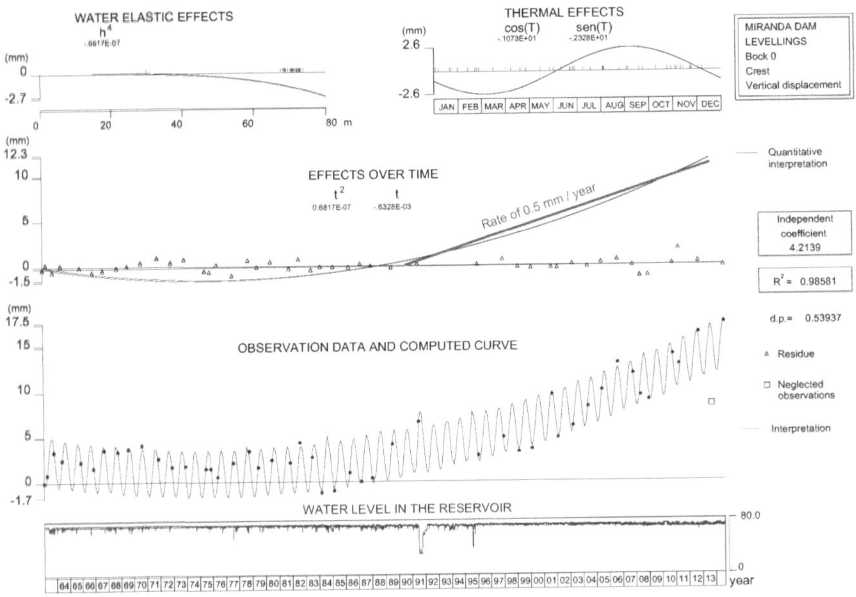

Figure 3. Miranda dam. Results of quantitative analysis of vertical displacements at the top of the central buttress, obtained from geodetic levelling, between 1963 and 2014

3. Laboratory Testing Campaign

3.1. *Samples Extracted from the Dam's Body*

The laboratory test campaign was performed on 114 cores extracted from the dam and on aggregates collected from a quarry, presumably the one used during the dam construction. Fig. 4 shows the locations on the dam's body where the cores were extracted and Table 4 details their number and size.

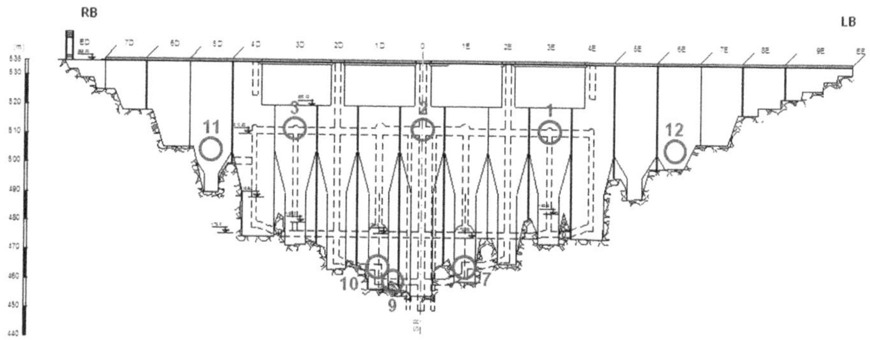

Figure 4. *Identification of the core extraction locations*

Element	Location	Cores Ø 100 mm	Cores Ø 142 mm	Cores Ø 191 mm	Cores Ø 216 mm
1E	7	2	7	0	4
1D	10	3	6	0	4
3E	1	2	6	1	4
0	2	2	6	4	2
3D	3	5	6	1	4
1D	9	2	6	0	5
5D	11*	3	6	0	5
6E	12 *	4	9	0	5

NOTE: * Cores extracted from outside the dam.

Table 4. *Cores extracted from the dam*

3.2. *Tests and Analysis*

3.2.1. *Methodology*

The test campaign comprised the petrographic and mineralogical analysis of the concrete, chemical analysis of the concrete (determination of the cement, sulfate, alkali and silica contents) and the aggregate (determination of the silica and alkali contents), evaluation of the aggregate potential alkali reactivity, evaluation of the potential for further expansion due to ASR and ISR, and the mechanical testing of the concrete (compressive and tensile strengths and stiffness damage test). However, due to the limited length of this paper not all results of the study (LNEC 2015) will be presented or discussed here.

3.2.2. *Petrographic Examination and Mineralogical Analysis*

The petrographic examination, made according to LNEC Specification E 388 (LNEC 1993a), was performed on 8 cores, one from each location, to confirm the consistency in the concrete's constituent materials throughout the structure, characterize the rock materials, identify the presence of potentially alkali-silica reactive constituents, and to observe any distress signs that allow to identify the mechanisms which might be contributing to the outward manifestations of damage. The samples were prepared according to LNEC Specification E 389 (LNEC 1993b). The petrographic examination of the rocks was made not only to identify alkali-reactive minerals and their distribution, but also the similarity between them and the aggregates used in the dam concrete. The examination was complemented with a mineralogical analysis performed by X-ray diffraction, according to a LNEC internal method.

3.2.3. *Chemical Analysis*

One of the factors contributing to the development of internal expansive reactions is the cement content, because the higher the cement content the higher is the alkali content of the concrete. Consequently, the cement content was determined to identify possible variations in the concrete composition throughout the sites sampled, as well as to correlate this variation with the intensity of the observed expansive phenomena. Soluble silica, coming from the aggregates, and alkalis, deriving from the cement and the aggregates, are elements that potentiate ASR; therefore, the silica content (soluble (ASTM 2007) and total (LNEC 1993c, IPQ 2014a) - determined by fusing the sample with lithium tetraborate), alkali content (soluble in acid (Arliguie & Hornain 2007), in water (IPQ 1976) and total (IPQ 2014a) - determined by fusing the sample with lithium tetraborate), as well as their solubilization evolution with time in alkaline solutions, were also determined. The chemical analysis was performed on concrete from 8 cores, one from each location

3.2.4. *Aggregate Potential Reactivity*

The aggregate potential alkali reactivity was assessed according to RILEM Recommended Test Methods AAR-3.1 (Nixon & Sims 2016) and AAR-4.1 (Nixon & Sims 2016).

3.2.5. *Concrete residual expansivity*

The potential for further expansion of the concrete due to ASR was determined on cylindrical specimens obtained from the cores and using the ASR accelerating conditions specified in RILEM Recommended Test Method AAR-3 (Nixon & Sims 2016). This test was performed to 48 specimens, six from each location, all having 142 mm in diameter and 284 mm in length. The residual expansion of the concrete, in regard to ISR, was assessed according to LPC Method no. 67 (Pavoine & Divet

2009) on 20 specimens, also covering all locations. The specimens had a diameter of 100 mm and a length of 200 mm.

3.2.6. *Concrete Ultimate Expansivity*

The aggregate residual alkali reactivity in the concrete was determined by testing specimens immersed in 1N sodium hydroxide solution at 38°C. The samples were monitored for uniaxial length change and mass change throughout the test period. This test was performed to 12 specimens that had been previously subjected to the concrete residual expansion test for a period of 392 days.

3.2.7. *Mechanical Testing*

The mechanical testing was made to assess the concrete general condition and to compare its characteristics in the different sites in the dam, serving also as an indicator of the existence of deleterious internal expansive phenomena. The concrete compressive and tensile splitting strengths were determined according to the Portuguese standards (IPQ 2009) and (IPQ 2011), respectively. The former was carried out on specimens having a diameter of 191 mm and 216 mm and a length/diameter ratio between 1.1 and 2.0; whilst the latter used cylindrical specimens having a diameter of 191 mm and 216 mm and a length/diameter ratio between 0.8 and 1.9. The stiffness damage test was performed according to a LNEC internal procedure, which derives from the test method proposed originally by Chrisp et al. in 1989 (Chrisp *et al.* 1993) and latter modified by (Smaoui *et al.* 2004), and comprises the test parameters set in (LNEC 1993d) and (IPQ 2014b).

4. Results and Discussion

4.1. *Petrographic Examination and Mineralogical Analysis*

The petrographic examination revealed that coarse and fine aggregate have a similar mineralogical composition, *i.e.* granitic *latu senso*, having alkali reactive constituents (*e.g.* strained and deformed quartz and myrmekites) and constituents that may release alkalis into the concrete pore solution (*e.g.* alkaline feldspars, muscovite and biotite). Signs of expansive reactions, like aggregates with borders alteration, fissures in the paste and in paste/aggregate interface and neoformation products, were observed in some samples.

4.2. *Chemical Analysis*

The values obtained for the concrete total alkali content (maximum 5.7%, minimum 5.1%, average 5.4%) are compatible with compositions incorporating granitic aggregates having alkaline minerals, thus corroborating the findings from the petrographic examination. All concrete samples have water-soluble alkali content greater than 2.3 kg/m^3. This indicates that the alkalis present in the concrete are sufficient to sustain ASR. Moreover, in many of the cores, the alkali content exceeds 3.0 kg/m^3 and this represents conditions that are highly favorable for ASR to occur given the potential for and existing alkali-silica reaction signs already observed in the structure. The chemical analysis performed to the aggregate, produced by crushing the rocks collected at the quarry, revealed a water-soluble alkali content varying between 0.018% and 0.027%, which confirms that they might also contribute with alkalis for the development or sustenance of ASR.

4.3. *Aggregate Potential Reactivity*

The currently suggested maximum expansion limits for concrete used in very large, long-service structures, like dams, are: AAR-4.1 – 0.02% at 15 weeks, and/or 0.03% at 20 weeks or longer (*e.g.* 0.03% limit at 1 year); AAR-3.1 – 0.03% at 1 year, and/or 0.04% at 2 years (Nixon & Sims 2016). Hence, the results obtained (Fig. 5) indicate that the aggregate has the potential to engage in deleterious ASR.

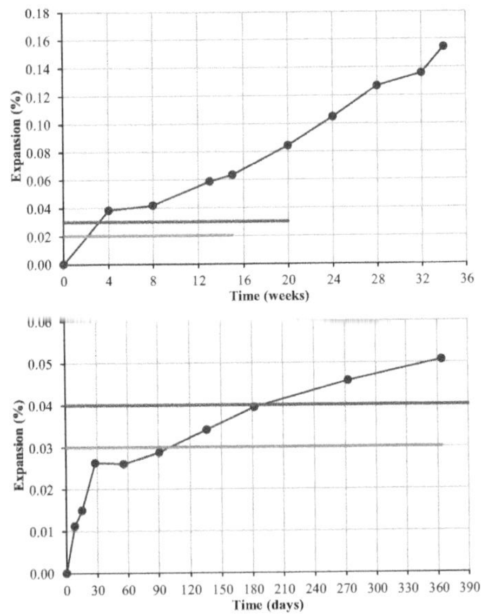

Figure 5. *Aggregate potential reactivity test results (left: AAR-4.1, right: AAR-3.1)*

4.4. *Concrete Residual Expansivity*

The average residual expansion of the cores, for each of the locations, is summarized in Table 5. In terms of the alkali-silica reaction, in the already assessed period, the residual expansion individual test results for all cores ranged from 0.001% to 0.008%, which indicates a negligible to low potential for continued expansion due to ASR. This test is currently ongoing with a group of specimens selected among those from the first test period. Concerning the internal sulfate reaction, the individual expansion values varied between 0.001% and 0.032%, meaning that there is also a negligible to low residual concrete expansion.

Location	Residual expansivity due to ASR					Residual expansivity due to ISR			
	Expansion at 392 days (%)		Expansion at 392 days ($x10^{-6}$)		Rate (μm/m/year)	Expansion (%)		Expansion ($x10^{-6}$)	
	Average	St. dev.	Average	St. dev.		Average	St. dev.	Average	St. dev.
7	0.002	0.001	25	9	5	0.011	0.014	114	145
10	0.004	0.002	36	17	20	-0.017	-	-167	-
1	0.005	0.001	54	13	29	0.013	0.000	130	5
2	0.005	0.001	48	8	16	0.003	0.009	34	86
3	0.005	0.002	47	20	22	0.007	0.006	71	59
9	0.004	0.001	35	10	12	-0.007	0.003	-71	34
11	0.003	0.001	26	12	14	0.013	0.016	130	158
12	0.003	0.003	31	31	18	0.008	0.007	80	75

Table 5. *Concrete residual expansivity test results*

4.5. *Concrete Ultimate Expansivity*

It is widely accepted that the maximum ASR expansion that a concrete member of a structure may suffer in the field can be obtained from the determination of the maximal expansion of concrete cores tested in humid air (>95% R.H.) at 38°C (IStructE 1992). However, such an assumption does not consider the alkali leaching occurring during the test and the consequent leveling-off of the expansion curve. Consequently, immersion tests in 1N NaOH solution can be useful to assess the residual reactivity of the aggregates present in the concrete under study. The results from these tests are normally assessed at 1 year; but, from the behavior obtained so far (Fig. 6), with specimens that exhibited a negligible to low residual concrete expansivity at the 38°C & >95% R.H. test, it is already possible to observe that a moderate residual reactivity still exists in the aggregate.

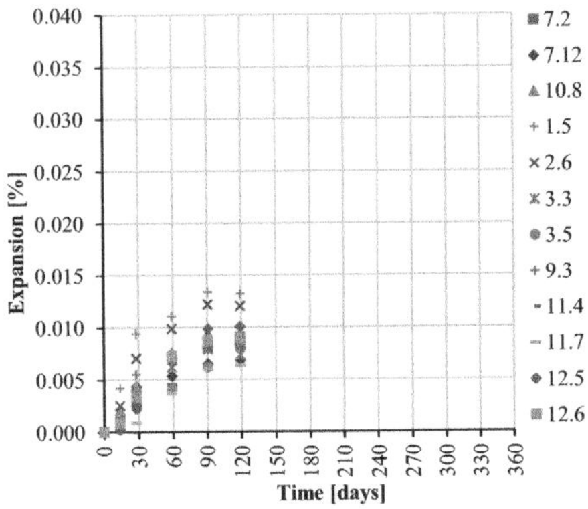

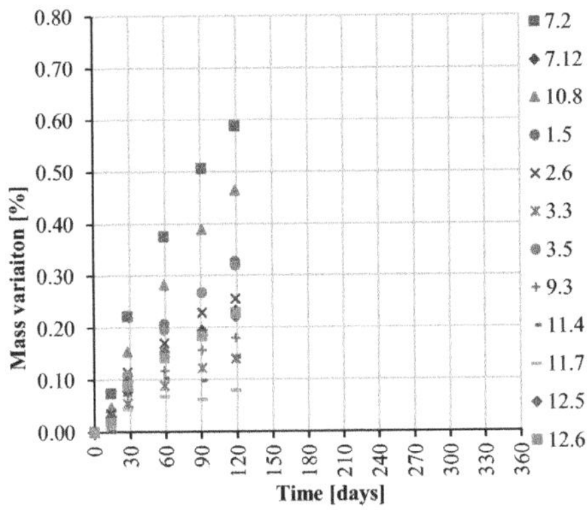

Figure 6. *Concrete ultimate expansivity test results*

4.6. *Mechanical Testing*

4.6.1. *Determination of Compressive Strength*

The results obtained, corrected for an equivalent strength of Ø150x300 mm cylinders, are presented in Fig. 7. The graph also displays the strength estimate made from the compressive strength of cores extracted on the dam during its construction (HED 1983), using the age coefficient defined in (MHOPT 1983) and the core shape correction factor set in (CS 1987, BSI 2007). The results show that there is a great variability on the condition of the concrete sampled throughout the structure, with the individual values ranging from 20.0 MPa (location 1) to 47.8 MPa (location 7). When comparing the experimental values with the correspondent estimated values for sound concrete, it is observed that the compressive strength is below the estimate only in one location. Overall, the sampled concrete exhibits an acceptable compressive strength.

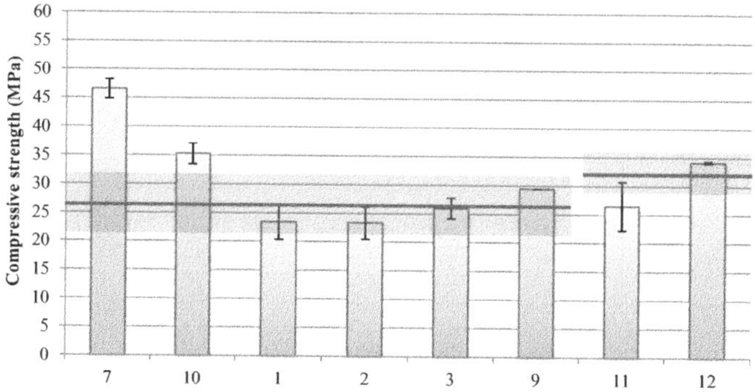

Figure 7. *Concrete compressive strength test results (in blue) and estimated values (in red)*

4.6.2. *Determination of tensile splitting strength*

The tensile splitting strength test results are presented in Table 6. The individual values obtained varied from 1.7 MPa (location 3) to 4.1 MPa (location 7). Even thought, the concrete from locations 11 and 12 has higher cement content than that from the other locations, the tensile splitting strength of that concrete was not the highest, suggesting that some internal damage could exist.

Location	Tensile splitting strength (MPa)		
	No. of specimens	Average	St. dev.
7	2	4.0	0.2
10	2	2.9	0.3
1	2	2.1	0.2
2	3	2.6	0.4
3	2	2.5	1.1
9	1	2.5	-
11	2	2.3	0.3
12	2	2.5	0.1

Table 6. *Concrete tensile splitting strength test results*

4.6.3. *Stiffness Damage Test*

The results obtained in the stiffness damage test are shown in Table 7 and Fig. 8, the latter also includes the estimate made from the compressive strength of cores extracted during the dam construction (LNEC 2015). When comparing the experimental values with the correspondent estimated values for sound concrete, it is observed that the stiffness is below what was expected in five locations. The corresponding values for accumulated final extension and dissipated energy confirm the existence of some internal damage in the concrete. The values obtained for one of the specimens from location 2 are indicative of a relevant internal damage.

Location	No. of specimens	Stiffness (GPa)		Accumulated final extension ($x10^{-6}$)		Dissipated energy (J/m^3)	
		Average	St. dev.	Average	St. dev.	Average	St. dev.
7	4	31.7	3.7	52	13	262	101
10	4	36.6	4.8	46	11	217	44
1	4	25.1	1.5	59	13	318	43
2	4	28.0	7.3	75	43	438	314
3	4	23.3	1.6	78	6	422	37
9	4	32.2	2.8	49	6	226	21
11	5	27.5	2.6	66	17	339	104
12	5	28.8	2.3	67	11	365	77

Table 7. *Stiffness damage test results*

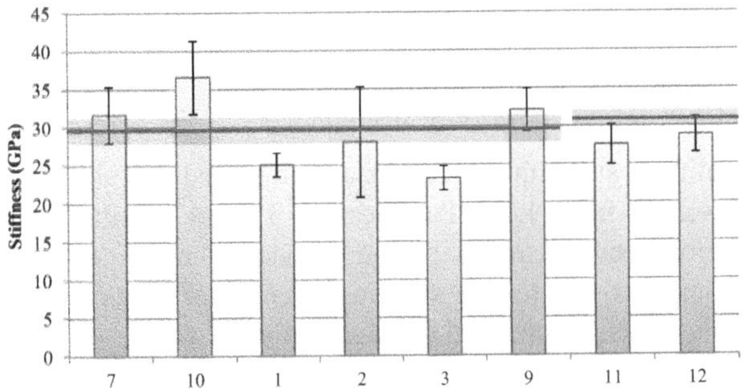

Figure 8. *Stiffness damage test results (in blue) and estimated values (in red)*

5. Conclusions

This paper analyzed and discussed very succinctly a subset of the data produced from the characterization and testing of the 114 cores extracted from the Miranda dam. This paper aimed at providing the scientific and technical community an example of a practical application of several test methods, current available, for the diagnosis and prognosis of ASR and ISR in large concrete structures.

The combination of the various test methods allowed determining that (i) the concrete is affected by both ASR and ISR (ii) the internal expansive reactions have not yet evolved to an extent so that they result in a relevant reduction of the concrete mechanical properties, (iii) the potential for continued deterioration due to ISR is negligible and due to ASR appears to be somewhat limited; it spite of that, a final conclusion will only be made once all the ongoing tests are completed.

The inspection results and the treatment of the monitoring data revealed that (i) the strains and displacements remain small, (ii) the magnitude of the swelling effects in the structure are still moderate and (iii) the map cracking on the dam surface has had a limited evolution in time.

Consequently, it can be concluded that, for Miranda dam, no problems related with serviceability and safety are expected in the near future.

A methodology, such as the one described in the paper, is thus useful for any long-life structure where concrete swelling deterioration mechanisms are observed visually or structurally.

Finally, it is important that the evolution of this pathology continues to be carefully accompanied. Moreover, the available data can be used, in the future, as input values on modeling for structural assessment.

6. References

Arliguie G., Hornain H., *Grandubé - grandeurs associeés à la durabilité des bétons*. Presses de l'Ecole Nationale des Ponts et Chaussées: Paris, 2007.

ASTM, "ASTM C289-07 Standard Test Method for Potential Alkali-Silica Reactivity of Aggregates (Chemical Method)". ASTM International: West Conshohocken, PA, 2007.

Batista A. L., Gomes J. P., "Characteristic behaviour of the Portuguese large concrete dams built with granite aggregates and affected by ASR", in *15th International Conference on Alkali-Aggregate Reaction – ICAAR 2016*. São Paulo, Brazil. 2016.

BSI, "BS EN 13791:2007 Assessment of in-situ compressive strength in structures and precast components". British Standards Institution (BSI): London, 2007.

CEN, "EN 14216:2015 Cement. Composition, specifications and conformity criteria for very low heat special cements". European Committee for Standardization (CEN): Brussels, 2015.

CEN, "EN 197-1:2011 Cement. Composition, specifications and conformity criteria for common cements". European Committee for Standardization (CEN): Brussels, 2011.

Chrisp T. M.; Waldron P., Wood J. G. M., "Development of a nondestructive test to quantify damage in deteriorated concrete". *Magazine of Concrete Research,* vol. 45, (165), pp. 247-256, 1993.

CS, "Concrete Society Technical Report No. 11 including Addendum (1987) - Concrete core testing for strength". The Concrete Society: London, 1987.

HED, "A betonagem da barragem de Miranda - Relatório final", Hidro-Elétrica do Douro, S.A.R.L., 1961.

HED, "Construção do Aproveitamento Hidroelétrico de Miranda - Douro Internacional. Relatório Final. Capítulo VIII - Estudo, fabrico e controle dos betões no estaleiro", Hidro-Elétrica do Douro, S.A.R.L., 1983.

IPQ, "NP 1382:1976 Aggregates for mortars and concretes. Determination of soluble alkali content. Flame spectrophotometry process (in Portuguese)". Instituto Português da Qualidade, I.P. (IPQ): Caparica, 1976.

IPQ, "NP EN 12390-13:2014 Testing hardened concrete. Part 13: Determination of secant modulus of elasticity in compression (in Portuguese)". Instituto Português da Qualidade, I.P. (IPQ): Caparica, 2014b.

IPQ, "NP EN 12390-6:2011 Testing hardened concrete. Part 6: Tensile splitting strength of test specimens (in Portuguese)". Instituto Português da Qualidade, I.P. (IPQ): Caparica, 2011.

IPQ, "NP EN 12504-1:2009 Testing concrete in structures. Part 1: Cored specimens. Taking, examining and testing in compression (in Portuguese)". Instituto Português da Qualidade, I.P. (IPQ): Caparica, 2009.

IPQ, "NP EN 196-2:2014 Method of testing cement. Part 2: Chemical analysis of cement (in Portuguese)". Instituto Português da Qualidade, I.P. (IPQ): Caparica, 2014a.

IStructE, "Structural effects of alkali-silica reaction: Technical guidance on the appraisal of existing structures (with Addendum from April 2010)". The Institution of Structural Engineers: London, 1992.

LNEC, "LNEC Specification E 388:1993 Concrete. Macro and microstructural analysis. Petrographic examination (in Portuguese)". Laboratório Nacional de Engenharia Civil, I. P. (LNEC): Lisboa, 1993a.

LNEC, "LNEC Specification E 389:1993 Concrete. Thin section preparation for microstructural analysis (in Portuguese)". Laboratório Nacional de Engenharia Civil, I. P. (LNEC): Lisboa, 1993b.

LNEC, "LNEC Specification E 397:1993 Hardened concrete. Determination of the modulus of elasticity of concrete in compression (in Portuguese)". Laboratório Nacional de Engenharia Civil, I. P. (LNEC): Lisboa, 1993d.

LNEC, "LNEC Specification E 407:1993 Cement. Determination of the soluble silica content. Molybdate spectrophotometric method (in Portuguese)". Laboratório Nacional de Engenharia Civil, I. P. (LNEC): Lisboa, 1993c.

LNEC, "Miranda dam - Assessment of the concrete degradation through the results obtained in the laboratory tests performed, between 2013 and 2015, on samples collected from the dam (in Portuguese)", LNEC - Laboratório Nacional de Engenharia Civil, I.P., Report 406/2015 – DM/NBPC, 2015.

MHOPT, "Decreto-Lei n.º 349-C/83 de 30 de Julho - Regulamento de Estruturas de Betão Armado e Pré-Esforçado". Ministério da Habitação, Obras Públicas e Transportes: Diário da República, I Série, Número 174, 1983.

Neville A. M., *Properties of Concrete*, 4th ed., John Wiley & Sons: New York, 1996.

Nixon P. J., Sims I., "RILEM Recommendations for the Prevention of Damage by Alkali-Aggregate Reactions in New Concrete Structures (State-of-the-Art Report of the RILEM Technical Committee 219-ACS)", in *RILEM State-of-the-Art Reports*. Springer: Dordrecht, 2016.

Pavoine M.A., Divet L., "Méthode d'essai des LPC n. 67. Réaction sulfatique interne au béton. Essai d'expansion résiduelle sur carotte de béton extraite de l'ouvrage", in Techniques et méthodes des laboratoires des ponts et chaussées. Laboratoire Central des Ponts et Chaussées (LCPC): Paris, 2009.

Smaoui N., Bérubé M.A., Fournier B., Bissonnette B., Durand B., "Evaluation of the expansion attained to date by concrete affected by alkali-silica reaction. Part I: Experimental study". *Canadian Journal of Civil Engineering,* vol. 31, (5), pp. 826-845, 2004.

Long-Term Behaviour of EDF Dams Regarding Concrete Swelling Structures

Synthesis of Monitoring Data Over Time

THIERRY GUILLOTEAU* — FRANÇOIS MARTINOT* — JÉRÔME SAUSSE**

** EDF/DTG*
21 avenue de l'Europe, BP41, 38040 Grenoble, France
thierry.guilloteau@edf.fr, francois.martinot@edf.fr

*** EDF/DTG*
18 avenue Raymond Poincaré, BP423, 19311 Brive La Gaillarde, France
jerome.sausse@edf.fr

ABSTRACT. *EDF operates156 monitored concrete dams. Two specific studies of this large panel were realized in 2007 and 2014 to identify the potential existence of swelling reactions, characterize it with explicit indicators and criteria and try to identify common influencing factors and trends. The main results were that at least 20% of them are potentially affected by swelling reactions, considering at least dams with a medium ratio higher than 2µm/m/year (only 5 dams have a medium ratio above 30 µm/m/year). All these dams were built between 1930 and 1960 (aged from 60 to 90 years), and are mainly gravity and single curvature arch dams. An attempt to reconstitute the total deformation was carried out, showing the importance to consider cumulated deformations for dam behaviour analyses.*

KEYWORDS: *dams, monitoring, swelling reaction, safety, surveillance.*

1. Introduction

This communication is based on 2 studies conducted on Edf monitored concrete dams, and regarding concrete swelling reactions:

– the first one in 2007 (Sausse et al. 2011), with the following objectives: identification of dams affected by a potentially significant swelling expansion, characterisation and inter-comparison of criteria linked to the monitored effects of swelling, identification of common trends;

– the second one in 2014 (Mauris et al. 2015), with the objectives of completing the first study on: quantification of amplitudes after first impoundment or after the detection of the phenomena, better understanding of evolution of monitored kinetics in time.

All information published here come from data acquired on monitoring devices installed on dams. As required by the French regulations, this information is edited and diagnosed periodically for each dam.

2. Dams and Data Selection – Methodology of Analysis

2.1. *Dams and Data Selection*

EDF is monitoring more than 300 dams of class A, B and C, from which 156 are concrete dams potentially sensitive to swelling reactions:

- 64 gravity dams;
- 50 arch dams;
- 20 gated dams;
- And 22 of other types.

Common devices for measuring strains and displacements (absolute and relative) on structures are:

- geodesy in X,Y plan and altimetry in Z axis;
- pendulums for displacements on X, Y and sometimes Z axis;
- embedded vibrating wire extensometers;
- surfacing long-base extensometers;
- crack/joint meters (in 1 or 3 dimensions).

Some dams have redundant measurements (for example: Z pendulum and altimetry), but in some cases there is no device for monitoring displacements or strains. It usually concerns small dams, just equipped with devices controlling leakages, and for which visual inspection is the best and economical way to detect concrete pathologies.

Pathology of concrete swelling generates expansion of the structures in all directions but it is mainly the irreversible displacement of dam crest upward that is the better indicator of a swelling reaction. In that case, monitoring devices informing on vertical strains or displacements are very useful.

In the particular case of arch dams, horizontal displacements may provide enough information to detect and quantify swelling reactions. Indeed, expansion of the arch introduces necessarily a displacement upstream (by the arc elongation). Nevertheless, upstream irreversible displacement is not necessarily the only consequence of a swelling reaction (e.g. rotation of the arch, irreversible movement of a bank…). In the case of gravity dams, swelling does not necessarily generate a horizontal upstream displacement, even if it is often observed. Therefore the installation of vertical displacement devices is recommended if swelling reaction is suspected.

2.2. *Methodology Followed by 2007 and 2014 Studies*

For the purpose of the first study, dams were selected only if a minimum data was available:

– for arch dams: strains or horizontal or vertical displacements in crest;

– for other types of concrete dams (gravity, gated, others): strains or horizontal or vertical displacements in crest.

Monitoring gives a possible access to different criteria quantifying the swelling reaction. Criteria considered in the 2007 study were based on the most recent data available and gave a picture of the current situation. These criteria were calculated using kinetics of expansion and translated in rates of expansion (Sausse et al. 2011):

– maximum and medium rate of swelling in vertical direction for all dams;

– maximum and medium rate of arc elongation, for arch dams, in horizontal direction;

– medium ratio of swelling along the upper arch for arch dams.

Computing criteria for swelling quantification could be both simple and tricky, because the results can differ considerably: depending on sensors (local like strain gauges) or long basis (like pendulums or extensometers), the part and the geometry of the structure, the available data (period of observation, accuracy and reliability, frequency of measurements...).

Among the 156 concrete dams, 93 had the minimum data required for the characterisation of a potential swelling reaction and the associated criteria, and 30 were identified as potentially affected by a swelling reaction (Figure 1). These results are presented and analysed in chapter 3.

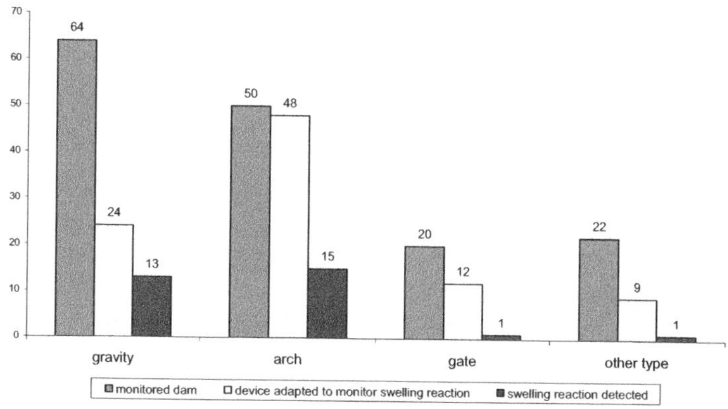

Figure 1. *Dams on which enough data was available for swelling identification and quantification, referring to 2007 study*

It was established that 20% of large concrete dams were concerned by swelling phenomena (including AAR and ISR). This ratio was a little bit increasing since 2007, considering new potential cases (especially small dams), but it remains under 30%. Portuguese experience (Batista et al. 2012) indicates 16 cases of large concrete dams affected by both AAR (i.e. 31% of the large concrete dams). In Brazil, 25 cases were reported in a 2016 publication (Nogueira Silva 2016).

The objectives of the 2014 study were to complete the "picture" given by the 2007 study: actual rates of swelling ratio and their kinetics are of primary interest but don't illustrate all the evolution and implications of these phenomena in time. Thus, in the second study, the addressed questions were to try to determine the total strains following the first impoundment or the first detection of signs of swelling.

Dams considered were the 30 dams identified in the 2007 study. The results and analyses are presented in chapter 4.

3. Monitored Ratio and Kinetics of Swelling Reactions on Concrete Dams Portfolio (2007 Study)

In this chapter, the results and analyses concern 30 dams for which the medium rate of swelling in vertical or along the arc of arch dams is greater than 2 μm/m/an.

It is the minimum value under which it is quite impossible to dissociate the internal phenomena of swelling reactions from external factors such as global climate heating. In many cases, and with such low rates, it is also difficult to confirm the nature of the swelling reaction (AAR, ISR,..) by laboratory analysis or other investigations.

3.1. *Indicators of Dams Affected of a Swelling Reaction*

3.1.1. *Rates of Swelling in Vertical Direction (All Types of Dams)*

Among the 54 dams equipped with vertical displacement devices, 20 show an irreversible expansion with a maximum rate varying from 2 to 70 µm/m/year. High levels of expansion (more than 30µm/m/year) are very rare and concern only 5 dams among the 150 studied (less than 5%). These dams are very well known (i.e. high level of monitoring, recent FEM stability studies, international publications…) and 2 of them have benefited of a specialized maintenance.

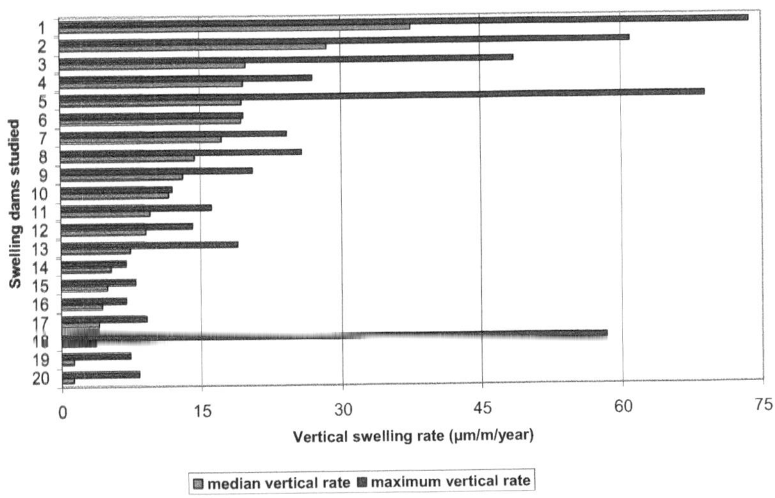

Figure 2. *Vertical swelling rates observed*

The distinction between maximal rate (localized on a part of the dam) and the medium rate (average of each rate calculated on each device) underlines some homogeneous swellings and heterogeneous swellings (e.g. only a specific part of dam is affected by swelling reaction). The dam n°18 on figure 7 shows the most heterogeneous swelling studied: only few concrete lifts are affected by ISR phenomena (Figure 3).

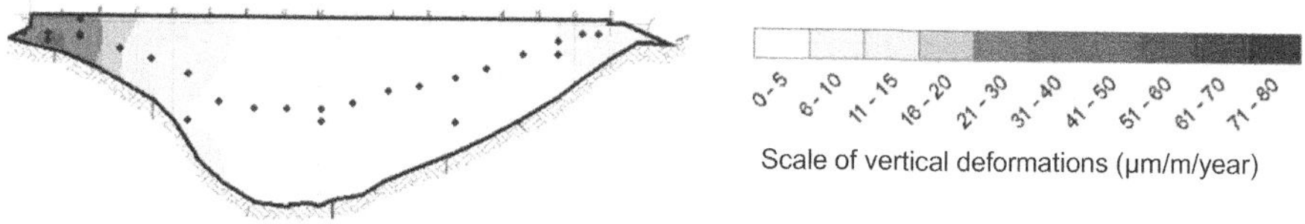

Figure 3. *Vertical swelling rates observed on dam n°18*

3.1.2. *Rates of Swelling Along the Crest Arc (Only Arch Dams)*

Irreversible movements towards the upstream can be the sign of swelling reaction of arch dams if this displacement is the consequence of an arc elongation. 15 arch dams show an arc elongation greater than 2µm/m/year (Figures 1 & 4).

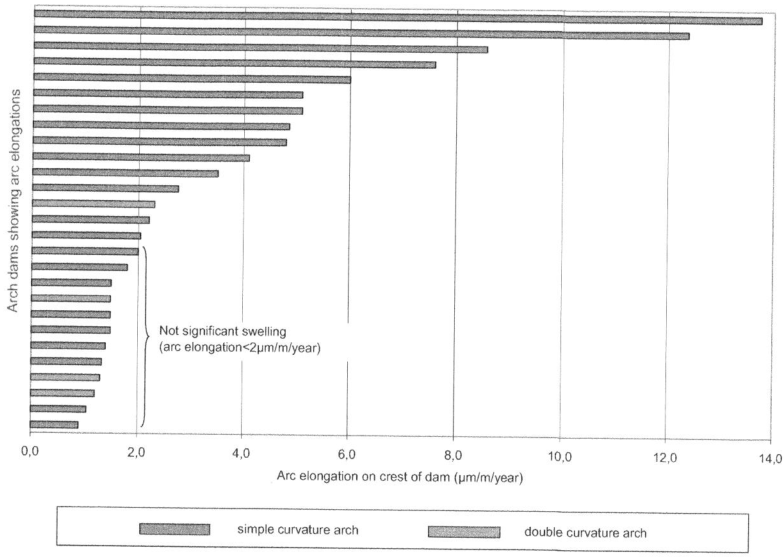

Figure 4. *Arc elongation on crest of arch dams*

Eight of them are also equipped with vertical displacement monitoring devices and for this panel, a relation can be established between vertical and horizontal strains (Figure 5). At first glance, the ratio of vertical strains versus horizontal strains is around 2. It depends on the capacity of deformation of abutments, and the part of compressive stresses induced by horizontal forces of swelling reaction. On 2 arch dams (in red on figure 5), the ratio is closed to 4, due to possible stiffness of abutments and higher compressive stresses.

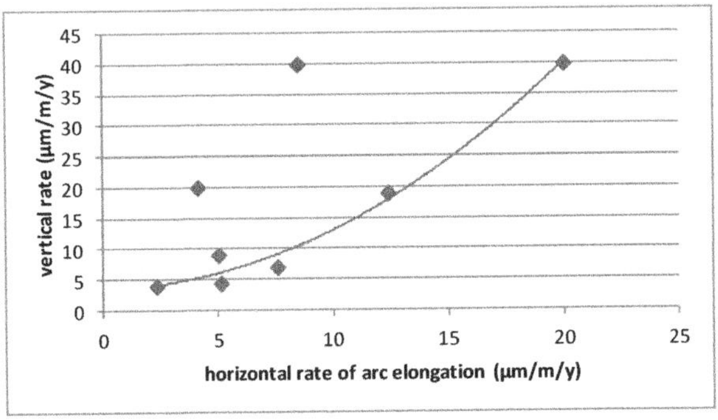

Figure 5. *Vertical rate versus horizontal rate of swelling (on 8 arch dams)*

3.2. *Common Factors Observed on Swelling Dams*

The exploitation of data extracted from the 150 concrete monitored dams allows an analysis of intrinsic criteria (e.g. geology, age, morphology of dam…) presenting a potential influence on swelling reactions.

3.2.1. *Types of Dams*

The identification of dams affected by swelling reactions underlines the arch and gravity dams as the more concerned types. The other types with thinner structures present higher reinforcement and stresses, limiting the effects of the expanding cement.

The analysis of swelling arch dams is very interesting because most of them are single curvature (cylindrical) arches (Figure 4). Indeed, it is observed that double curvature arches generally show irreversible downstream displacements due to a more important part of creep of these thin structures (due to higher internal stresses). Therefore, it can be argued that more important stresses in double curvature arches may confine swelling effects which are principally observed on cylindrical arches.

Swelling effects on large gravity dams are also interesting considering the role of joints. Effectively, gravity dams show horizontal displacements towards the upstream, when the number of contraction joints per meter is limited or when their joints are closed. At the opposite, when joints are numerous and opened, horizontal displacements due to swelling reaction is lower and could be oriented towards the downstream, in some cases. This is also observed just after a slot cutting operation (e.g. Chambon dam). Therefore, with the ageing of gravity dams affected by swelling pathology, contraction joints tend to close gradually, inducing a displacement of the crest towards the upstream.

3.2.2. *Age and Building Period of Dams*

The number of dams showing a vertical expansion is maximal on the decade 1951-1960, but it is also the period when the number of construction was the most important in France. The period between 1930 and 1950 is also remarkable because 65% (12 of 42 dams built) of monitored dams built in this decade show a vertical irreversible movement. After 1960, no case of swelling is registered on the 50 dams built.

Figure 6. *Dams identified with swelling, and building dates*

The poor quality of cements produced during and just after the second world war, and the difficulties of supply during this period could explain this important percentage of swelling dams aged 60 to 90 years.

Following the 60s and the identification of AAR pathology, professionals (cement manufacturers, project managers, owners…) took care of the formulation of cement (i.e. rate of alkali) and its reactive potential.

3.2.3. *Geology and Dam Localization*

The localization of swelling dams in France underlines a central zone where most cases are observed (Figure 7). This part of France is the mountainous "Massif Central" whose geology shows a predominant part of igneous and magmatic rocks, able to liberate alkali. In this part of France, about 50% of dams equipped with vertical displacements devices show irreversible movements attributed to chemical swelling reactions. In the other regions, this part of swelling dams is reduced to 25%. The pertinence of this geographic (and geologic) factor (BHATT 2015) is also confirmed by the analysis of the age of dams situated in Massif Central: the ratio of dams built during the critical period (i.e. 1930-1950) is quite the same as shown in the other regions of France.

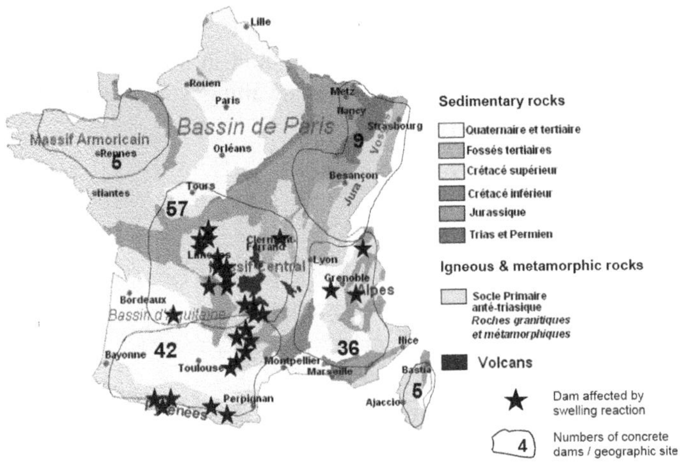

Figure 7. *Localization of swelling dams*

4. Estimation of Total Effects and Trends on Dams Potentially Affected by Swelling Reactions (2014 Study)

In the previous paragraph, we characterized and compared the present activity of swelling reactions measured at the time being on dams. However, this picture does not give information on the total effects of this phenomenon since its initiation. Indeed, dams affected by the most active swelling reaction were well identified in 2007, but it was also important for EDF to know which had suffered the main cumulated deformations.

Knowing the ages of the 30 dams studied in 2007 (with a swelling ratio greater than 2 μm/m/year), it is easy to give a first assess of total strain: age (years) x swelling rate (μm/m/year). Nevertheless, this approach is not allowable because of the non-linear kinetics of swelling reactions, well known referring to laboratory experiences (LARIVE 2000). It was also necessary to take into account the part of shrinkage and creep, and all causes of delayed start of swelling effects.

4.1. *Modelling Swelling Kinetics since First Detection*

The difficulty of the 2014 study was the lack of data, during the first years and decades of life of dams. Indeed, a large majority of dams studied were well instrumented after the possible beginning of the swelling reaction. So it was tried to model the probable behaviour since the beginning of reactions, for all the dams studied.

However, embedded extensometers data was only available on 8 arch dams, at least since their first impoundment and over 40 years. These measurements showed that strain kinetics was well fitted by a polynomial law of time "t", that ranges from f(t) to $f(t^{2.5})$, when applied since the date of the probable beginning of reaction. This date of reaction's detection is unique to each dam: sometimes first swelling clues are highlighted just after impoundment, sometimes they are disclosed later since swelling effects are masked by shrinkage and creep (shown on dams n°2 and 6, Figure 8).

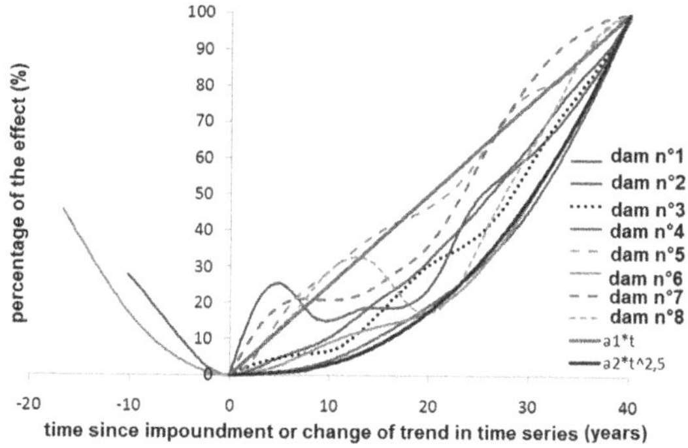

Figure 8. *Strain kinetics modeled on 8 dams using monitored data from embedded extensometers*

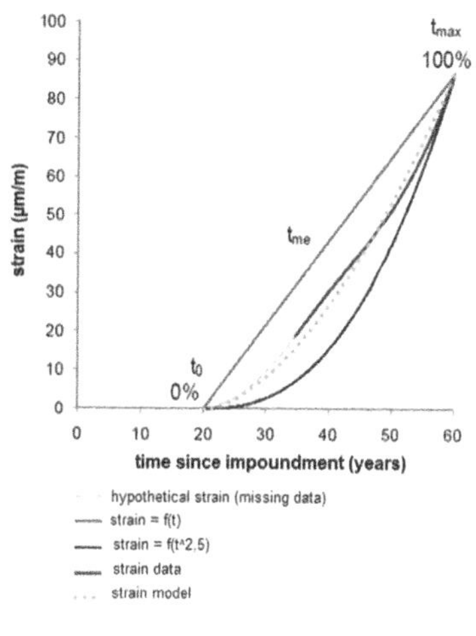

Figure 9. *Fitting of the monitoring data with $f(t^x)$ and $(1 \leq x \leq 2.5)$*

Based on this statistical model (using a polynomial law of time fitted between t and $t^{2.5}$), whole time series were rebuilt to estimate cumulated effects. Thus, tangential strains along the crest for arch dams and vertical expansion for gravity dams were determined since the estimation of the beginning of reaction to now.

This model appears like an approximation of the "initiation" part of the theoretical "S" curve (Larive 2000) of swelling kinetics.

4.2. *Results: Quantification of Total Strains since the Date of First Detection*

After modeling, we obtain total estimated strain values in vertical and horizontal directions (arc elongations for arch dams). Shapes of kinetics are generally better explained with a linear law of time (t) for vertical strains, and with a polynomial function (with a degree upper than 2), for arc elongations (Figure 10).

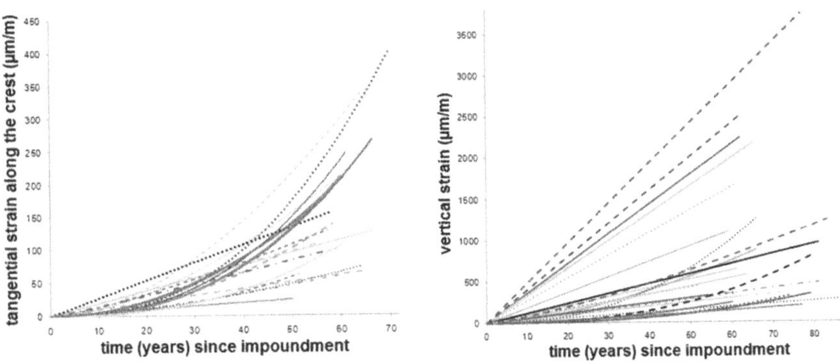

Figure 10. *Tangential and vertical strains modeled since first impoundment on 30 dams*

Cumulated arc elongations should not exceed 250 µm/m. They are lower than cumulated vertical strain values because of the absence of cracks in this direction, and a part of swelling effects resulting in local compressions.

Cumulated vertical strain values could vary between 200 and 3500 µm/m. Among 30 dams studied, only 7 showed deformations higher than 1000 µm/m. These values, of free vertical strains, include necessarily a part of cracks (Amberg 2011). All these dams, and particularly those with estimated cumulated vertical strain values upper than 1000 µm/m have a reinforced monitoring system, 3D FEM studies (Bourdarot 2010 and Grimal 2007) and sometimes remedial works adapted to the pathology and its negative effects on safety.

All these results are coherent with others cases of swelling dams, published worldwide (Charlwood 2009 and Curtis 2000). Nevertheless, these strain amplitudes must be interpreted with caution since whole time series are rebuilt with heterogeneous confidence levels (shrinkage and creep, missing data with war, monitoring devices, visual clues).

4.3. *Comparison of Total Swelling Strains with Seasonal Strains on Arch Dams*

Considering swelling arch dams, the assessment of the importance of the amplitude of swelling reactions can be considered by comparison to seasonal thermal strains, determined by applying a classic statistical HST (Ferry 1958) analysis (Figure 11).

Therefore, we should consider this criterion: when total swelling strain becomes higher than seasonal strain, we should consider the need of the adaptation of monitoring to the pathology and its consequences, and to determine potential stresses with a specific 3D model, taking into account strain measurements.

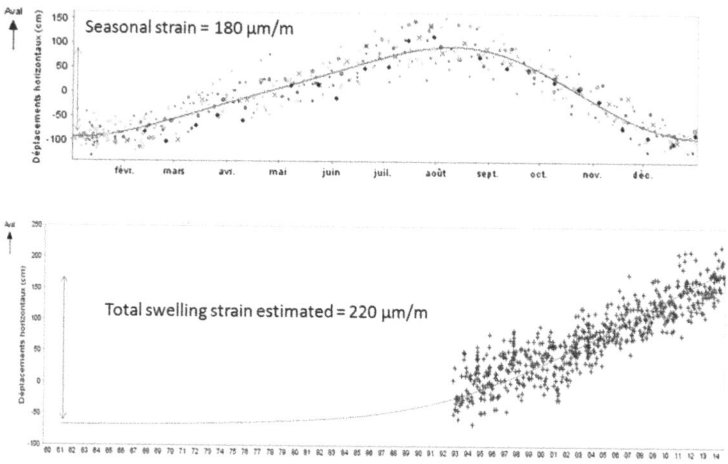

Figure 11. *Comparison of total arc elongation strain with the seasonal deformation*

5. Conclusion

Swelling of concrete is generally found on concrete dams built during the period 1930-1960. It is easily detected by standard monitoring devices installed in gravity and arch dams. Its characterisation is often limited to the publication of the annual kinetics of strain, and this value is frequentally used for comparisons betwwen dams and to estimate the severity of the phenomena. This ratio must be considered with caution and indications on where, how and with which data it is computed is necessary.

For an appropriate and valid evaluation of the kinetics of swelling on a dam, it is recommended to:

– favorize measurements in directions where the phenomenon can freely expand and generate maximum displacements and strain, generally following a vertical orientation;

– adapt the monitoring system to its type and geometry, to its different structures (central and abutments blocks, spillways, ...) and to the possible heterogeneity of concrete;

– perform regular and accurate measurements, to quantify the progress of evolutions.

Nevertheless, the determination of current kinetics is not enough in itself to assess the consequences in time of swelling. Visual inspection provides an essential source of information to complete the analysis based on monitoring data. But, a simple and cautious estimation of the cumulated amplitudes of deformation is also useful. It can be easily done using a polynomial modelisation and an assumption on the date of initiation of the measured evolutions.

Considering all this information, and the fact that swelling pathologies have no immediate or brutal consequences, dam owners have the possibility to plan with anticipation investigations, FEM studies and maintenance with confidence.

6. References

Amberg, F. (2011) *Performance of Dams affected by expanding concrete.* Dams and Reservoirs under Changing Challenges – Schleiss & Boes (Eds).

Batista, A. et al. (2012*). Practical assessment of the structural effects of swelling processes and updated inventory of the affected Portuguese concrete dams.* 54º Congresso Brasileiro do concreto.

Bhatt, S. (2015). *Effects of Alkali Silica/Aggregate Reaction on Concrete Structures in Bundelkhand Region, Central India,* Engineering Geology for Society and Territory - Volume 6. pp 195-201.

Bourdarot, E. (2007). *Expansion Issues in EDF Dams.* Granada: Chemical expansion of concrete in dams.

Bourdarot, E. et al. (2010). *A review of continuum damage modelling for dam analysis,* European Journal of Environmental and Civil Engineering.

Charlwood, R. (2009). *AAR in Dams and Hydroelectric Plants.* Short Course on Management of Alkali Aggregate Affected Structures: Analysis, Performance & Prediction.

Curtis, D. (2000). "A review and analysis of AAR-effects in arch dams". *Proceedings of the 11th International Conference on Alkali-Aggregate Reaction (ICAAR)*, Quebec, Canada, June 11 16, 2000, pp. 1273-1282.

Ferry, S. (1958). "Méthodes d'analyse et de surveillance des déplacements observés par le moyen de pendules dans les barrages". In: *VIth International congress on large dams.* New York; 1958. p. 1179–201 [in French].

Grimal, E. (2007). "Caractérisation des effets du gonflement provoqué par la réaction alcali-silice sur le comportement mécanique d'une structure en béton". Université de Toulouse: Thèse de doctorat/PhD dissertation.

Larive, C. (2000.). *The rôle of water in alkali silica reaction.* Quebec: 11th conference on alkali silicate reaction/CIRAG ICAAR.

Mauris, F. (2015). "Synthesis of hydraulic structures behavior: lessons learned from monitored dams of EDF in France LTBD 2015", *3rd International Workshop on Long-Term Behaviour and Environmentally Friendly Rehabilitation Technologies of Dams* 17 – 19 October 2015 – Nanjing, China.

Nogueira Silva et al. (2016). *Alkali-Aggregate Reaction - Brazil Experience*, EJGE, p. 6879-6892.

Sausse, J. et al. (2011). *Diagnosis of dams affected by swelling reactions: lessons learned from 150 monitored concrete dams in france.* Lisbon 6th International conference on dam engineering.

Important Lessons Learnt from the Proper Surveillance of Swelling Concrete

LOUIS CHRISTIAN HATTINGH* — CHRIS OOSTHUIZEN** — ILIDIO TEMBE*** — C.N. MAHLABELA****

* *Hattingh Anderson Associates CC*
Pretoria, South Africa
halh@icon.co.za

** *DS3Consultants (Pty) Ltd*
Pretoria, South Africa
coosthuizenc@gmail.com

*** *Hidroeléctrica de Cahora Bassa*
Songo, Mozambique
ilidio.tembe@hcb.co.mz

**** *Department of Water and Sanitation*
Pretoria, South Africa
MahlabelaC@dwa.gov.za

ABSTRACT. *When dealing with swelling concrete in dams the main questions that need to be answered pertains to the safety of the structure and the possible failure mechanisms. To answer these there are normally a number of uncertainties that should be considered. These include the current extent of reaction/swelling, future prediction of reaction/swelling, 3D swelling behaviour, as well as the impact of openings on swelling behaviour and changes in stress profile. This paper intends to provide some answers to these uncertainties and highlight the importance of a properly designed surveillance system with the necessary redundancy. The surveillance strategies currently in use in the dam fraternity are also discussed using a number of case histories of dams in Southern Africa.*

KEYWORDS: *surveillance, swelling of concrete, ASR, impact of openings, impact of confinement, 3D behaviour, redundancy.*

1. Introduction

When dealing with swelling concrete in dams (whether it is arch, gravity or buttress structures), the main questions that need to be answered pertains to the safety of the structure and the possible failure mechanisms. To answer these questions there are normally a number of uncertainties that should be considered. These include:

- the current extent of reaction/swelling;
- the rate and prediction of future reaction/swelling (in other words when if at all will the swelling reaction stop if at all);
- 3D swelling behaviour of the structure;
- the impact of openings on swelling behaviour; and
- changes in stress profile within the structure.

This paper will make use of the definitions of surveillance and monitoring as contained in the latest ICOLD bulletins on surveillance (ICOLD: 2000 & 2014). Surveillance is therefore defined as consisting of the following four aspects:

- manual monitoring;
- automated monitoring;
- visual inspections; and
- analyses (checking and testing).

Typically the surveillance system of a new dam is not specifically designed to monitor the behaviour of any swelling of concrete. The concrete mixes of new dams are designed not have any significant swelling due to any chemical reaction (whether it is alkali silica reaction or any other adverse swelling reaction). It is important to note that for the purposes of this paper the term swelling concrete would be used to describe swelling of concrete due to any chemical reaction.

Therefore when it becomes apparent that some form of swelling due to chemical reaction is taking place at a dam one has to make use of the existing surveillance system. In many cases the results available from the existing surveillance system does not provide the required answers to the above-mentioned uncertainties. In many cases additional monitoring systems and inspections are implemented.

A proper surveillance system of a dam should monitor both loads and responses and for an arch dam typically includes the following:

- loads:
    - hydrostatic (water levels);
    - temperature (air & concrete);

  - foundation piezometric pressures;
- responses (both static and dynamic):
  - ground motion
  - displacements & rotations (relative and absolute);
  - strains;
  - stress;
  - seepage;
  - pore pressure; and
  - dynamic responses.

The monitoring system should also comprise sufficient redundancy for most of the responses of the dam. Not only are displacements, rotations and strains some of the most common responses used to shine some light on the uncertainties mentioned above that needs to be addressed, it is these responses that in general are the easiest to attain the required redundancy.

However, for the large majority of dams subjected to swelling concrete issues, the monitoring is normally restricted to 1D displacement measurements. Typical examples for arch dams would be vertical precision levelling of the crest and if one is lucky perhaps an additional single pendulum measuring the horizontal displacement of the centre cantilever.

In the following sections some case histories of dams in Southern Africa will be discussed. The intention is not only to provide some answers to these uncertainties but will highlight the importance of a properly designed surveillance system. The necessary redundancy and also question some of the surveillance strategies currently in use in the dam fraternity when dealing with swelling concrete will be addressed.

2. Cahora Bassa Dam

Cahora Bassa Dam, a 170 m high double curvature concrete arch dam, is located on the Zambezi River close to Songo in the Tete Province of Mozambique (see Figure 1). Construction of the dam that forms part of a 2 075 MW hydro electrical scheme, was finished in 1976. The rest of the scheme consists of an underground powerhouse on the right flank complete with upstream intakes and penstocks, 5 power units of 415 MW/480 MVA each as well as downstream surge chambers and tail race tunnels.

An extensive monitoring system provided during construction and subsequently upgraded, monitors either the loads on the dam or its static and/or dynamic response.

Redundancy was provided for the hydrostatic and temperature loads and more importantly for the displacement and strain responses of the dam structure (Hattingh et al, 2014).

Figure 1. *Downstream view of Cahora Bassa Dam*

The strain in the dam is monitored by a number of different instrument types and includes:

– Precision levelling (relative vertical displacements on the crest and the horizontal galleries the results of which is used to determine vertical strain);

– Rod extensometers (relative displacement the results of which is used to determine vertical strain);

– Deformeters (crack gauges) over the vertical construction joints in the galleries (relative displacement the results of which is used to determine vertical strain);

– Carlson strain gauges installed in 1D, 2D as well as 3D clusters (measuring strain in different directions and also includes in a single "no-stress" strain gauge installed in a vertical direction at each installation); and

– Creep cells (strains).

Upon close examination of the monitoring results, signs of alkali silica reaction (ASR) were soon evident after its completion in the early 1980s. In 2013 detail behaviour analysis made some interesting conclusions with regards to the strain responses observed in the main body of the dam by the monitoring system. Especially with regards to the impact of openings on the strain behaviour of the concrete in the vicinity of openings (including galleries) as well as the impact of confinement on strain (Hattingh et al, 2014). These are very important considerations when interpreting monitoring results of structures subjected to swelling concrete. More specific details are discussed in the next few paragraphs.

2.1. *Impact of Openings on the Strain Behaviour of Swelling Concrete*

A number of additional rod extensometer installations have been performed since 1998 (see Figure 2). In 1998, 5 vertical rod extensometers were installed into the foundation (EF 1 to 5). It is important to note these installations measure the displacements of the foundation as well as a small section of concrete as they have been installed in the peripheral gallery. During the second installation in 2006, 6 rod extensometers were installed to monitor the swelling at the centre of the dam (EB 203, EB 224, EB 240, EB 271, EB 296 and EB 326). In addition 2 short rod extensometers were also installed in Blocks 1-3 and 14-16 only in the concrete section below the peripheral gallery to determine the swelling effect of concrete on the readings of the rod extensometers installed in the foundation (EB 171 and EB 225). Finally, 3 horizontal rod extensometers (with varying lengths) were installed on the right flank as well as the left flank into the foundation at RL 296 m from the peripheral gallery during 2008 and early 2009 respectively (EBME and EBMD).

The annual vertical swelling strain rate of the top part in the centre of the dam (above RL 296 m - the part where the effect of gravity is at its minimum – therefore the closest to the "free" strain rate) is around 30 micro strain/year (see Figure 3). The vertical strain rate, as expected, decreases with depth with the lowest strain rates of less than 2 micro strain/year observed between RL 203.5 m to RL 171.96 m.

Of particular interest is the vertical swelling strain rates of 19 and 23 micro strain/year observed at the installations between the peripheral gallery and the rock interface (EB 171 and EB 225 respectively in Figure 6) when comparing these with strain rates observed at the different levels in the centre of the dam. For example EB 171 (which is not vertically constrained due to the gallery opening above it) an annual strain rate in the concrete between RL 166.5 m and RL 171 m of 19 micro strain/year is observed compared to less than 2 micro strain/year observed between RL 203.5 m to RL 171.96 m where the concrete is mainly constrained in a vertical direction.

This provides a clear indication of the effect of gallery openings or any other openings for that matter on the local strain behaviour of concrete. This must always be kept in mind not only when analysing the results of the rod extensometers, the precision levelling as well as the Carlson strain gauges but all other instruments measuring displacement, strains or stresses.

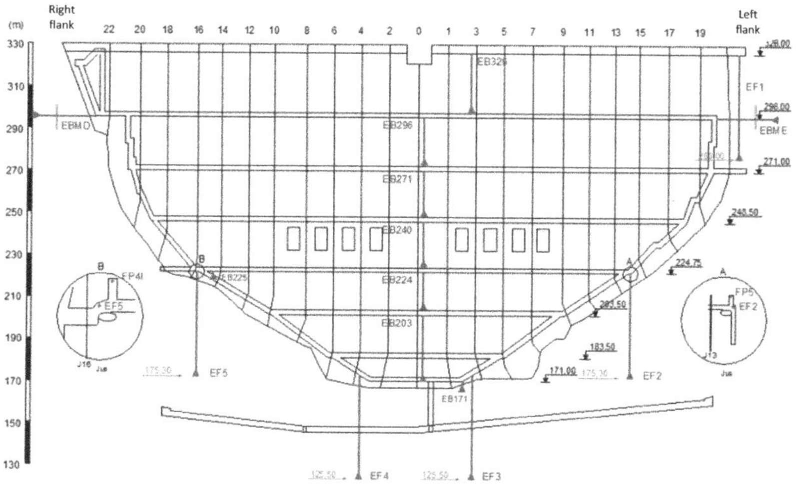

Figure 2. *Cahora Bassa Dam: Layout of the rod extensometers*

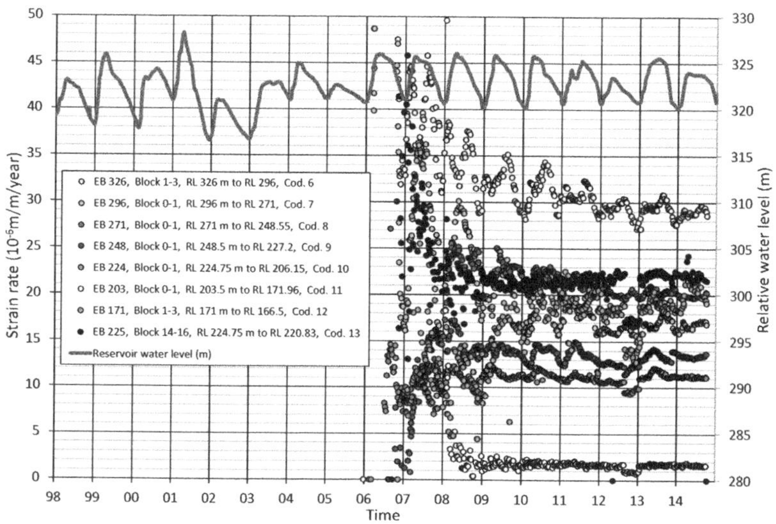

Figure 3. *Cahora Bassa Dam: Rod extensometers: Annual strain rate of the concrete*

2.2. *Impact of Confinement on Strain*

The Carlson strain gauges have been installed in groups of 1 (1D), 5 (2D in tangential direction) and 9 (3D) instruments each respectively during construction (see Figure 4 for the layout and Figure 5 for the typical strain directions for each installation). Each of these installations includes a single "no-stress" strain gauge installed in a vertical direction.

From the "no stress" strain gauges close to the top of the wall annual vertical swelling strain rates of between 20 and 35 micro strain are evident (similarly to the rod extensometer results, this is the part where the effect of gravity is at its minimum – therefore the closest to the "free" strain rate – one could conclude a minimum "free" strain rate) (see Figure 6). This is similar to what is observed from the rod extensometer results in the same area.

When analysing the average "no stress" strain rates, it is evident that the average "stress free" swelling strain rates across the dam are approximately 25 micro strain/year (see Table 1). The effect of confinement/stress is also evident through the dam section with the average radial strain rates of around 25 micro strain/year close to both the downstream and upstream faces decreasing to a round only 9 micro strain/year at the centre as well as decrease of the "no stress" strain rates from 25 to 16 micro strain/year in the centre of the section. It is important to note that in general there is no significant change in the vertical strain rates through the section of the dam.

The apparent "stress free" strain rates as expected decrease with depth to around 20 micro strain/year at level RL 203 m. This most probably could be attributed to the fact that these "stress free" strain gauges are not completely stress free due to the installation methodology and possibly, to a lesser degree, the decrease in concrete temperature in the central part of the dam body with a decrease in relative level (3 to 4 degrees from top to bottom) and its influence on the rate of swelling.

2.3. *Importance of Redundancy*

These important conclusions about the behaviour of swelling concrete and the impact of openings on the strain behaviour of the concrete especially in the vicinity of openings (including galleries) as well as the impact of confinement on strain most probably would not have been possible without the redundancy in the original displacement and strain monitoring system of Cahora Bassa Dam. Value for money when considering the relative small size of the investment in the monitoring system compared to the overall investment. The attention to detail during the installation of the instruments and the dedicated and careful operation of the instruments added value to the redundancy in the monitoring. The impact on decision-making regarding the impact of swelling concrete on the safety of the dam is clearly evident.

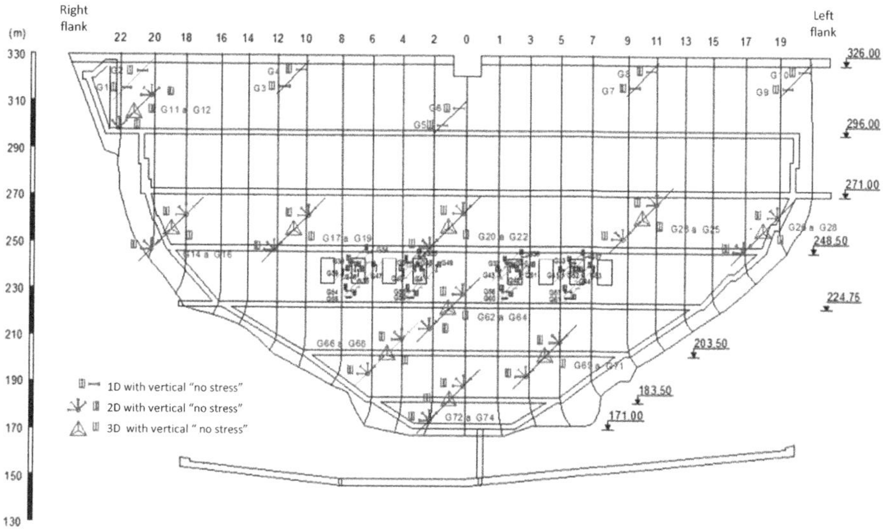

Figure 4. *Cahora Bassa Dam: Layout of Carlson strain gauges*

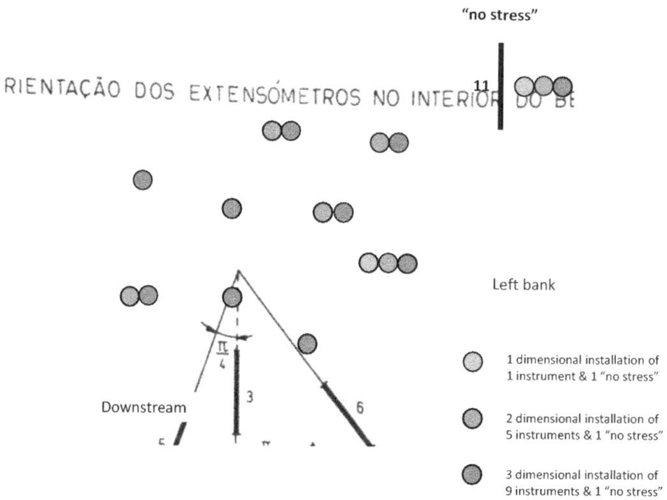

Figure 5. *Cahora Bassa Dam: Carlson strain gauges: strain directions for the different installations*

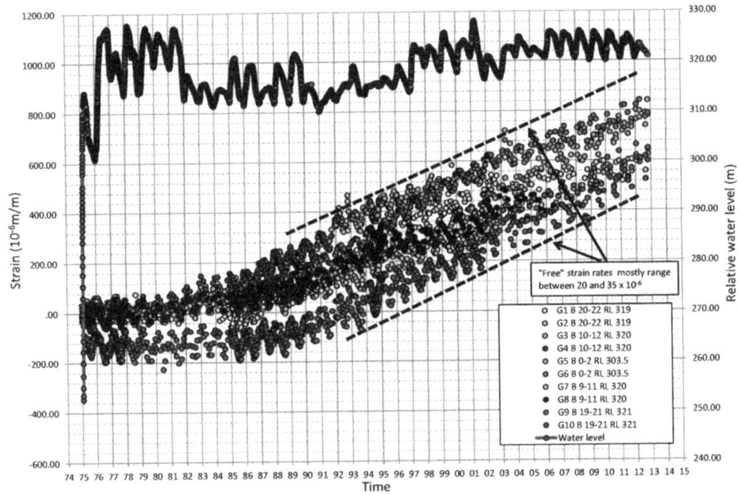

Figure 6. *Cahora Bassa Dam: Carlson strain gauges: "Stress free" strain at the top of the dam wall above RL 300 m*

Strain direction (instrument position – see Figure 8)	Average strain rate (micro strain/year)		
	1 m from downstream face	Centre of block	1 m from upstream face
Vertical (3)	16.7	17.0	15.6
Upstream/downstream (Radial) (2)	25.8	9.33	26.5
"No stress" (11)	24.0	16.5	25.7

Table 1. *Cahora Bassa Dam: Carlson strain gauges: Average strain rates excluding those around the spillway openings*

3. Kouga Dam

The 69 m high Kouga Dam (a double curvature arch dam) is located on the Kougha River in South Africa (see Figure 7 for a downstream view). The dam has been extensively monitored during and after its completion in 1969. Initially the dam reacted as expected but since 1972, it became clear that some form of concrete swelling was occurring. In addition to this, some inelastic movement of the right flank followed a few years later that complicated the behaviour model of the structure even further. The static monitoring system of Kouga Dam kept on expanding since the early signs of swelling became evident. Initially only clinometers and pendulum clinometers as well as a geodetic network of targets on the downstream face of the dam wall were used.

3D-crack gauges across major joints and major cracks on the non-overspill crest were added in 1991 and the geodetic surveys were improved with amongst others, precise traversing in the gallery and along the cat walks. This was followed by sliding micrometers and later Trivec's were installed in the cored boreholes. The original 3D-crack gauges were replaced by upgraded 3D-crack gauges measuring translations and rotations. The latest additions in 2011 are real-time automatic 3D-crack gauges in the gallery and a GPS system at four survey beacons on the crest. After the initial use of a non-permanent 3D ambient vibration measurement system in 2010 at the dam (Moyo et al, 2013), the installation of a permanent 3D ambient vibration measurement (AVM) system was completed in 2012.

Figure 7. *Downstream view of Kouga Dam (previously named Paul Sauer Dam)*

The importance of not only focussing on the behaviour of the centre cantilever of an arch dam as well as the importance 3D-crack meters and monitoring redundancy in the proper interpretation of the behaviour of Kouga Dam are highlighted in the following paragraphs (based on Hattingh et al, 2012a & 2012b).

3.1. *3D long-Term Displacement Behaviour*

From the results of 6-monthly geodetic surveys the following are evident (see Figures 8 and 9):

– The rate of the permanent vertical displacement due to swelling can be grouped into four distinct periods, up to 1972, from 1972 to 1983, from 1983 to 2000 and since 2000. the highest rate of vertical displacement due to swelling was evident from 1972 to 1983 and rate of displacement due to swelling has decreased in the subsequent periods. Virtually no permanent displacement is evident since 2000;

– also clearly evident from the vertical displacement is the inelastic deformation (due to drying shrinkage) of the upper part of the wall during low water levels;

– the maximum permanent horizontal displacement is taking place at the quarter points of the dam similar to what is evident at other arch dams with swelling concrete (charlwood, 2009). very little or no permanent horizontal displacement is evident at the centre of the dam;

– no decrease in the permanent horizontal displacement due to swelling at the quarter points is evident since detail monitoring started during the early 1970's. the influence of the water level on the horizontal behaviour of the dam is also clearly evident;

– upon closer examination some inelastic downstream displacement after periods of extended lower water levels is also evident most probably due to drying shrinkage. it must be mentioned that similar inelastic deformations are also observed after prolonged periods of low water levels in several other concrete arch dams in south africa (hattingh et al, 2011); and

– it is interesting to note that elges et al in 1995 postulated that the swelling would for all practical purposes cease from about 2000 onwards. although this postulation appears true for the vertical swelling, the horizontal swelling has continued unabated since. in many cases where only the vertical swelling is monitored the postulation that the swelling has stopped is clearly not correct. this phenomenon is also evident for example at bimont dam (goguel, 2009) where horizontal swelling was still evident more than 25 years after the vertical swelling appeared to stop. it is important to note that recent research highlighted the very long term expansion due to swelling as a result of alkali supply from various sources, alkali recycling as gel transformation and alkali release from aggregates.

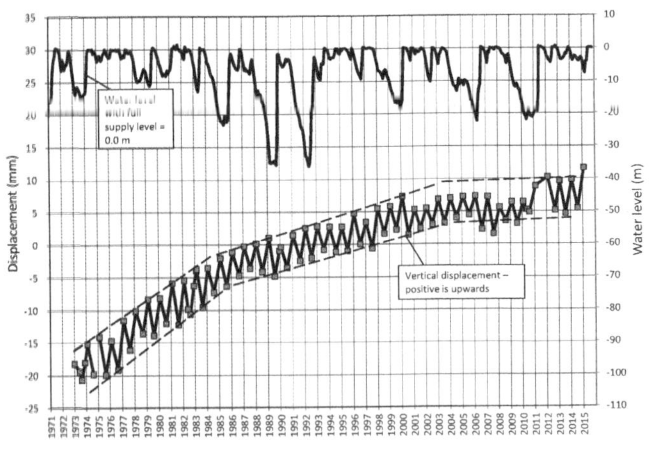

Figure 8. *Kouga Dam: Vertical displacement at the centre of the dam 4 m below full supply level*

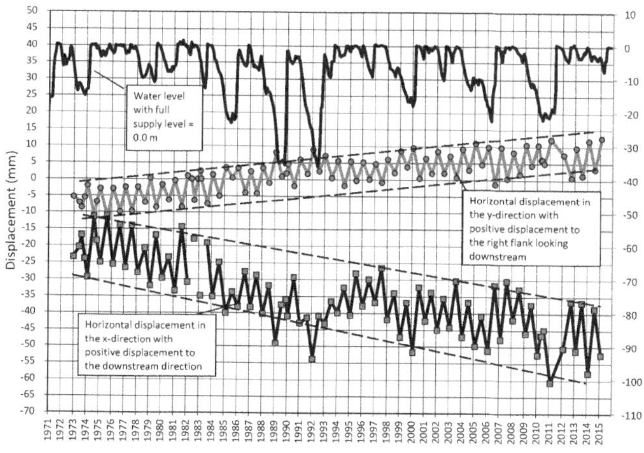

Figure 9. *Kouga Dam: Horizontal displacement at a quarter point of the dam on the right flank (looking downstream) 4 m below full supply level*

3.2. *Proper Use of 3D-Crack Meters in Monitoring Redundancy*

It has been observed by the authors that in the majority of cases where 3D-crack gauges are used to monitor relative displacements and rotations of constructions joints, the results of these instruments are not properly interpreted. In fact in many cases the results are normally disregarded because it appears not to make any sense. From the results of the 3D-crack gauges Kouga Dam the following have been observed (Hattingh et al, 2012a & 2012b):

– seasonal cyclical tangential displacements are evident on the flanks and are decreasing in magnitude towards the centre of the wall. the magnitude of the tangential displacement on the left flank is of a larger magnitude than what is observed on the right flank;

– in the vicinity of the quarter points of the wall the tangential displacement start to be directly related to water level with minimal seasonal cyclical displacements evident. the maximum water related tangential displacements are evident at the deepest section just to the left of the centre of the wall;

– some permanent vertical displacements are evident especially on the flanks indicating differential vertical displacement between the blocks most probability due to the vertical swelling of the blocks. seasonal cyclical vertical displacements are also evident on the flanks and are decreasing in magnitude to-wards the centre of the wall;

– these tangential as well as vertical displacement observations confirm slight asymmetric primary arch action in the spillway section related to the water level (at lower water levels the primary arch action never happens above the water level), cantilever action of the blocks on the flanks as well as additional "stiffness" in the left flank arch section created by the auxiliary chute spillways.

Several ambient vibration measurements have been carried out in 2010 at similar relatively low water levels during late summer and late winter to determine the natural frequencies of the dam. Eight natural frequency modes were identified. One of these was an "independent" vertical mode of vibration of a few isolated blocks on the left flank that only occurred during the winter measurements (Moyo et al, 2013). Initially this "independent" vertical mode of vibration was doubted until the dynamic results were compared with the static results of the 3D-crack gauges. This comparison confirmed that the same conclusions can be reached from the results of both the 3D-crack gauges as well as the ambient vibration measurement (including the cantilevering effect of individual blocks in the upper part of the dam wall).

4. Concluding Remarks

Using the strain results of the different monitoring systems a quite clear picture of the strain behaviour of Cahora Bassa Dam is evident especially as a result of the swelling due to ASR. Of particular importance is the observed impact of openings on the strain behaviour of the concrete especially in the vicinity of the galleries as well as the impact of confinement on strain (strain rates significantly decrease with an increase of confinement).

These important conclusions on the swelling behaviour of the concrete would not have been possible if only one particular instrument type was used for monitoring strain. It highlights the importance of having redundancy measurements with different types of instruments that can provide similar results is stressed, even if some manipulation may be required to ensure confidence in the measurements. In the case of Cahora Bassa Dam it was possible to compare all the results by converting them to strain rates from displacements (precision levelling and rod extensometers) and strains respectively.

The strain rates of the different monitoring systems correlate extremely well. In particular, when considering the impact of gallery openings on the actual strain rates and the position of the different monitoring points relative to the different galleries. It is therefore highly recommended that all these aspects be taken into account when interpreting monitoring results as well as considering different types of remedial solutions and their long-term impacts on the behaviour of the overall structure (for example with slot cutting).

From the Kouga Dam example (and numerous others as well) it is clearly evident that only monitoring the behaviour of the central cantilever or vertical displacement of the crest of an arch dam will in many instances not provide you with correct interpretation of the behaviour of the structure to draw conclusions on the current as well as future swelling behaviour of the concrete.

The importance of having sufficient redundancy in a monitoring system is also highlighted by the correct interpretation of unusual behaviour of some of the blocks of Kouga Dam by comparing the static 3D-crack gauge results or the dynamic ambient vibration results. Without this redundancy uncertainty would have existed regarding these observations and they in most cases would have been disregarded due to the unusual nature.

5. Acknowledgements

The authors wish to express their gratitude to all other present and past members of the Dam Safety section of Hidroeléctrica de Cahora Bassa (HCB) as well as all present and past members of the Subdirectorate: Dam Surveillance of the Department of Water and Sanitation of South Africa that contributed in some or other way to the development of the above-mentioned paper. The opinions expressed are those of the authors and do not necessarily reflect the views of either of the dam owners.

6. References

Charlwood, R.G. AAR in Dams and Hydroelectric Plants. Management of Alkali Aggregate Affected Structures: Analysis, Performance & Prediction. Short Course held in Paris, France. 2009.

Elges, H., Geertsema, A., Lecocq, P., Oosthuizen, C. Detection, Monitoring, and Modelling of Alkali-Aggregate Reaction in Kouga Dam (South Africa). *Alkali-Aggregate Reactions in Hydroelectric Plants and Dams*. Proceedings of USCOLD 2nd International Conference held in Chattanooga, USA. 1995.

Gogual, B. Arch Dam Case Studies. Management of Alkali Aggregate Affected Structures: Analysis, Performance & Prediction. Short Course held in Paris, France, 2009.

Hattingh, L.C., Oosthuizen, C. Unpredictable behaviour of a large arch dam in South Africa. Dams and Reservoirs under changing Challenges. Proceedings of the ICOLD Symposium held in Lucern, Switzerland. 2011.

Hattingh, L.C., Oosthuizen, C. Rehabilitation of a large arch dam: Importance of dam safety surveillance. Published as part of the Proceedings of the *3rd International Conference on Concrete Repair, Rehabilitation and Retrofitting* held in Cape Town, South Africa, 2012a.

Hattingh, L Oosthuizen, C. "Unusual Behaviour of a Large Arch Dam in South Africa as a Result of Swelling Due To Chemical Reaction". Published as part of proceedings of the ICOLD Conference held in Kyoto, Japan, 2012b.

Hattingh, L, Tembe, I, Carvalho, E, Oosthuizen, C. "3D Swelling Due To Chemical Reaction – The Cahora Bassa Dam Experience". Published as part of proceedings of the SANCOLD Conference held in Johannesburg, South Africa, 2014.

ICOLD. *Automated Dam Monitoring Systems*. Bulletin 118. ICOLD, Paris, France. 2000.

ICOLD. *Dam Surveillance Guide*. Bulletin 158. ICOLD, Paris, France. 2014

Moyo, P., Oosthuizen, C. Hattingh, L.C. "Ambient Vibration Measurements at Kouga Dams...Getting Much More Information Than Expected". Published as part of the Proceedings of the Symposium titled *Changing Times: Infrastructure Development to Infrastructure Management* held during the ICOLD Annual meeting in Seattle, USA. 2013.

Index of Authors